Geomorphological Hazards in High Mountain Areas

The GeoJournal Library

Volume 46

Managing Editors: Herman van der Wusten, University of Amsterdam,
The Netherlands
Olga Gritsai, Russian Academy of Sciences, Moscow,
Russia

Former Series Editor:
Wolf Tietze, Helmstedt, Germany

Editorial Board: Paul Claval, France
R. G. Crane, U.S.A.
Yehuda Gradus, Israel
Risto Laulajainen, Sweden
Gerd Lüttig, Germany
Walther Manshard, Germany
Osamu Nishikawa, Japan
Peter Tyson, South Africa

The titles published in this series are listed at the end of this volume.

Geomorphological Hazards in High Mountain Areas

edited by

JAN KALVODA

Charles University,
Faculty of Science,
Prague, Czech Republic

and

CHARLES L. ROSENFELD

Oregon State University,
Corvallis, Oregon, U.S.A.

SPRINGER SCIENCE+BUSINESS MEDIA, B.V.

A C.I.P. Catalogue record for this book is available from the Library of Congress.

ISBN 978-94-010-6200-8   ISBN 978-94-011-5228-0 (eBook)
DOI 10.1007/978-94-011-5228-0

CONTENTS

vi CONTENTS

DEDICATION TO PROFESSOR CLIFFORD EMBLETON

There are some men
who should have mountains
to bear their names to time.
         -Leonard Cohen

This book brings to the community of science a current view of hazards associated with the geomorphic processes in the mountain environment. Clifford Embleton, Professor of Geography at King's College, London, dedicated his life and considerable talents to geomorphology and natural hazards research, especially in the glacial and periglacial domain. Among his important works are *Glacial and Periglacial Geomorphology,* with C.A.M. King, published in 1968 and expanded in 1975, *Geomorphology of Europe* in 1984, and over 75 articles for scientific journals. Clifford was a highly organized, meticulous, quiet contributor to many of the major initiatives of geomorphology in the twenty-first century, including NASA's 'Geomorphology from Space', and the Study Group on Rapid Onset Geomorphological Processes of the International Geographical Union. We are particularly indebted to Professor Embleton as the founder of the IGU's Commission on Natural Hazards Studies, which largely through his vision has produced this volume.

Clifford Embleton was first and foremost a true field geomorphologist, a geographer who absorbed his landscapes through the soles of his boots'. We, his friends and colleagues, dedicate this work to his memory, knowing that his spirit leads us further into the mountain environments that he loved. Clifford is survived by his wife, Dr. Christine Embleton-Hamann, and three sons.

Charles L. Rosenfeld
Chairman, IGU Commission
on Natural Hazards Studies

PROLOGUE

High mountains are often perceived as a landscape of dreams and rugged beauty with wildly exalted rock masses humbly bowing their napes to the law of extinction. They are also secrets uplifted from the depths of the planet, sharp waste lands battered by gales, frost and burning sunshine, an ocean of stone cyclops lathered by glacial masses. Every stone in the high mountains is evidence of the history of these scars of the Earth. The fine cutting of fissures healed by quartz, dark nooks of glens and dazzling vertical walls - everything is full of tension, unrest and expectation. The certainty of disintegration of seemingly intact rock plates and points of ridges, rocky slopes and, in the valleys, stony carpets woven from the wrecks of courage and pride that shoot up to the clouds, all compose an immutable rhythm. The history of stone, a second of which equal as a generation of man, was fascinating in its austerity both at time of its formation within the Earth's core and at the time when it was first touched by gusts of wind, beams of light and the clinging of water drops, ice crystals and lichens. All aspects of very high mountains bear clear testimony to an incessant cascade of energy, spilling from one landform to another. Bygone landscapes are reflected in the present ones, manifesting changes and mixing rise and fall, fertility and death in fantastic chaos. Since long ago, humans have been measuring their curiosity, apprehension and faith by the unreal curves of high mountains ridges and summits. They were crossed by pilgrims and merchants, and mountain tribes learned to live in narrow valleys, intermountain basins and on mountain slopes. But only in the present second of the mountain history have strange people, of often very lonely regions, come thus far, driven by vain and incomprehensible reasons, to climb mountain summits and to explore the high mountain nature. Some of them pay with their lives for their audacity, but nothing changes their determination.

High mountain regions have become a symbol of inaccessibility and of almost insurmountable obstacles, a scene of mountaineering and scientific adventures. The high mountain environment is at the same time transforming into an inexhaustible laboratory, in which the multiple and dynamic natural processes continue in the sort of extreme conditions that are so well liked by Earth sciences' students. It is certainly useful to marshal the research on high mountain regions into the rational structure of natural sciences. But if we pass through the long and often bitter course of reasons and facts, we often discover that logic and abstraction are bound, if not conditioned, by a subtle thread of intuition and fantasy. Only with their help is it possible to grasp the substance, if not directly the spirit, of the landforms, events and phenomena studied by Earth sciences. In spite of the volume of findings during the last ten years, this look at the high mountains nature is still well-founded. One of the wonders found in the extreme conditions of high mountain deserts is the ecstasy of the unknown. In the chasm of our own ignorance and weariness we are forced to wander through the monolithic temples of mountains, admiring the profuse riches of landforms, events and phenomena. To see certainly does not mean to understand. The deeper we penetrate into the natural processes of high mountains, the richer is the mosaic of their structure and the more difficult their explication.

The research profession has very strict and purposeful criteria for theoretical approaches and experimental methods of research work. Maybe for that reason we pay very
little attention to the aesthetic values of the world in which we live, to efforts of mutual
understanding and more to the search for and the evaluation of the beauty of the laws
of natural phenomena and processes which we try to grasp, at least to a certain degree.
They are definitely essential aspects of science and university work, which are the most
important aspects from the human viewpoint. At the same time, they may be suspension bridges on the way to understanding such mysterious features of nature that we are
yet not able to even name.

Geomorphological hazards are part of a larger group of natural hazards, including
for instance floods, soil or water quality risks due to agricultural technologies, deserti-
fication, sudden weather changes and wildfires. Natural hazards studies are also
essential for a better understanding of ecosystem protection and the relationship
between the environment and human health. The protection of people from natural
risks in high mountains areas requires an understanding of the nature of hazards phenomena. Therefore, research into basic geomorphological hazards in high mountains
could be understood not only as a set of case studies but, first of all, as a rare opportu-
nity for the preparation of theoretical models and for the understanding of the general
architecture of the origin of natural disasters. At present, a lot of international research
projects are focused on determining the extent to which the geodynamic processes affecting the Earth's surface constrain land use and engineering works. It is also important
to estimate the range of the effects of anthropogenous activities on the rate of natural
geodynamic processes. The monograph "Geomorphological Hazards in High Mountains Areas" has been prepared within the long-term research programme of the
International Geographical Union Commission "Natural Hazards Studies"
(CONAHA). Noted specialists were asked to contribute manuscripts to the general
theme given by the title of this GeoJournal Library volume. This was a sort of a symposium by correspondence, which resulted in the selection of the most appropriate
manuscripts for our volume. It is interesting that given this freestyle correspondence,
together with the orientation and the experience of different authors, a certain concen-
sus in opinions on a set of basic geomorphological hazards in high mountains areas
has developed. The manuscripts represented in the volume include selected regional
examples of research of volcanic and post-volcanic activity, active fault zones, earthquakes, as well as of landslides, rockfalls, avalanches, creep movements, intensive
weathering, erosion of soils and parent material, glacier thawing and surging, debris
flows, mud torrents, floods and man-made landforms changes.
Besides the co-editor Charles L. Rosenfeld, chairman of CONAHA, I was helped
during the compilation of the volume by several colleagues. Co-editorial work was
partly done during my research stay in Centre d'Etudes et Recherches Ecogéographi-
ques (Université Louis Pasteur, Strasbourg) in 1996, which has for several years been
the seat of the European Center on Geomorphological Hazards. The camera-ready
version of the manuscripts was prepared at the Department of Physical Geography and
Geoecology, Faculty of Science, Charles University in Prague, with the help of Dr.
Helena Švachová and Mgr. Jakub Langhammer. We would like to thank Charles University and the Department of Culture and Science of the Czech Ministry of Foreign

Affairs for their support. We are naturally very grateful to all the authors of the manuscripts included in this monographical volume, as their research endeavours were decisive for the realization of the volume.

The GeoJournal Library volume "Geomorphological Hazards in High Mountains Areas" has been prepared with a view to the 28th International Geographical Congress at the Hague (1996) and the 4th International Conference on Geomorphology at Bologna (1997). We would like to dedicate it to the 650th anniversary of Charles University in Prague, founded on 7th of April 1348.

Jan Kalvoda
Faculty of Science,
Charles University, Prague

SEDIMENTOLOGY AND CLAST ORIENTATION OF DEPOSITS PRODUCED BY GLACIAL-LAKE OUTBURST FLOODS IN THE MOUNT EVEREST REGION, NEPAL

DANIEL A. CENDERELLI, ELLEN E. WOHL

1. Introduction

Climatic warming during the last 100 to 150 years has caused many alpine glaciers to thin and retreat (Grove, 1990; O'Connor and Costa, 1993; Evans and Clague, 1994). This glacial downwasting has disturbed geomorphic systems in mountainous environments, increasing the risk of geologic and hydrologic hazards such as glacial-lake outburst floods, landslides, and glacier avalanches (O'Connor and Costa, 1993; Evans and Clague, 1994). Glacial-lake outbursts are perhaps the most devastating of these processes because they can travel tens to hundreds of kilometers downstream dramatically modifying channel morphology and, in some cases, causing loss of human life and destruction of property. The devastating effects of glacial-lake outburst floods have been documented in the Andes (Lliboutry et al., 1977), Himalayas (Hewitt, 1982; Fushimi et al., 1985; Ives, 1986; Vuichard and Zimmermann, 1986, 1987), Canadian Rockies (Clarke, 1982; Blown and Church, 1985; Desloges and Church, 1992; Evans and Clague, 1994), Swiss Alps (Haeberli, 1983; Haeberli et al., 1989), and Cascades (O'Connor et al., in press).

Despite the recognized hazards and impacts of glacial-lake outburst floods in alpine regions, few studies have evaluated the morphology, sedimentology, and clast-orientation (fabric) of deposits along the flood route. Recent studies have described transformations from debris flow to hyperconcentrated flow to streamflow during a single flow event because of changes in sediment concentration as sediment is mobilized and incorporated into the floodwaters or deposited and diluted with additional floodwaters along the flood route (Pierson and Scott, 1985; Smith, 1986; Wells and Harvey, 1987). Characterizing the morphology, sedimentology, and fabric of glacial-lake outburst flood deposits is necessary to properly identify the type(s) of flow process, to understand coarse sediment transport and deposition, and to recognize older flood deposits.

The Mount Everest region (Figure 1) recently experienced two glacial-lake outburst floods; one in 1977 and one in 1985. The 1977 and 1985 glacial-lake outburst floods produced an assortment of erosional and depositional features along the outburst-flood routes. Of particular interest in this study are the large bar complexes deposited in expanding reaches along the outburst-flood routes. Deposit morphology, sedimentology, and fabric criteria were used to infer flow processes along the outburst-flood routes. The deposits were either derived from a flow process intermediate between hyperconcentrated-flow and water-flood processes or from water-flood processes. Hyperconcentrated flow represents a flow process intermediate between debris flow and water flow. Because sediment samples were not collected during the outburst floods,

1

J. Kalvoda and C.L. Rosenfeld (eds.), Geomorphological Hazards in High Mountain Areas, 1-26.
© 1998 Kluwer Academic Publishers.

rheologic criteria such as sediment concentration, bulk density, and yield strength could not be used to classify the outburst-flood deposits. However, studies by Pierson and Scott (1985), Smith (1986), Wells and Harvey (1987), and Costa (1988) have provided researchers with sedimentologic and fabric criteria of deposits to infer flow processes during a flow event (Table 1).

Table 1. Sedimentologic characteristics and descriptions of hyperconcentrated-flow and water-flood deposits from various studies.

Study - deposit type	Mean particle size (mm)	Sorting	Stratification	Grading	Clast long-axis orientation to flow direction	Imbrication
Pierson and Scott (1985)[a] hyperconcentrated flow	0.35-0.55	very poor (2.0-2.4ϕ)	none or faint bedding	none	not applicable	not applicable
Smith (1986) hyperconcentrated flow	> 64	poor	none or faint bedding	variable; commonly normal	transverse or parallel[b]	poor
water flood	not given	better than hypercon centrated	none, horizontal, cross-bedding	variable	transverse	strong
Wells and Harvey (1987) - hyperconcentrated flow	60-230	moderate; matrix is very poor	weak to moderate	none	not given	not given
water flood	> 50	moderate to poor; matrix very poor	none, front to tail fining	not applicable	transverse	moderate to strong
Costa (1988) hyperconcentrated flow	0.5-2, pred. coarse sand	poor (1.1-1.6ϕ)	none or weak	reverse or normal	not given	weak
water flood	not given	well to poor	none, horizontal, cross-bedding	not given	not given	moderate to strong

a. Did not discuss water-flood deposits.

b. Transverse orientations for clasts greater than 150 mm. Parallel orientations for clasts less than 150 mm.

Each of these studies included descriptions of debris-flow deposits, but debris-flow criteria were not included in Table 1 because debris-flow processes did not occur along the flood routes. Examination of Table 1 indicates that the criteria used to distinguish hyperconcentrated-flow deposits from water-flood deposits differs between studies.

These differences reflect our limited understanding of the hydraulic conditions and depositional mechanics associated with coarse sediment-laden flows, as well as the similar nature of deposits formed by hyperconcentrated-flow and water-flood processes.

Along the upper 16 km of the 1985 outburst-flood route (Figure 1, reaches L1, L2, L3, and L4), deposits are thicker, reverse graded, more poorly sorted, more coarsely skewed, and display weaker imbrication than coarse fluvial deposits further downstream, but do not convincingly display the sedimentologic and fabric characteristics of hyperconcentrated-flow deposits described by Pierson and Scott (1985), Smith (1986), Wells and Harvey (1987), and Costa (1988).

Therefore, the deposits along the upper 16 km of the 1985 outburst flood were probably formed in a flow transitional or approaching hyperconcentrated conditions and are classified as transitional-flow deposits. Specifically, deposits in reaches L1, L2, L3, and L4 are inferred to have been formed by transitional-flow processes and deposits in reaches L5, L6, L7, L8, N1, N2, and N3 by water-flood processes. To quantify and support our observations in the field, we conducted sedimentologic and fabric analyses on 123 and 43 deposits, respectively.

2. Physical setting and glacial-lake outburst flood hydrology

The Mt. Everest region is located in eastern Nepal (Figure 1). The area is underlain primarily by Precambrian gneisses and granites (Vuichard, 1986), lies within the High Himalayas, and is characterized by extremely high relief. The four major valleys in the area, Bhoti Kosi, Dudh Kosi, Imja Khola, and Khumbu Khola, are deeply incised and their valley floors can be 4000 m lower than the surrounding mountains. Above an elevation of 3400 to 3800 m, the valleys are distinctly U-shaped from Pleistocene glaciations (Iwata, 1976; Fushimi, 1977, 1978) and below this elevation the valleys are narrow and deeply incised. Alpine glaciers are present at elevations above 5000 meters and have been, for the most part, in retreat since their Little Ice Age maximum positions, creating numerous moraine- and glacier-dammed lakes (Mayewski and Jeschke, 1979; Fushimi et al., 1985).

Hydrology in the study area is strongly influenced by monsoonal precipitation and late spring/early summer snowmelt. Eighty percent of the total annual precipitation occurs between June and September (Ageta, 1976). However, the rain-shadow effect created by the high mountain topography reduces the intensity and amount of monsoonal precipitation with increasing elevation (Ageta, 1976; Zimmermann et al., 1986). For example, from June to September in 1974 the total precipitation was 428 mm at 4400 m, 685 mm at 3900 m, and 1100 mm at 2700 m (Ageta, 1976).

On September 3, 1977, a series of ice-cored moraine-dammed lakes below the Nare Glacier failed, sending a flood surge down the Nare, Imja, and Dudh valleys (Buchroithner et al., 1982; Fushimi et al., 1985; Zimmermann et al., 1986)(Figure 1). This flood caused extensive erosion and deposition for 35 km downstream from the source area, destabilized valley side slopes, and destroyed bridges and trails, disrupting travel in the region (Zimmermann et al., 1986; Ives, 1986). Based on a field survey, the volume of water released by the lake was estimated to be 500,000 m^3 (Fushimi et al., 1985). Preliminary step-backwater modeling indicates that the peak discharge was 1400 m^3/s approximately nine kilometers downstream from the lake (Figure 1. reaches N1 and N2). Because the 1985 glacial-lake outburst flood was at least two times greater than the 1977 glacial-lake outburst flood (see discussion below), geomorphic evidence of the 1977 glacial-lake outburst flood was either removed or altered by the 1985 glacial-lake outburst flood downstream from the confluence of the Bhoti and Dudh Kosi (Figure 1).

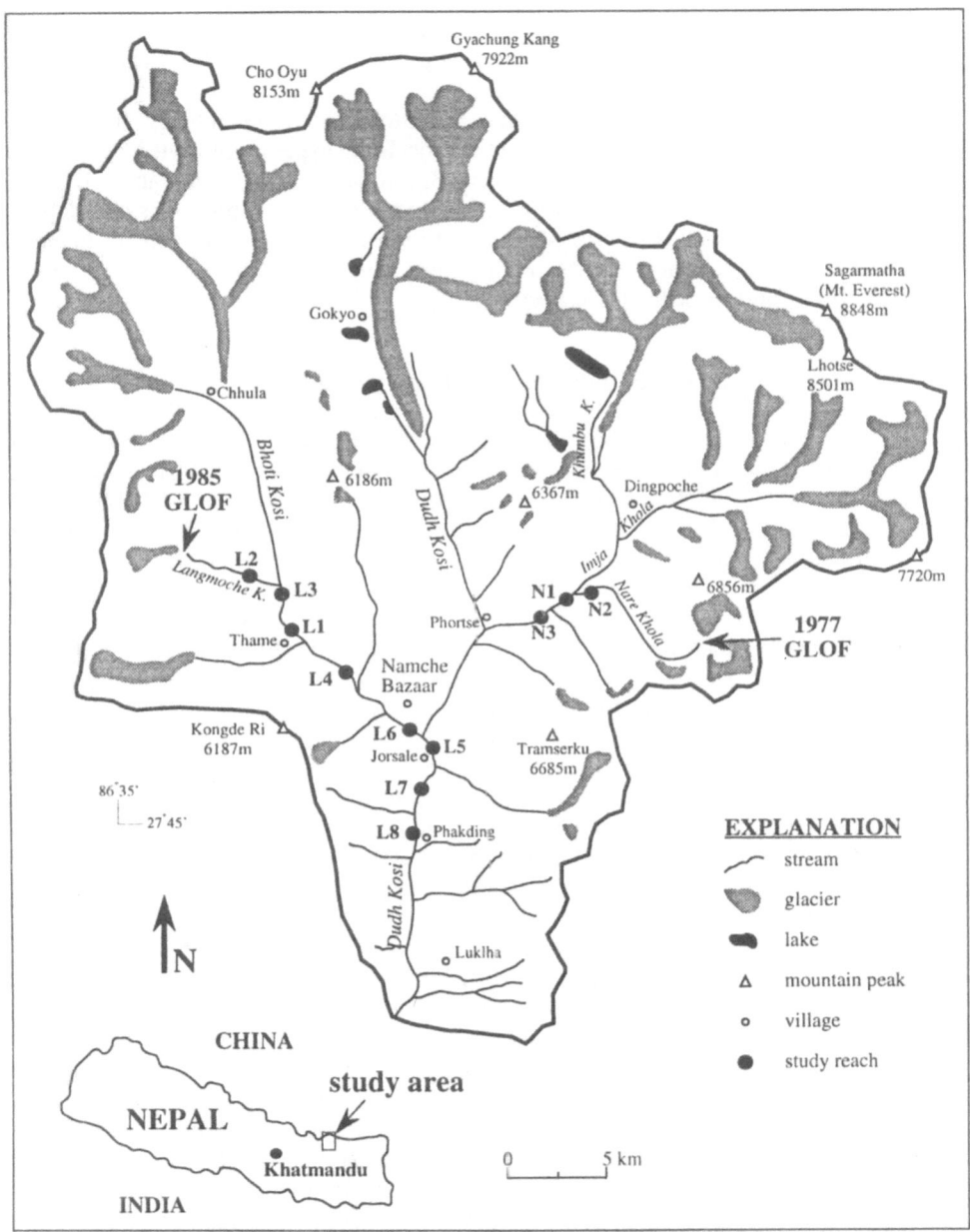

Figure 1. Map of the study area showing locations of reaches studied along the 1977 and 1985 glacial-lake outburst floods (GLOF). N designates 1977 study reaches and L designates 1985 study reaches.

Figure 2. Downstream view of depositional features at reach L1. Deposition occurs across the entire valley bottom and consists of multiple longitudinal and transverse bars. Valley width is between 150 and 200 m.

On August 4, 1985, a moraine-dammed lake located below the Langmoche glacier failed when an ice avalanche from the glacier fell into the lake, triggering a surge wave that breached the moraine (Ives, 1986; Vuichard and Zimmermann, 1986, 1987)(Figure 1). The volume of water released was estimated to be between 500,000 and 1,000,000 m³ (Ives, 1986; Vuichard and Zimmermann, 1986, 1987). The resulting flood caused considerable erosion and deposition for 40 km along the Langmoche Khola, Bhoti Kosi, and Dudh Kosi (Figure 1). Erosion was most extensive in narrow, steep reaches where the channel boundaries consisted of glacio-fluvial sediment, and deposition was prevalent where the valley widened and was less steep. Vuichard and Zimmermann (1987) estimated that 3,300,000 m³ of sediment were eroded and deposited along the flood route; approximately 70 percent of this total occurred in the first 16 km of the flood route. This flood destroyed several bridges, tens of houses, and a nearly completed hydroelectric power plant (Ives, 1986; Vuichard and Zimmermann, 1987). Preliminary step-backwater modeling indicates that the peak discharge was 2600 m³/s approximately seven kilometers downstream from the lake (reach L2) and attenuated to a peak discharge of 1400 m³/s at a distance of 27 km from the lake (reach L8)(Figure 1). The peak discharges of the outburst floods were between 5 and 10 times greater than the typical annual peak discharges from snowmelt floods.

3. Deposit characteristics

Eleven depositional reaches were selected for detailed study; three reaches along the 1977 outburst-flood route and eight reaches along the 1985 outburst-flood route (Figure 1). Reach parameters for the study sites have been derived from field measurements and maps (Table 2).

Figure 3. Downstream view of reach L5. Channel is bounded by bedrock along right margin and terraces along left margin. Sheet deposits (right bank, foreground) and longitudinal bars (left bank, background) are deposited approximately four to five meters above the channel on glaciofluvial terrace surfaces. Valley width is 70 m. Note the people crossing the suspension bridge.

Deposition along the 1977 and 1985 glacial-lake outburst flood routes was dominated by coarse sediment. Deposition occurred across the entire valley bottom in wider and less steep valley segments (Table 1) where flow energy was reduced and flow was diverging. In general, the deposits are clast supported, poorly to moderately imbricated, moderately to very poorly sorted, structureless, and comprised primarily of cobbles and boulders with variable long-axis orientations. The largest clast transported by the 1977 outburst flood had an intermediate diameter of 2.3 m and the largest clast transported by the 1985 outburst flood had an intermediate diameter of 2.7 m.

3.1. DEPOSIT MORPHOLOGY

Valley morphology influenced depositional features. Depositional reaches L1, L2, L3, L4, N1, and N2 are located in wide, expanding reaches (valley widths greater than 100 m) immediately downstream from narrow, constricted reaches (widths less than 30 m). Reaches L3 and N1 are also located immediately downstream from the confluence where the upstream outburst-flood tributary enters the main channel. In contrast, reaches L5, L6, L7, L8, and N3 are located in relatively narrow expanded reaches (valley widths less than 70 m) immediately downstream from narrower, more constricted reaches (valley widths less than 30 m). Study reach characteristics are summarized in Table 2.

Table 2. Summary of reach parameters. N designates reaches along the 1977 outburst-flood route. L designates reaches along the 1985 outburst-flood route. Refer to Figure 1 for reach locations.

Study reach	Distance from lake (m)	Average gradient	Average width (m)	Length (m)[a]	Predominant flow process
L2	7125	0.035	130	250	transitional
L3	9350	0.056	110	270	transitional
L1	10900	0.060	160	600	transitional
L4	15600	0.045	140	200	transitional
L6	20710	0.049	50	120	water flood
L5	22140	0.032	70	280	water flood
L7	24665	0.026	60	330	water flood
L8	26700	0.024	70	530	water flood
N2	8175	0.060	100	250	water flood
N1	8600	0.050	120	200	water flood
N3	11450	0.055	50	170	water flood

a. The length between the furthest upstream sampled deposit and the furthest downstream sampled deposit

Depositional features in reaches L1, L2, L3, L4, N1, and N2 consist primarily of multiple, linear and curvilinear longitudinal and transverse bars that are clast-supported and parallel or subparallel to valley alignment (Figure 2). These deposits overlie older pre-outburst flood glacio-fluvial terraces and are six to eight meters above the present-day channel. Longitudinal bars are long (20-80 m), narrow (4-6 m), and 0.5 to 3.0 m thick. Transverse bars are long (50-100 m), wide (10-30 m), and 0.5 to 3.0 m thick. The longitudinal and transverse bars have steep-sided flanks and fronts with nearly flat to slightly convex surfaces. For the most part, the largest clasts are located near the front of the deposits. Isolated boulder clusters are occasionally present on the surface of the transverse bars. The longitudinal and transverse bars are separated by chute-channels (2-4 m below the bar crest) displaying step-pool morphology (Figure 2). Similar channels between bar deposits have been described by Stewart and LaMarche (1967) and Martini (1977). The chute channels represent zones where flow was converging between the longitudinal bars.

As pointed out by Costa (1988), hyperconcentrated-flow deposits and water-flood deposits have similar morphologic features, making them difficult to distinguish from each other using morphologic criteria alone. The transitional-flow deposits in reaches L1, L2, L3, and L4 have morphologic features similar to the water-flood deposits in reaches N1 and N2 with the exception that the transitional-flow deposits are reverse graded or lack grading and appear massive. In contrast, the water-flood deposits are massive and lack grading.

Depositional features in reaches L5, L6, L7, L8, and N3 consist primarily of single, longitudinal bars that are clast-supported and parallel or subparallel to valley alignment (Figure 3). Also occurring in these reaches, although not as extensive, are thin sheet deposits that are clast-supported and oriented parallel to flow direction and valley alignment. Both of these deposit types were deposited four to six meters above the channel on older pre-outburst flood glacio-fluvial terrace surfaces along the valley margins. The longitudinal bars in these reaches are long (5-30 m), narrow (1-3 m), and 0.5 to 1.0 m thick. The sheet deposits are long (5-30 m), wide (5-15 m), and up to 0.5 m thick.

3.2. SEDIMENTOLOGY

Surface particle counts using a grid-method (modified from Wolman, 1954) were conducted on the uppermost segments of deposits to determine the particle-size distribution of the glacial-lake outbursts during peak flow. The lengths of the intermediate axes of 100 clasts were measured at 123 sampling localities. Additionally, the lengths of the intermediate axes of the ten largest clasts within and in close proximity to the sampling grid were measured. Cumulative particle-size distributions (Figures 4 and 5) of the deposits indicate they are unimodal, but asymmetrical. Transitional-flow deposits (reaches L1, L2, L3, and L4) have less-steep curves than water-flood deposits (reaches L5, L6, L7, L8, N1, N2, and N3), indicating they are more poorly sorted (Figures 4 and 5). The plots also indicate there is considerable variability in the size of particles deposited along a given reach.

Particle-size statistics of mean size, sorting, skewness, and kurtosis were calculated for the deposits using the inclusive graphic method (Folk and Ward, 1957). The results of these analyses are summarized in Tables 3 and 4. The particle-size statistics indicate that the transitional-flow deposits (reaches L1, L2, L3, and L4) are clast-supported, poorly to very poorly sorted, coarsely- to strongly coarsely-skewed, and generally leptokurtic (Table 3a). The mean particle-sizes of the deposits are highly variable, ranging from 39 to 642 mm (Table 3a). The mean size of the ten largest clasts, D_m, in the deposits also reflects this variability, ranging from 323 to 1661 mm (Table 3a). The amount of fine-grained sediment (particle diameters less than 64 mm) in the transitional-flow deposits is variable and ranges from 0 to 52 percent (Table 3a).

The particle-size statistics indicate that the water-flood deposits (reaches L5, L6, L7, L8, N1, N2, and N3) are moderately to very poorly sorted, non-skewed to strongly coarsely-skewed, and generally leptokurtic (Tables 3b and 4). In the deposits, the mean particle-size ranges from 48 to 636 mm, the mean of the 10 largest clasts ranges from 267 to 1405 mm, and the amount of fine-grained sediment ranges from 0 to 41 percent (Tables 3b and 4). Examination of individual values, as well as reach averages, shows that transitional-flow deposits are more poorly sorted and coarsely skewed and have a higher percentage of fine-grained sediment than water-flood deposits (compare Table 3a to Tables 3b and 4). An analysis of variance (level of significance, $\alpha=0.05$) indicates that sorting, skewness, and percentage of fine-grained material in transitional-flow deposits are statistically different than sorting, skewness, and percentage of fine-grained in water-flood deposits, with sorting showing the strongest difference between the two deposits.

Table 3a. Summary of sedimentologic characteristics of transitional-flow deposits along the 1985 glacial-lake outburst flood route. Reaches are ordered upstream (L2) to downstream (L4) and locations are shown in Fig. 1.

Reach, sample no.	Percent finer than 64 mm	d_{50} (mm)	Mean of 10 largest clasts (mm)	Mean size (mm)	Sorting (φ units)	Skew-ness	Kurt-osis	Dist. from lake (m)
L2, X1S1F1	28.7	134	763	90	-2.82	-0.401	1.266	7000
L2, X1S2	18.6	155	555	141	-1.55	-0.311	1.395	7000
L2, X2S1	22.3	147	453	120	-1.80	-0.404	1.459	7050
L2, X2S2	48.0	74	1661	77	-3.53	-0.064	0.767	7050
L2, X2SHF1	43.0	74	323	39	-5.55	-0.553	1.721	7050
L2, X3S1F1	21.0	191	962	133	-2.45	-0.463	1.895	7130
L2, X3S2	21.2	181	797	141	-1.94	-0.417	1.405	7130
L2, X4S1	15.2	357	1148	261	-2.26	-0.545	1.775	7200
L2, X4S2	24.5	191	667	133	-2.41	-0.479	1.404	7200
L2, X4SHF1	36.5	83	649	68	-2.06	-0.347	1.698	7200
L2, X5S1	12.6	461	1118	320	-2.06	-0.605	1.961	7250
L2, X5S2	37.9	158	1284	61	-3.76	-0.479	0.724	7250
average	**27.5**	**184**	**865**	**132**	**-2.68**	**-0.422**	**1.456**	**7125**
std. dev.	**11.4**	**116**	**384**	**83**	**1.12**	**0.142**	**0.397**	
L3, X1S1	24.0	160	652	135	-1.82	-0.321	1.204	9200
L3, X3S1F1	18.5	298	1216	214	-2.34	-0.499	1.932	9310
L3, X3S2	5.8	326	1107	299	-1.18	-0.241	1.122	9310
L3, X4S1	34.6	187	997	109	-3.16	-0.404	0.906	9370
L3, X4S2	52.0	58	1372	69	-3.24	0.045	1.120	9370
L3, X4S3	10.6	352	791	300	-1.49	-0.380	1.512	9370
L3, X4SHF1	26.2	137	1427	138	-1.98	-0.002	1.223	9370
L3, X5S1F1	25.7	214	863	126	-2.70	-0.495	1.252	9410
L3, X5S2	30.3	205	748	116	-2.60	-0.480	0.915	9410
L3, X6S1F1	20.0	396	1124	224	-2.68	-0.578	2.134	9470
average	**24.8**	**233**	**1030**	**173**	**-2.32**	**-0.336**	**1.332**	**9350**
std. dev.	**12.9**	**106**	**265**	**81**	**0.69**	**0.212**	**0.410**	
L1, X19SHF1	21.2	221	746	98	-2.92	-0.603	1.801	10600
L1, X19S1F1	24.2	211	765	136	-2.22	-0.588	1.553	10600
L1, X18S2	37.0	98	673	76	-1.84	-0.387	1.132	10650
L1, X18S1	25.0	174	934	153	-2.13	-0.140	1.012	10650
L1, X15S2F1	35.0	135	901	67	-3.36	-0.462	0.865	10840
L1, X15S1	34.7	139	843	119	-2.49	-0.237	0.989	10840
L1, X14S2F1	20.0	326	996	183	-2.40	-0.666	1.543	10920
L1, X14S1	38.0	226	1089	62	-3.80	-0.645	0.658	10920
L1, X12SHF1	10.1	304	967	257	-1.36	-0.290	1.323	11050
L1, X12S2F2	17.0	239	1014	203	-2.15	-0.405	1.410	11050
L1, X12S1F1	5.1	776	1517	638	-1.61	-0.597	2.788	11050
L1, X11S4	21.0	231	956	127	-2.75	-0.529	1.783	11090
L1, X11S3F2	24.0	331	1438	128	-3.91	-0.522	1.475	11090
L1, X11S2F1	18.0	324	722	220	-2.21	-0.609	1.787	11090
L1, X11S1	8.0	501	1114	434	-1.93	-0.433	2.111	11090
L1, X10SHF1	25.0	194	1172	82	-3.04	-0.602	1.968	11140
L1, X10S3	27.0	410	1328	131	-3.61	-0.648	1.143	11200
L1, X10S2	21.0	304	1019	103	-3.13	-0.721	2.155	11200
L1, X10S1F1	12.0	393	1633	355	-1.49	-0.258	1.154	11200
average	**22.3**	**292**	**1043**	**188**	**-2.55**	**-0.492**	**1.508**	**10900**
std. dev.	**9.5**	**156**	**272**	**146**	**0.78**	**0.166**	**0.526**	
L4, X9S2F1	14.4	357	996	267	-2.16	-0.556	2.195	15500
L4, X9S1	17.1	276	968	212	-2.29	-0.464	1.978	15500
L4, X8SHF1	27.9	139	700	124	-1.62	-0.170	0.853	15550
L4, X8S2	47.6	67	291	62	-1.33	-0.157	1.053	15550
L4, X8S1	12.5	158	493	154	-1.46	-0.266	2.164	15550
L4, X7S2	28.2	474	1092	126	-3.62	-0.732	0.945	15600
L4, X7S1F1	0.0	676	1235	642	-0.95	-0.100	1.026	15600
L4, X6S3	34.6	402	974	117	-3.53	-0.716	0.722	15650
L4, X6S2F1	15.5	440	1144	315	-2.32	-0.555	3.121	15650
L4, X6S1	37.1	100	834	78	-2.54	-0.317	1.205	15650
L4, X5S2	24.3	362	1172	153	-3.00	-0.615	1.307	15700
L4, X5S1	25.5	388	1347	226	-2.61	-0.517	0.980	15700
average	**23.7**	**320**	**937**	**206**	**-2.29**	**-0.430**	**1.462**	**15600**
std. dev.	**12.8**	**179**	**311**	**157**	**0.84**	**0.221**	**0.733**	

Table 3b. Summary of sedimentologic characteristics of water-flood deposits along the 1985 glacial-lake outburst flood route. Reaches are ordered upstream (L6) to downstream (L8) and locations are shown in Fig. 1.

Reach, sample no.	Percent finer than 64 mm	d_{50} (mm)	Mean of 10 largest clasts (mm)	Mean size (mm)	Sorting (ϕ units)	Skew-ness	Kurt-osis	Dist. from lake (m)
L6, X6S1	27.4	99	512	104	-2.13	-0.129	2.175	20650
L6, X5S1	8.5	161	462	153	-0.98	-0.126	1.112	20670
L6, X4S2	4.8	175	525	179	-0.94	0.085	1.122	20700
L6, X4S1F1	1.8	491	953	443	-0.95	-0.229	1.087	20700
L6, X2S2	23.8	130	498	105	-2.10	-0.394	2.035	20770
L6, X2S1F1	3.9	338	870	343	-0.98	-0.015	1.693	20770
average	**11.7**	**232**	**637**	**221**	**-1.35**	**-0.135**	**1.537**	**20710**
std. dev.	**11.0**	**152**	**215**	**140**	**0.59**	**0.167**	**0.497**	
L5, X10S2	20.6	136	563	120	-1.82	-0.345	1.796	22000
L5, X10S1	21.0	184	922	187	-1.89	-0.076	0.777	22000
L5, X9S1F1	17.8	207	611	169	-1.39	-0.386	1.053	22030
L5, X7S1F1	3.9	362	790	317	-1.09	-0.282	1.083	22120
L5, X6SHF2	11.4	181	754	173	-1.26	-0.072	1.401	22160
L5, X6SHF1	9.4	274	670	247	-1.33	-0.283	1.147	22160
L5, X6S1	15.1	175	650	166	-2.00	-0.181	2.422	22160
L5, X5S2	16.0	416	1001	379	-2.65	-0.248	1.188	22210
L5, X5S1F1	7.8	331	1405	365	-1.60	-0.015	0.958	22210
L5, X4S3F2	41.0	76	791	75	-1.66	-0.133	1.740	22280
L5, X4S2F1	16.8	149	743	169	-1.82	-0.072	1.254	22280
L5, X4S1	22.4	153	898	176	-1.93	0.018	0.976	22280
average	**16.9**	**220**	**816**	**212**	**-1.70**	**-0.173**	**1.316**	**22140**
std. dev.	**9.4**	**103**	**226**	**95**	**0.42**	**0.134**	**0.462**	
L7, X9S1	25.7	150	528	124	-1.71	-0.378	1.129	24500
L7, X9S2	3.8	699	1151	636	-1.18	-0.320	1.694	24500
L7, X8S1	30.1	128	685	108	-1.74	-0.188	1.050	24530
L7, X8S2F1	18.8	244	786	195	-2.09	-0.428	1.259	24530
L7, X7S1F1	18.3	168	683	150	-1.71	-0.310	1.370	24580
L7, X7SHF1	9.6	311	923	246	-1.48	-0.276	0.917	24580
L7, X5S1	37.7	91	618	65	-2.38	-0.348	1.456	24650
L7, X5SHF1	10.7	320	846	278	-1.45	-0.271	1.260	24650
L7, X4S1F1	13.3	135	478	143	-1.13	0.028	1.050	24700
L7, X3S1	11.8	335	1178	310	-2.35	-0.303	1.145	24740
L7, X2S1F1	26.9	124	469	108	-1.35	-0.216	0.899	24830
average	**18.8**	**246**	**759**	**215**	**-1.69**	**-0.274**	**1.203**	**24665**
std. dev.	**10.3**	**174**	**247**	**160**	**0.43**	**0.121**	**0.239**	
L8, X2SHF1	18.1	362	814	232	-2.02	-0.489	1.459	26320
L8, X6S1F1	10.7	158	669	147	-1.00	-0.167	1.501	26470
L8, X6S2	27.9	446	1037	185	-2.87	-0.675	1.035	26470
L8, X8S1	14.3	218	501	181	-1.91	-0.502	1.950	26550
L8, X8S2F1	27.6	158	395	60	-2.89	-0.709	1.344	26550
L8, X9S1	27.2	96	267	48	-2.43	-0.643	2.072	26620
L8, X9S2	21.4	128	379	122	-1.54	-0.304	1.683	26620
L8, X10S1F1	24.5	199	603	146	-1.73	-0.349	0.942	26660
L8, X10S2F2	10.7	146	370	143	-0.90	-0.103	1.226	26660
L8, X10S3F3	8.6	133	348	132	-0.85	-0.041	1.100	26660
L8, X11S1F1	19.1	147	601	150	-1.56	-0.064	1.307	26700
L8, X13SHF1	23.3	191	739	106	-2.78	-0.487	1.622	26740
L8, X13SHF2	10.7	172	891	226	-1.76	0.243	0.797	26740
L8, X14S1	24.0	112	385	105	-1.36	-0.177	1.422	26770
L8, X15S1F1	10.8	252	947	242	-1.86	-0.267	1.392	26810
L8, X15S2F2	14.4	177	941	181	-1.80	-0.090	1.689	26810
L8, X16S1F1	6.9	653	1095	536	-1.67	-0.582	2.730	26850
average	**17.7**	**220**	**646**	**173**	**-1.82**	**-0.318**	**1.487**	**26700**
std. dev.	**7.3**	**143**	**270**	**108**	**0.63**	**0.268**	**0.465**	

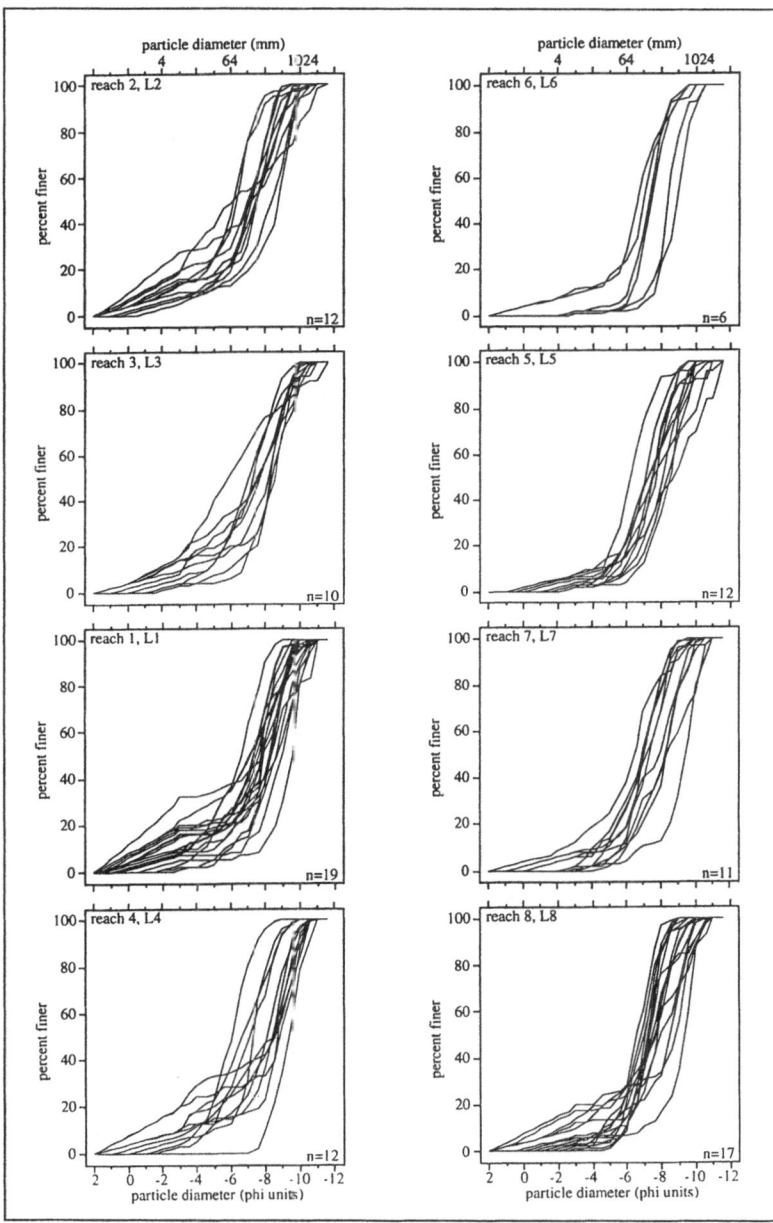

Figure 4. Particle-size distribution curves of deposits along the 1985 outburst-flood route. Samples in reaches L2, L3, L1, and L4 are transitional-flow deposits. Samples in reaches L6, L5, L7, and L8 are water-flood deposits.

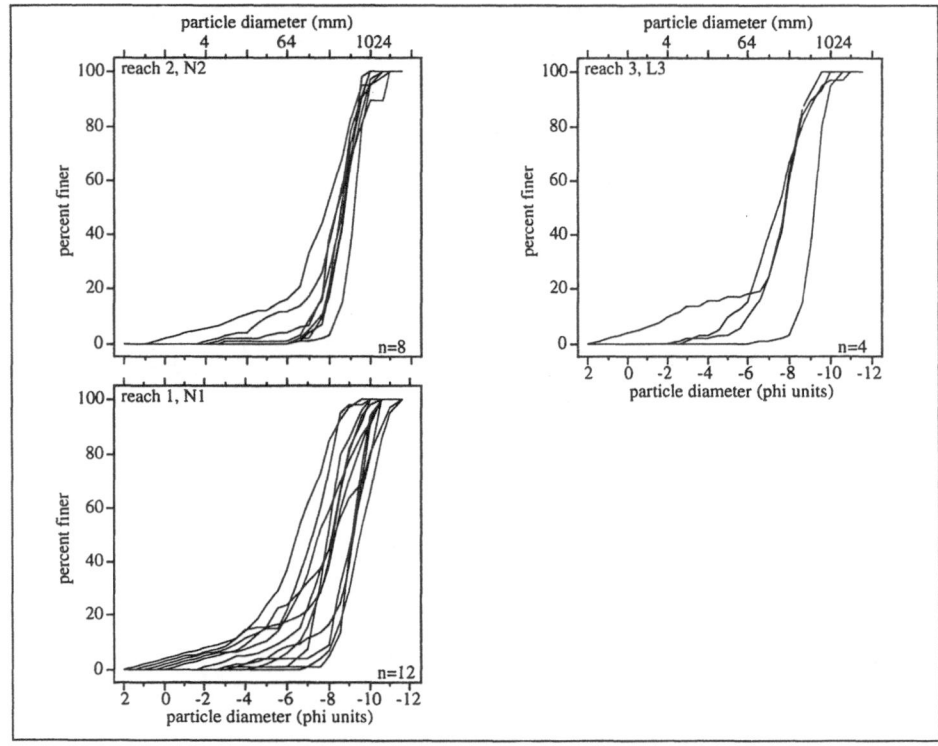

Figure 5. Particle-size distribution curves of water-flood deposits along the 1977 outburst-flood route.

These differences are illustrated in bivariate plots of sorting, skewness, and percentage of fine-grained sediment versus mean-particle size (Figure 6). There is some overlap between the transitional-flow deposits and the water-flood deposits and this may indicate variations in flow conditions at a given reach or the similar nature of water-flood and transitional-flow conditions. Hyperconcentrated-flow deposits have been described as being more poorly sorted than water-flood deposits (Smith, 1986; Costa, 1988). Whether the flow was hyperconcentrated in the upstream reaches of the 1985 outburst-flood is uncertain. However, the sedimentologic evidence suggests that floodwaters were at least sediment-laden and approaching hyperconcentrated conditions. The deposits are, therefore, classified as forming from transitional-flow processes.

In transitional-flow conditions, turbulence is reduced, but remains an important sediment-support mechanism during transport, along with dispersive stresses and buoyancy (Smith, 1986; Costa, 1988). Because of these sediment-support mechanisms during transport, deposition occurs rapidly through grain-by-grain accumulations from both tractive and suspension processes (Smith, 1986; Costa, 1988). The poorly sorted nature and high percentage of fine-grained material at the surface of the transitional-flow deposits (Figure 6) indicate that the sediment concentration profile was becoming more uniform as cobbles and boulders were being transported above the bed, probably by buoyancy and dispersive pressures from particle collisions, just prior to their rapid deposition.

Table 4. Summary of sedimentologic characteristics of water-flood deposits along the 1977 glacial-lake outburst flood route. Reaches are ordered upstream (N2) to downstream (N3) and locations are shown in Fig. 1.

Reach, sample no.	Percent finer than 64 mm	d_{50} (mm)	Mean of 10 largest clasts (mm)	Mean size (mm)	Sorting (ϕ units)	Skew- ness	Kurt- osis	Dist. from lake (m)
N2, X3S1	11.5	338	999	332	-1.75	-0.150	1.505	8000
N2, X6S1	16.0	234	639	201	-1.97	-0.388	1.570	8125
N2, X8S1	0.0	380	870	385	-0.81	0.030	1.123	8150
N2, X8S2F1	3.8	407	795	370	-0.99	-0.281	1.112	8150
N2, X9S1	0.0	419	905	423	-0.82	-0.038	0.976	8200
N2, X9S2F1	0.9	335	868	345	-0.87	-0.001	1.025	8200
N2, X10S1F1	0.0	425	873	404	-0.73	-0.257	1.052	8250
N2, X10S2	0.0	588	1119	577	-0.56	-0.119	1.076	8250
average	**4.0**	**391**	**884**	**380**	**-1.06**	**-0.151**	**1.180**	**8175**
std. dev.	**6.3**	**101**	**140**	**105**	**0.51**	**0.148**	**0.226**	
N1, X7S1	17.7	296	991	184	-2.37	-0.596	2.401	8500
N1, X7S2	21.9	148	558	132	-1.70	-0.394	1.502	8500
N1, X7SHF1	23.8	313	1423	230	-2.56	-0.314	0.939	8500
N1, X6S1	8.7	592	1145	514	-1.25	-0.441	1.916	8550
N1, X6S2F1	4.0	580	1232	576	-0.98	-0.115	0.903	8550
N1, X5S1F1	1.0	729	1470	739	-1.01	-0.026	0.929	8575
N1, X5S2	11.4	282	727	254	-1.50	-0.299	1.360	8575
N1, X4S1F1	1.0	247	759	260	-0.86	0.107	1.093	8650
N1, X4S2	36.8	89	699	75	-2.09	-0.292	1.241	8650
N1, X3S1	5.8	320	1079	320	-1.31	-0.068	1.007	8700
N1, X3S2F1	0.0	560	1056	581	-0.60	0.038	1.060	8700
N1, X3SHF1	19.1	182	942	188	-2.14	-0.141	1.314	8700
average	**12.6**	**362**	**1007**	**338**	**-1.53**	**-0.212**	**1.305**	**8600**
std. dev.	**11.4**	**203**	**287**	**211**	**0.64**	**0.211**	**0.452**	
N3, X11S1	35.9	111	384	84	-1.85	-0.399	1.080	11380
N3, X10S1	18.3	211	560	127	-2.23	-0.612	2.717	11420
N3, X8SHF1	9.4	223	665	203	-1.14	-0.150	1.281	11500
N3, X7S1	15.2	169	889	170	-1.44	-0.057	1.023	11550
average	**19.7**	**179**	**624**	**146**	**-1.66**	**-0.305**	**1.525**	**11450**
std. dev.	**11.4**	**50**	**211**	**52**	**0.47**	**0.251**	**0.802**	

The reverse grading present in many of the transitional-flow deposits is additional evidence that dispersive pressures and buoyancy were lifting cobbles and boulders above the bed. In water floods, turbulence is the principal sediment support-mechanism and particles are deposited individually through grain-by-grain tractive processes (Smith, 1986; Costa, 1988). As a result, water-flood deposits are typically better sorted and have less fine-grained sediment along their surface (Figure 6). These characteristics reflect the nonuniform sediment concentration profiles of water floods.

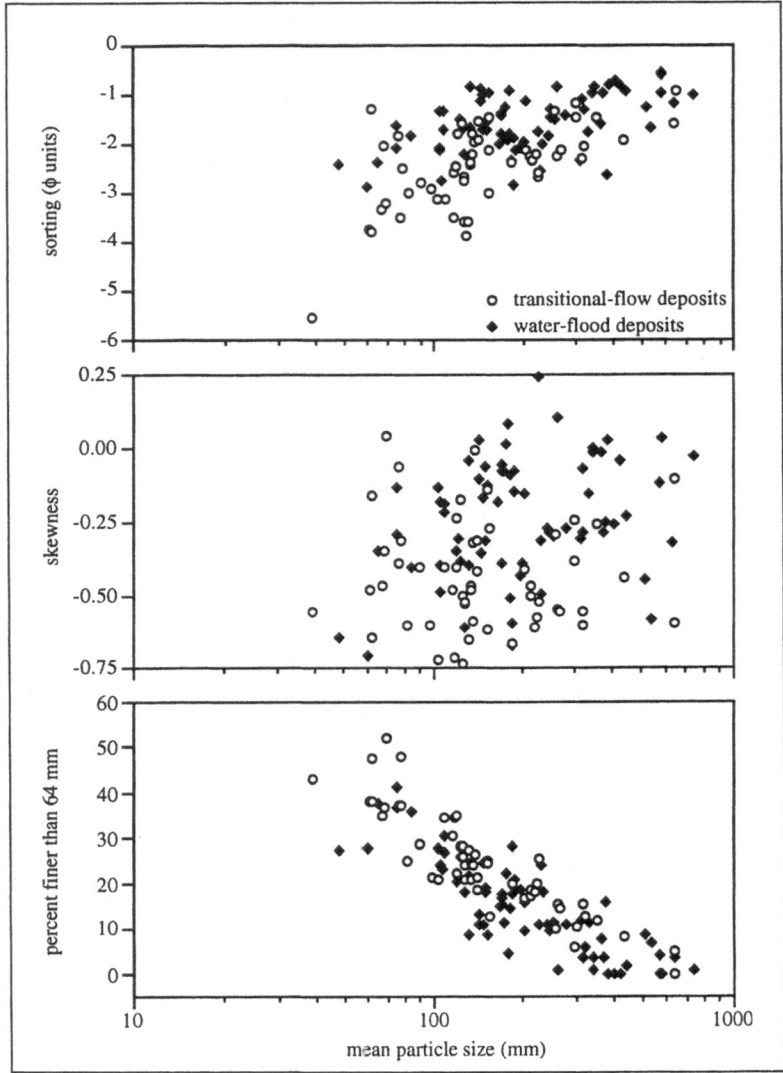

Figure 6. Plots of sorting, skewness, and percentage of fine-grained sediments versus mean grain size that distinguish water-flood deposits (closed triangles) from transitional-flow deposits (open circles).

Examination of various deposit sedimentologic characteristics along the 1985 outburst-flood route illustrates how flow conditions changed from transitional flow to water flood with increasing distance from the glacial lake (Table 3, Figure 7). The four upstream reaches (L1, L2, L3, L4) have similar average reach values for sorting, skewness, percentage of fine-grained sediment, and the mean size of the ten largest clasts, as do the four downstream reaches (L5, L6, L7, and L8). The average reach values indicate that the deposits are becoming better sorted, less coarsely skewed, and have less fine-grained material at their surface with distance from the lake (Figure 7).

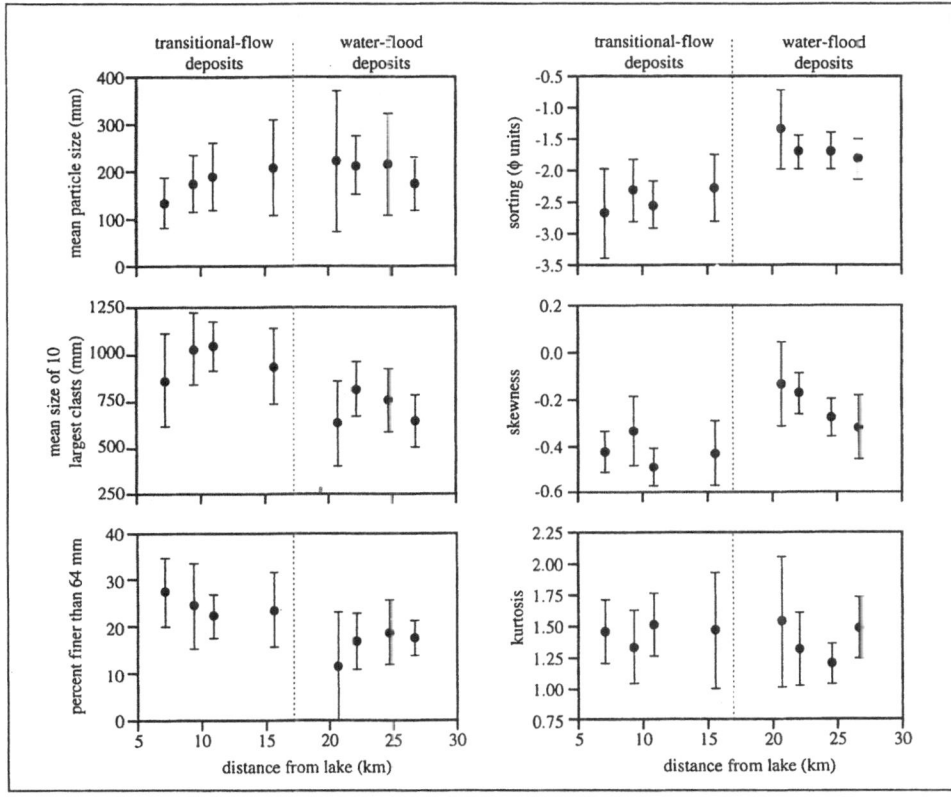

Figure 7. Downstream changes in sedimentologic characteristics of deposits along the 1985 outburst-flood route. Reach order from the lake is L2, L3, L1, L4, L6, L5, L7, and L8. Average reach values and 95 percent confidence intervals of the sedimentologic parameters are plotted for each reach.

As indicated by the 95 percent confidence intervals, there is considerable variability for a given sedimentologic parameter at a particular reach. This variability may reflect the complexity of flow hydraulics and depositional mechanics at a given reach. Additionally, the higher variability of deposit sedimentologic parameters in reaches L2, L4, and L6 may indicate the occurrence of both transitional-flow and water-flood conditions in these reaches. The average reach values for mean-particle size and kurtosis did not vary substantially along the outburst-flood route (Figure 7). However, the mean size of the ten largest clasts decreases with distance from the lake, indicating that flow competence of the outburst-flood decreased as it moved downstream (Figure 7). Preliminary step-backwater analyses indicate that the outburst-flood attenuated by almost 50 percent from the furthest upstream reach (L2) to the furthest downstream reach (L8). This supports the sedimentologic evidence that flow competence was decreasing in the downstream direction.

3.3. CLAST ORIENTATION

The orientation of clasts in a deposit is often a useful property for understanding depositional mechanics and distinguishing deposits formed from different flow processes. The long-axis orientation (fabric) of clasts in the deposits was measured at 43 sites. The azimuth (trend) and dip (plunge) of 25 clasts, larger than 64 mm in diameter and with long-axis to intermediate-axis ratios of greater than 1.5, were measured to assess the preferred long-axis orientation of clasts in the deposits. The shape of the particles measured fall along the boundary that delineates blade- and rod-shapes (Zingg, 1935) and does not vary between sampling sites.

The azimuth and dip of the clasts measured were plotted on lower-hemisphere, equal-area stereonets with contour intervals of two standard deviations (Figures 8a, 8b, and 9). The plots have been rotated so that the flow direction is toward the bottom of the page. In the transitional-flow deposits the contoured diagrams show weak to strongly developed clast long-axis orientations that dip upstream, but have varying azimuths among individual sites with respect to flow direction (Figure 8a). The water-flood deposits show weak to strongly developed clast long-axis orientations that, for the most part, dip upstream and have varying azimuths among individual sites with respect to flow direction, although oblique orientations seem more prevalent (Figures 8b and 9). The contoured diagrams for both types of deposits have primarily girdle distributions (orientations plot along the outer margin) and secondarily cluster distributions (Figures 8a, 8b, and 9). Girdle distributions in fluvial deposits have been attributed to particles being deposited rapidly (Carling, 1987). The girdle distributions in the transitional-flow and water-flood deposits indicate that deposition was rapid for both flow processes.

The eigenvalue method, a three-dimensional statistical analysis, was used to analyze the preferred orientation or spatial distribution of clasts in the deposits. From this analysis, three eigenvectors (V_1, V_2, and V_3) and corresponding eigenvalues (λ_1, λ_2, and λ_3) are calculated (Anderson and Stephens, 1972; Woodcock, 1977). The largest eigenvector, V_1, represents the mean-orientation of maximum clustering of clast long axes. The smallest eigenvector, V_3, represents the mean-orientation of minimum clustering of clast short axes, and is perpendicular to V_1 or the pole to the plane of the long axes (Mills, 1984). The orientation of V_3 can be used to assess the orientation and dip of the a:b plane defined by the long-axis orientations. S_1 and S_3 are normalized eigenvalues and measure the degree of clustering or fabric strength about their respective eigenvectors (Anderson and Stephens, 1972; Mark, 1973; Woodcock, 1977). Values of S_1 and S_3 were tested to determine if V_1 and V_3 were statistically different ($\alpha=0.10$) from the expected fabric strength values of a random sample of axes drawn from a uniform population (Anderson and Stephens, 1972). The results of the eigenvalue analysis are summarized in Table 5.

The mean orientation of V_1 and degree of clustering of the data, S1, reflect the patterns of the contoured diagrams (compare Table 5 to Figures 8a, 8b, and 9). Of the 27 water-flood deposits, 18 had statistically significant V_1 orientations and 19 had statistically significant V_3 orientations (Table 5). Of the 18 transitional-flow deposits, 11 had statistically significant V_1 orientations and 12 had statistically significant V_3 orientations. The results of the eigenvalue analysis indicate that there are no distinct

Table 5. Summary of eigenvalue analysis of clast fabric for the transitional-flow deposits (reaches L1, L2, L3, L4) and water-flood deposits (reaches L5, L6, L7, L8, N1, N2).

Reach, sample no.	Dist. from lake (m)	flow dir./ dep. slope (°)	V_1 az./dip (°)	S_1	V_3 az./dip (°)	S_3	dip of a:b plane (°)	General azimuth of V_1 and dip direction of a:b plane
L2, X1S1F1	7000	090/06	208/03	0.528	097/83	0.137	07	T, U
L2, X3S1F1	7130	076/06	128/20	0.440	247/52	0.232	38	R, R
L3, X3S1F1	9310	205/05	007/17	0.511	145/68	0.181	22	P, U
L3, X5S1F1	9410	215/09	040/15	0.546	172/69	0.223	21	P, R
L3, X6S1F1	9470	180/06	048/13	0.453	182/72	0.156	18	R, U
L1, X19S1F1	10600	214/04	104/01	0.577	195/65	0.128	25	T, U
L1, X15S2F1	10840	158/05	232/35	0.481	095/46	0.194	44	R, R
L1, X14S2F1	10920	165/04	268/17	0.545	167/31	0.105	59	T, U
L1, X12S2F2	11050	150/06	027/20	0.544	147/55	0.089	35	O, U
L1, X12S1F1	11050	135/02	312/26	0.585	153/63	0.127	27	P, U
L1, X11S3F2	11090	150/04	310/23	0.558	117/66	0.181	24	P, U
L1, X11S2F1	11090	160/05	293/29	0.514	155/53	0.145	37	O, U
L1, X10S1F1	11200	155/04	013/23	0.491	152/61	0.142	29	R, U
L4, X9S2F1	15500	115/06	200/05	0.501	102/58	0.154	32	T, U
L4, X7S1F1	15600	145/05	341/39	0.498	123/45	0.158	45	P, U
L4, X6S2F1	15650	135/05	214/24	0.418	073/60	0.266	30	R, R
L6, X4S1F1	20700	120/08	025/38	0.454	239/47	0.227	43	R, R
L6, X2S1F1	20770	105/05	343/24	0.489	120/58	0.225	32	R, R
L5, X9S1F1	22030	230/02	345/14	0.566	235/53	0.154	37	T, U
L5, X7S1F1	22120	225/-2	152/09	0.506	269/71	0.187	19	T, R
L5, X5S1F1	22210	240/06	116/12	0.434	223/54	0.195	36	R, R
L5, X4S3F2	22280	215/03	352/25	0.476	204/61	0.153	29	R, U
L5, X4S2F1	22280	200/00	338/18	0.505	115/66	0.136	24	O, U
L7, X8S2F1	24530	225/04	199/05	0.496	071/81	0.142	09	P, D
L7, X7S1F1	24580	240/03	115/05	0.523	218/68	0.138	22	O, U
L7, X4S1F1	24700	215/03	287/14	0.482	167/65	0.117	25	R, U
L7, X2S1F1	24830	240/01	100/16	0.631	289/74	0.122	16	O, U
L8, X6S1F1	26470	185/01	010/23	0.568	187/67	0.083	23	P, U
L8, X8S2F1	26550	155/03	243/06	0.510	136/71	0.134	19	T, U
L8, X10S1F1	26660	170/02	128/00	0.540	218/69	0.125	21	O, U
L8, X10S2F2	26660	170/03	312/08	0.499	201/69	0.087	21	O, U
L8, X10S3F3	26660	170/03	276/05	0.519	178/57	0.075	33	T, U
L8, X11S1F1	26700	165/02	288/09	0.523	133/80	0.077	20	O, U
L8, X15S1F1	26810	225/02	018/15	0.605	255/64	0.145	26	P, U
L8, X15S2F2	26810	225/02	007/25	0.480	233/56	0.155	34	R, U
L8, X16S1F1	26850	250/01	025/09	0.516	277/62	0.199	28	O, R
N2, X8S2F1	8150	245/06	306/02	0.491	194/86	0.205	04	R, R
N2, X9S2F1	8200	245/11	150/12	0.404	315/78	0.239	12	R, R
N2, X10S1F1	8250	235/07	163/22	0.556	261/19	0.185	71	T, R
N1, X6S2F1	8550	220/06	071/32	0.579	205/48	0.086	42	O, U
N1, X5S1F1	8575	220/06	134/47	0.501	286/39	0.182	51	T, U
N1, X4S1F1	8650	220/04	080/03	0.534	343/68	0.180	22	O, D
N1, X3S2F1	8700	230/09	277/13	0.465	171/50	0.167	40	R, U

For $\alpha=0.10$ and n=25, a S_1 value greater than 0.496 indicates that V_1 has a statistically significant preferred orientation and a S_3 value less than 0.184 indicates that V_3 has a statistically significant preferred orientation. Larger values of S_1 and smaller values of S_3 indicate greater fabric strength.
P=Parallel, indicates V_1 is within 30° of flow direction. O=oblique, indicates V_1 is between 30 and 60° of flow direction. T=transverse, indicates V_1 is between 60 and 90° of flow direction. U=upstream dip direction of a:b plane. D=downstream dip direction of a:b plane. R=random or no statistically significant preferred orientation.

differences in fabric strength between transitional-flow deposits and water-flood deposits (Table 5).

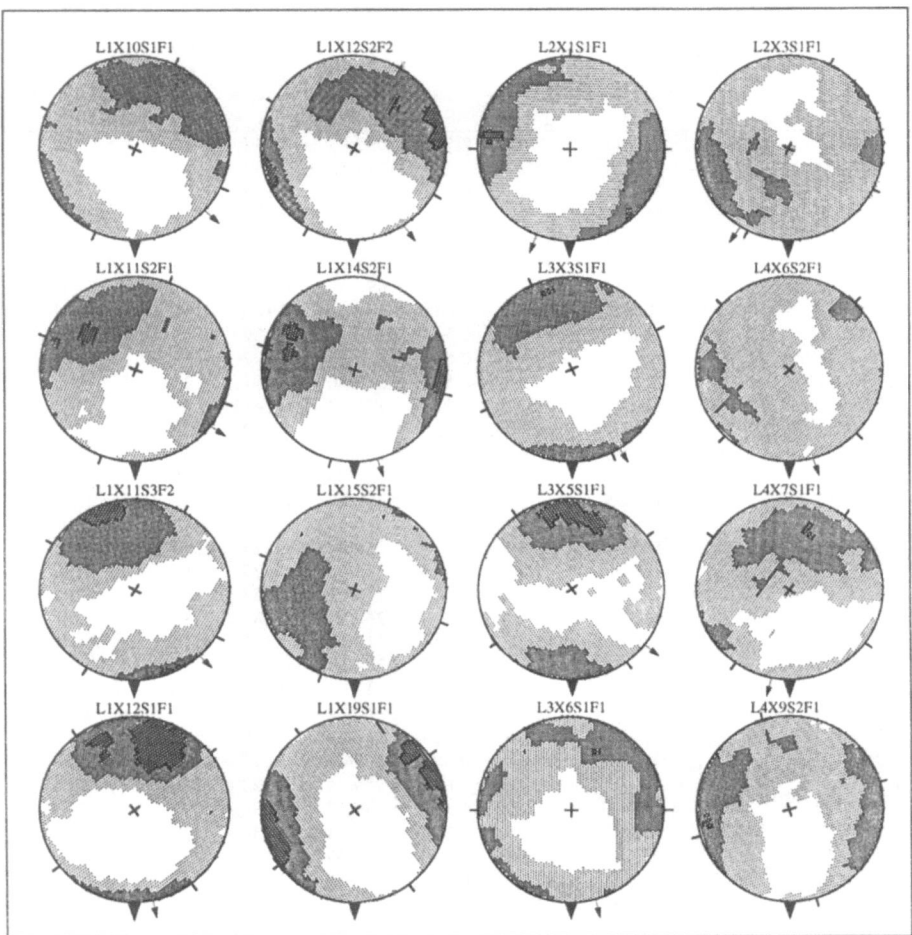

Figure 8a. Lower hemisphere, equal-area stereonets of clast long-axis orientations measured in transitional-flow deposits at reaches L1, L2, L3, and L4 of the 1985 GLOF. The diagrams are rotated so that flow direction (large arrow on equator) is toward the bottom of the page. The smaller arrow on the equator delineates the valley trend if it differs from the flow direction. The contoured diagrams have intervals of two standard deviations and the darkest shaded pattern indicates areas of highest concentration of clast long-axis orientations, whereas the lighter shaded patterns indicate areas of lower concentration.

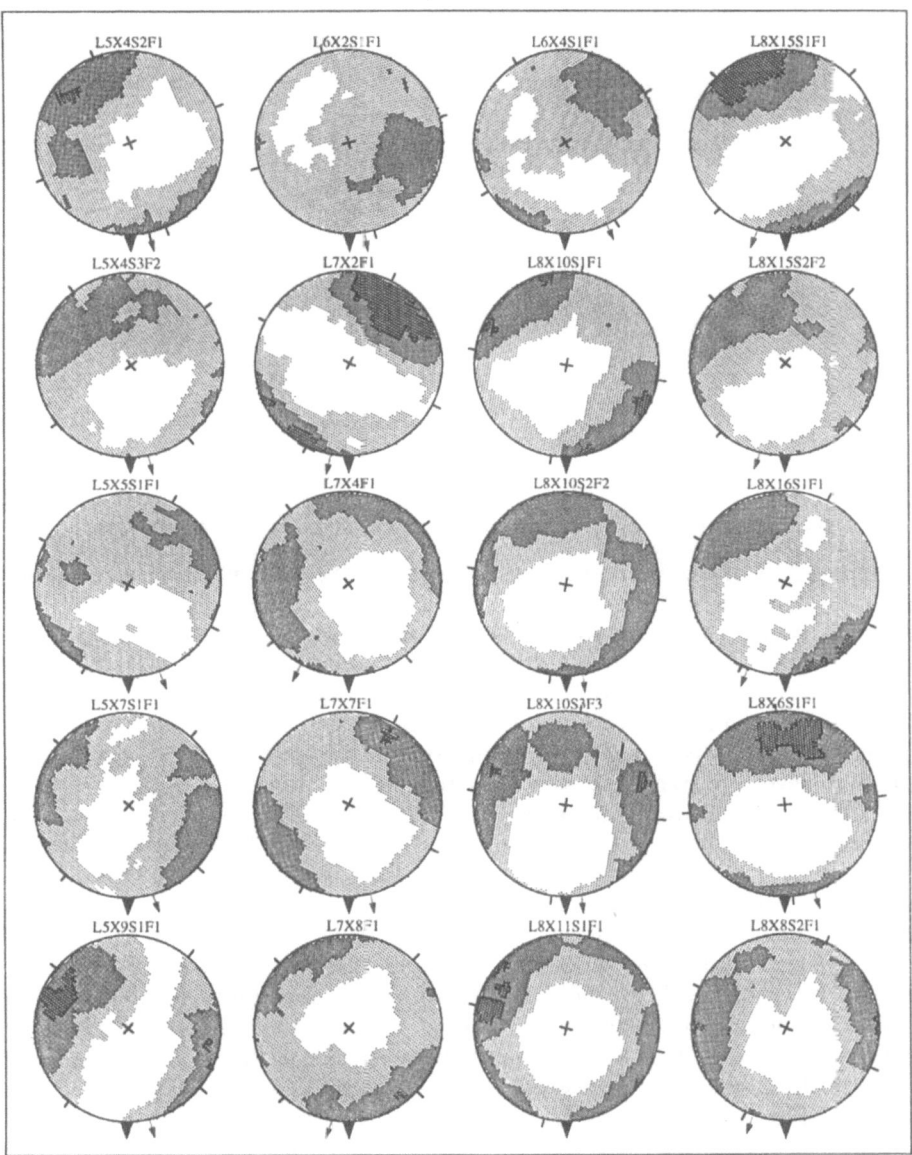

Figure 8b. Lower hemisphere, equal-area stereonets of clast long-axis orientations measured in water-flood deposits at reaches L5, L6, L7, and L8 of the 1985 GLOF. Refer to Figure 8a for descriptions of the diagrams.

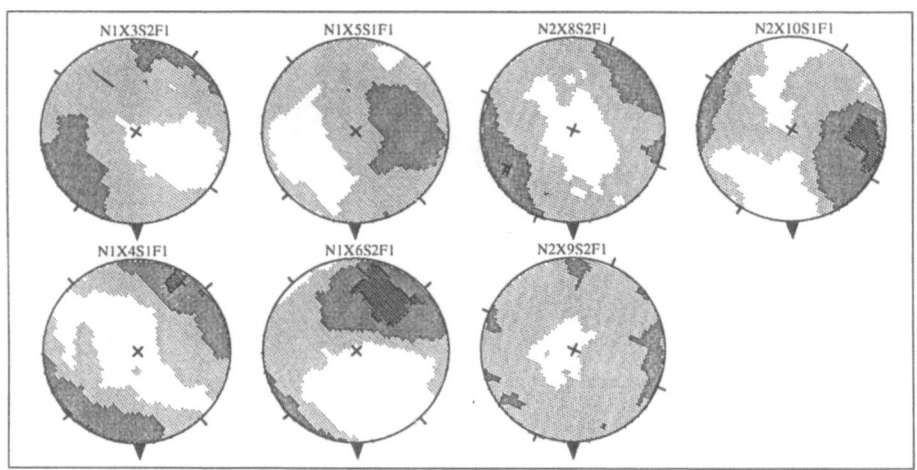

Figure 9. Lower hemisphere, equal-area stereonets of clast long-axis orientations measured in water-flood deposits at reaches N1 and N2 of the 1977 GLOF. Refer to Figure 8a for descriptions of the diagrams.

Examination of the V_1 orientations with respect to flow direction in the transitional-flow and water-flood deposits illustrates the complexity of depositional mechanics in both flow processes because of turbulent flow conditions (Figure 10a, Table 5). In the water-flood deposits, three have V_1s that are parallel to flow direction, nine have V_1s that are oblique to flow direction, six have V_1s that are transverse to flow direction, and nine have V_1s with random orientations. In the transitional-flow deposits, five have V_1s parallel to flow direction, two have V_1s oblique to flow direction, four have V_1s transverse to flow direction, and five have V_1s with random orientations. In other words, the majority of the statistically significant long-axis orientations in water-flood deposits are oblique and transverse to flow direction, whereas the majority of the statistically significant long-axis orientations in transitional-flow deposits are parallel and transverse with respect to flow direction (Table 10a). The transverse orientations indicate that the particles were rolling or sliding along the channel bottom prior to deposition and the parallel orientations indicate the particles were not in contact with the channel bottom prior to deposition (Johannson, 1963). The oblique orientations with respect to flow direction may indicate particles being propelled from the main concentration of flow into a zone of flow separation where the longitudinal bar was being formed. The notably small number of oblique orientations in the transitional-flow deposits may be evidence that secondary turbulence was being dampened because of increased sediment concentration (Table 10a). The large number of parallel orientations in the transitional-flow deposits is additional evidence that dispersive pressures and buoyancy were lifting and transporting cobbles and boulders above the bed just prior to deposition. The random long-axis clast orientations in the water-flood and transitional-flow deposits probably reflect the interaction of multiple vortices during deposition.

Steeply dipping clasts in deposits have been interpreted as forming in high velocity depositional environments (Jarrett and Costa, 1986; Carling, 1987). In the transitional-flow deposits, all of the statistically significant V_3 orientations are in the direction of flow, indicating the clast a:b planes are oriented upstream with an average dip of 30 degrees (Table 4). In the water-flood deposits, 17 of the 19 a:b planes have an average upstream dip of 27 degrees (Table 4). Figure 10b illustrates that most of the a:b plane dips for both type of deposits are between 20 and 40 degrees. The girdle distributions and steep dip of clasts in the transitonal-flow and water-flood deposits (Figures 8a, 8b, 9, and 10b) indicate that deposition was rapid and occurred at high velocities (Carling, 1987).

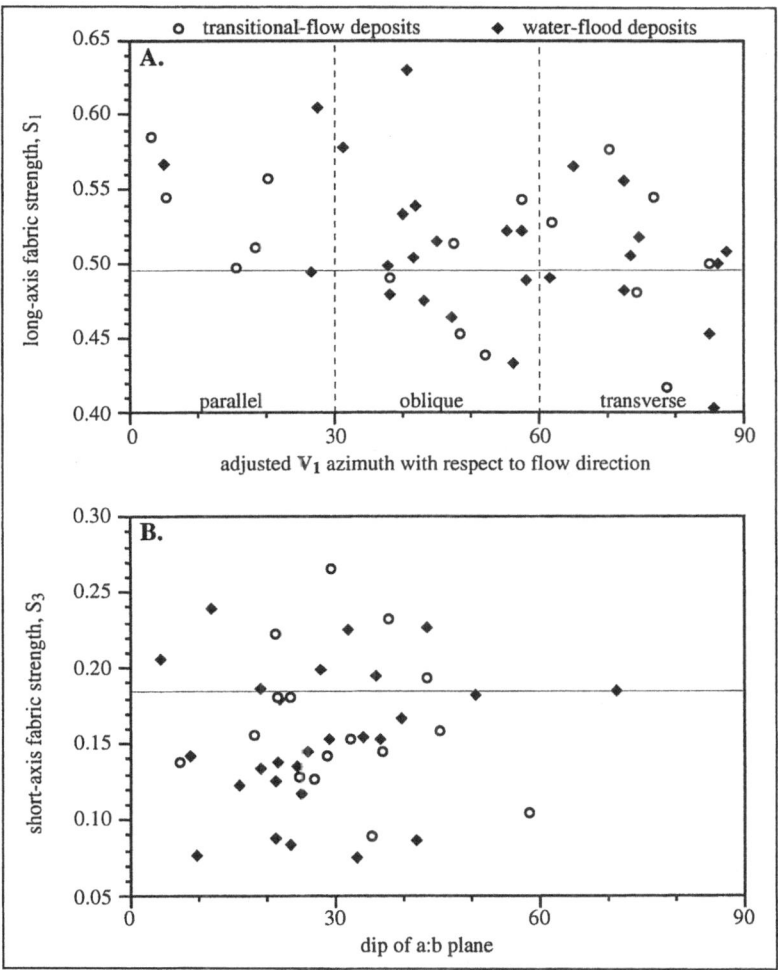

Figure 10. a) Plot of long-axis fabric strength, S_1, versus the adjusted long-axis orientation, V_1, with respect to flow direction of water-flood deposits (closed triangles) and transitional-flow deposits (open circles). Parallel indicates that V_1 is within 30° of flow direction, oblique indicates that V_1 is between 30° and 60° of flow direction, and transverse indicates that V_1 is between 60° and 90° of flow direction. Points plotting above S_1=0.496 have a statistically significant preferred V_1 orientation. b) Plot of short-axis fabric strength, S_3, versus dip of a b plane of water-flood deposits (closed triangles) and transitional-flow deposits (open circles). Points plotting below S_3=0.184 have a statistically significant preferred V_3 orientation.

4. Discussion

Changes in flow processes along a flood route can occur in a variety of geomorphic settings, but they have been primarily documented and described in mountainous environments (Krumbein, 1942; Stewart and Lamarche, 1967; Scott and Gravlee, 1968; Pierson and Scott, 1985; Smith, 1986; Wells and Harvey, 1987; O'Connor *et al.*, in press). Flow transformations from debris flow to hyperconcentrated flow to water flood occur when the flow is diluted by the addition of water and/or when sediment concentration is reduced because of deposition (Pierson and Scott, 1985; Smith, 1986). Transformations from water flow to hyperconcentrated flow to debris flow occur when large quantities of sediment are eroded by and incorporated into the flow (Pierson and Scott, 1985).

At elevations above 3400 m in the Mt. Everest region, channels and slopes consist primarily glacial and glacio-fluvial sediment. The large volumes of water released by the lakes quickly became sediment-laden floodwaters as the flood surge incorporated glacial and glacio-fluvial sediment into the flow. Along the first 16 km of the 1985 outburst-flood route (reaches L1, L2, L3, and L4), the sediment-laden floodwaters transformed into transitional flow or a flow approaching hyperconcentrated conditions because of further erosion and incorporation of glacio-fluvial sediment into the flow. The most intense erosion occurred in narrow, steep reaches. At these locations, valley side slopes of glacio-fluvial sediment were severely undercut and destabilized by the flow surge. Sixteen kilometers from the lake the transitional flow changed back into a water-flood after extensive deposition in a wider, less steep reach reduced the sediment concentration of the flow. Immediately downstream from this reach, the flow entered a narrow, unglaciated bedrock gorge with limited amounts of sediment available for erosion and incorporation into the floodwaters. The 1977 outburst flood was a sediment-laden water flood along its entire flood route and did not undergo flow transformations even though the geomorphic conditions in the upper reaches (N1 and N2) of the 1977 outburst flood are similar to those in the upper reaches of the 1985 outburst flood. It appears that the sediment laden water-flood did not or could not mobilize and incorporate enough sediment to alter its flow conditions.

As discussed by Costa and Jarrett (1981), Gallino and Pierson (1985), and Costa (1988), incorrectly identifying or not recognizing the flow process will result in erroneous and inaccurate flow reconstructions and discharge estimates. Costa and Jarrett (1981) provide several examples of discharges being substantially overestimated because debris flows were interpreted as being water floods. Although coarse sediment-laden floodwaters approaching hyperconcentration are still thought to be Newtonian fluids (Costa, 1988; Carling, 1989), their hydraulics are poorly understood because of the reduction in turbulence and increased importance of dispersive stresses and buoyancy. Thus, application of hydraulic equations derived for clear water flows is questionable (Costa, 1988). Several studies of sediment-laden flows, often interpreted to be flows intermediate between debris flow and water flood, have described deposits generated by these flows to have surface elevations that coincide with or are slightly higher than the projected maximum water surface elevation (Stewart and LaMarche, 1967; Scott and Gravlee, 1968; Costa, 1984; Jarrett and Costa, 1986; Carling, 1987, 1989). In our study, preliminary results of step-backwater modeling generated water-surface profiles that closely matched the upper surface elevations of several deposits in the modeled reach. These results suggest that the upper surface of deposits generated by coarse, sediment-laden water floods or flows aproaching hyperconcentration may define the maximum water-surface elevation of the flow. However, this relationship is tenuous because of our limited understanding of the flow hydraulics and depositional mechanics associated with coarse, sediment-laden flows.

5. Summary

The results of this study indicate that differences in morphologic, sedimentologic, and fabric characteristics of deposits along outburst-flood routes can be used to recognize and identify changes in flow processes and to provide insights into transport and depositional mechanics. Because of the large amounts of sediment available in channels at higher elevations, the large volumes of water released by the lakes became sediment-laden as the floodwaters eroded and incorporated glacial and glacio-fluvial sediment from valley side slopes. Deposition occurred across the entire valley bottom along wider and less steep valley segments where flow energy was reduced and flow was diverging. Depositional reaches with valley widths greater than 100 m are characterized by multiple longitudinal and transverse bars, and chute channels. In reaches with valley widths less than 70 m, deposition consisted of single, longitudinal bars or thin sheet deposits.

Deposits produced by the 1977 glacial-lake outburst flood are interpreted to have formed from water-flood processes. The 1985 outburst flood deposits are interpreted to have formed by both transitional-flow and water-flood processes. Transitional-flow deposits are predominant along the first 16 km of the 1985 outburst-flood route. Transitional-flow deposits are typically reverse graded and are more poorly sorted, coarsely-skewed, and have a higher percentage of fine-grained sediment than water-flood deposits. Fabric analysis indicates that the orientation of clast long axes in water-flood deposits is typically oblique and transverse to flow direction, whereas the majority of the clast long-axis orientations in transitional-flow deposits are parallel and transverse with respect to flow direction. The girdle distributions and steep a:b plane dips of clasts in both the transitional-flow and water-flood deposits are evidence that deposition was rapid and occurred at high velocities.

Sorting, skewness, and percentage of fine-grained sediment in the transitional-flow deposits are statistically different from those in water-flood deposits. Deposit sorting is the best discriminator between transitional-flow and water-flood deposits. The overlap in values for sorting, skewness, and percentage of fine-grained sediment between the two types of deposits reflects the similar nature of water-flood and transitional-flow processes. Differences in sedimentologic and fabric characteristics of deposits at a given depositional reach (regardless of the flow process) reflect the complexity and variation of flow hydraulics and depositional mechanics that occur along a reach.

Acknowledgements

Funding for this study was provided by National Science Foundation Grant No. CMS-9320876.

References

1. Ageta, Y. (1976) Characteristics of precipitation during monsoon season in Khumbu Himal, *J. of the Japanese Soc. of Snow and Ice*, 38, 84-88.

2. Anderson, T.W., and Stephens, M.A. (1972) Tests for randomness of directions against equatorial and bimodal alternatives, *Biometrika*, 59, 613-622.

3. Blown, I., and Church, M. (1985) Catastrophic lake drainage within the Homathko River basin, British Columbia, *Canadian Geotechnical J.*, 22, 551-563.

4. Buchroithner, M.F., Jentsch, G., and Wanivenhaus, B. (1982) Monitoring of recent geological events in the Khumbu area (Himalaya, Nepal) by digital processing of Landsat MSS data, *Rock Mechanics*, 15, 181-197.

5. Carling, P.A. (1987) Hydrodynamic interpretation of a boulder berm and associated debris-torrent deposits, *Geomorphology*, 1, 53-67.

6. Carling, P.A. (1989) Hydrodynamic models of boulder berm deposition, *Geomorphology*, 2, 319-340.

7. Clarke, G.K.C. (1982) Glacier outburst floods from "Hazard Lake", Yukon Territory, and the problem of flood magnitude prediction, *J. of Glaciology*, 28, 3-21.

8. Costa, J.E. (1984) Physical geomorphology of debris flows, in Costa, J.E., and Fleisher, J.P., eds., *Developments and applications of geomorphology*, New York, Springer-Verlag, 268-317.

9. Costa, J.E. (1988) Rheologic, geomorphic, and sedimentologic differentiation of water floods, hyperconcentrated flows, and debris flows, in Baker, V.R., Kochel, R.C., and Patton, P.C., eds., *Flood Geomorphology*, New York, John Wiley and Sons, 113-122.

10. Costa, J.E., and Jarrett, R.D. (1981) Debris flows in small mountain stream channels of Colorado and their hydrologic implications, *Bulletin of the Engineering Geologists*, 18, 309-322.

11. Desloges, J.R., and Church, M. (1992) Geomorphic implications of glacier outburst flooding: Noeick River valley, British Columbia, *Canadian J. of Earth Sciences*, 29, 551-564.

12. Evans, S.G., and Clague, J.J. (1994) Recent climatic change and catastrophic geomorphic processes in mountain environments, *Geomorphology*, 10, 107-128.

13. Folk, R.L., and Ward, W.C. (1957) Brazos River bar: A study in the significance of grain size parameters, *J. of Sedimentary Petrology*, 27, 3-26.

14. Fushimi, H. (1977) Glaciations in the Khumbu Himal (1), *J. of the Japanese Soc. of Snow and Ice*, 39, 60-67.

15. Fushimi, H. (1978) Glaciations in the Khumbu Himal (2), *J. of the Japanese Soc. of Snow and Ice*, 40, 71-77.

16. Fushimi, H., Ikegami, K., and Higuchi, K. (1985) Nepal case study: Catastrophic floods, in Young, G.J., ed., *Techniques for prediction of runoff from glacierized areas*, IAHS pub. 149, 125-130.

17. Gallino, G.L., and Pierson, T.C. (1985) Polallie Creek debris flow and susequent dam-break flood of 1980, East Fork Hood River Basin, Oregon, *USGS Water-Supply Paper 2273*, 22.

18. Grove, J.M. (1990) *The Little Ice Age*, Reprinted by Routledge, New York and London, 498.

19. Haeberli, W. (1983) Frequency and characteristics of glacier floods in the Swiss Alps, *Annals of Glaciology*, 4, 85-90.

20. Haeberli, W., Alean, J.C., Müller, P., and Funk, M. (1989) Assessing risks from glacier hazards in high mountain regions: Some experiences in the Swiss Alps, *Annals of Glaciology*, 13, 96-102.

21. Hewitt, K. (1982) Natural dams and outburst floods of the Karakoram Himalaya, in Glen, J.W., ed., *Hydrological aspects of alpine and high-mountain areas*, IAHS pub. 138, 259-269.

22. Ives, J.D. (1986) Glacial lake outburst floods and risk engineering in the Himalaya: A review of the Langmoche Disaster, Khumbu Himal, 4 August 1985, *ICIMOD Occasional Paper 5*, 42.

23. Iwata, S. (1976) Late Pleistocene and Holocene moraines in the Sagamartha (Everest) region, Khumbu Himal, *J. of the Japanese Soc. of Snow and Ice*, 38, 109-114.

24. Jarrett, R.D., and Costa J.E. (1986) Hydrology, geomorphology, and dam-break modeling of the July 15, 1982 Lawn Lake Dam and Cascade Lake Dam Failures, Larimer County, Colorado, *USGS Professional Paper 1369*, 78.

25. Johansson, C.E., (1963) Orientation of pebbles in running water. A laboratory study, *Geografiska Annaler*, 45, 85-112.

26. Krumbein, W.C. (1942) Flood deposits of Arroyo Seco, Los Angeles County, California, *GSA Bulletin*, 53, 1355-1402.

27. Lliboutry, L., Arnao, B.M., Pautre, A., and Schneider, B. (1977) Glaciological problems set by the control of dangerous lakes in the Cordillera Blanca, Peru. I. Historical failures of morainic dams, their causes and prevention, *J. of Glaciology*, 18, 239-254.

28. Mark, D.M. (1973) Analysis of axial orientation data, including till fabrics, *GSA Bulletin*, 84, 1369-1374.

29. Martini, I.P. (1977) Gravelly flood deposits of Irvine Creek, Ontario, Canada, *Sedimentology*, 24, 603-622.

30. Mayewski, P.A., and Jeschke, P.A. (1979) Himalayan and trans-Himalayan glacier fluctuations since AD 1812, *Arctic and Alpine Research*, 11, 267-287.

31. Mills, H.H. (1984) Clast orientation in Mount St. Helens debris-flow deposits, North Fork Toutle River, Washington, *J. of Sedimentary Petrology*, 54, 625-634.

32. O'Connor, J.E., and Costa, J.E. (1993) Geologic and hydrologic hazards in glacierized basins in North America resulting from 19th and 20th century global warming, *Natural Hazards*, 8, 121-140.

33. O'Connor, J.E., Hardison, J.H., and Costa, J.E. (in press) Debris flows from moraine-dammed lakes in the Three Sisters and Mt. Jefferson Wilderness areas, Oregon, *USGS Water-Supply Paper*.

34. Pierson, T.C., and Scott, K.M. (1985) Downstream dilution of a lahar: Transition from debris flow to hyperconcentrated streamflow, *Water Resources Research*, 21, 1511-1524.

35. Scott, K.M., and Gravlee, G.C. (1968) Flood surge on the Rubicon River, California-Hydrology, hydraulics and boulder transport, *USGS Professional Paper 422M*, 1-38.

36. Smith, G.A. (1986) Coarse-grained nonmarine volcaniclastic sediment: Terminology and depositional processes, *GSA Bulletin*, 97, 1-10.

37. Stewart, J.H., and LaMarche, V.C. (1967) Erosion and deposition produced by the flood of December 1964 on Coffee Creek Trinity County, California, *USGS Professional Paper 422K*, 1-22.

38. Vuichard, D. (1986) Geological and petrographical investigations for the mountain hazards project, Khumbu Himal, Nepal, *Mountain Research and Development*, 6, 41-52.

39. Vuichard, D., and Zimmermann, M. (1986) The Langmoche flash-flood, Khumbu Himal, *Mountain Research and Development*, 6, 90-94.

40. Vuichard, D., and Zimmermann, M. (1987) The 1985 catastrophic drainage of a moraine-dammed lake, Khumbu Himal, Nepal: Cause and consequences, *Mountain Research and Development*, 7, 91-110.

41. Wells, S.G., and Harvey, A.M. (1987) Sedimentologic and geomorphic variations in storm-generated alluvial fans, Howgill Fells, northwest England, *GSA Bulletin*, 98, 182-198.

42. Wolman, M.G. (1954) A method of sampling coarse river bed material, *Transaction, American Geophysical Union*, 35, 951-956.

43. Woodcock, N.H. (1977) Specification of fabric shapes using an eigenvalue method, *GSA Bulletin*, 88, 1231-1236.

44. Zimmermann, M., Bichsel, M., Kienholz, H. (1986) Mountain hazards mapping in the Khumbu Himal, Nepal, with Prototype Map, scale 1:50,000, *Mountain Research and Development*, 6, 29-40.

45. Zingg, T. (1935) Beitrag zur Schotteranalyse, *Schweizerische, Mineralogische, und Petrographische Mettelungen*, 15, 38-140.

Authors

Daniel A. Cenderelli, Ellen E. Wohl
Colorado State University
Department of Earth Resources
Ft. Collins, Colorado 80525
U.S.A.

CATASTROPHIC FLOOD FLUSHING OF SEDIMENT, WESTERN HIMALAYA, PAKISTAN

J F. SHRODER, JR., M. P. BISHOP, R. SCHEPPY

1. Introduction

Catastrophic flooding in the Himalaya is an all too common phenomenon (Hewitt 1968 ; Goudie et al., 1984) that produces considerable hazard to residents. In recent years we have recognized that such massive movement of water is also a significant factor in eroding sediment from areas of active tectonism. In the Nanga Parbat project, in which a large international group of scientists is focusing on the relations between rapid uplift and deep denudation, we are attempting to map and measure erosion produced by slope failures, glaciers, rivers, and catastrophic flooding throughout Quaternary time. This paper includes information on these processes from both the Nanga Parbat region as well as from Hunza where collateral data have been collected (Figure 1). In this paper we differentiate floods from within glaciers, floods from behind glacier dams, floods from behind slope failure dams, and debris-flow floods induced by torrential rains.

Mountain landscapes are high energy environments where geomorphic processes tend to operate at higher magnitude and moderate to higher frequency rates than most other places. The globally maximal relief of the Himalaya and the Karakoram is associated with some of the highest recorded rates of regional denudation for large drainage basins. Regional denudation values exceed $1 - 3$ mm yr^{-1} and local values can be much higher (Hewitt, 1972; Ferguson, 1984). These high denudation rates imply hazard in that large magnitude floods carrying massive debris loads are a source of great danger to life and property. From another perspective, sediment transfer efficiency from slope failures to glaciers, and from glaciers to rivers, or from either slope failures or glaciers into catastrophic floods, is also a partial measure of natural hazard (Shroder, et al., 1996). Where high efficiency prevails, because of limited debris supply, hazards will not be common, but where high-magnitude and low-to-high frequency processes remove sediment quickly, hazard potentials will rise. For this reason we have established some transfer efficiencies for parts of the geomorphic-process cascade. Other things being equal, more efficient sediment transfer from slope failures to river- and ice-transporting mechanisms, coupled with higher hazard, occurs in valleys that have: (1) high maximum relief; (2) steep slopes; (3) large rivers or glaciers; (4) foliation or structural planes that closely parallel and daylight in canyon walls; (5) high precipitation, and (6) high seismicity. More efficient sediment transfer from glaciers to rivers, coupled with higher hazard, also occurs in those valleys where: (1) a steep glacier terminus zone increases transfer energetics; (2) the subglacial tunnel migrates laterally and removes moraine; (3) glacier outburst floods occur; (4) high glacier meltwater discharge occurs; (5) a narrow valley occurs in the

J. Kalvoda and C.L. Rosenfeld (eds.), Geomorphological Hazards in High Mountain Areas, 27-48.
© 1998 *Kluwer Academic Publishers.*

glacier terminus zone which restricts the moraine and facilitates its removal through concentrating erosion in a limited area.

Annual river discharges in mountain environments are commonly compressed in time and space by torrential rains, steep channels, natural dam bursts and other factors that concentrate flow. Hewitt (1972) noted that for the Himalayan Indus, where 91% of the sediment yield is carried in just over two summer months, flows in excess of the partial duration series base (i.e. the annual maximum flow with a recurrance interval of 1.5 yr) transport about 25% of the total load. The Indus river stems from an enormous drainage basin, which is known to have had legions of catastrophic outburst floods from slope failure and glacier dams across one or another of its many tributaries for centuries (Mason, 1929; Hewitt, 1968; Goudie et al., 1984). Unknown, however, has been the magnitude, frequency, and effect of extreme events in very many small basins or in restricted areas of these mountain environments. Furthermore, comparison of such limited mountain regions to small basins in lowland areas has been hampered by a lack of data. Hewitt (1972) felt that averaging out of extreme events for the whole of the Indus basin would not produce results typical for the smaller tributaries. This paper concentrates on damming events and catastrophic floods in several restricted basins in the Nanga Parbat area, and in a small part of the greater Hunza region. Repetative, moderate to higher frequency events are noted as characteristic of several of these areas.

The many large glaciers that descend from the heights of Nanga Parbat and its peripheral ranges produce considerable meltwater that is variously ponded in supra-, en-, and sub-glacial positions, as well as behind ice dams. These waters are held by the ice different lengths of time before breaking out in catastrophic floods that modify the landscape and transport considerable volumes of sediment. Erosional destruction of trails, roads, buildings, canals, and fields is a commonplace hazard. Deposition of gravel deposits and large and small boulder lobes in channels, and over the surfaces of debris fans and fields, occurs in the lower discharge flood events. The highest discharges carry large volumes of coarse bouldery sediment down tributaries and into the Indus where much of the load moves out of the Himalayan mountain system and into the lowlands. Tarbella dam and reservoir in the foothills (Figure 1) now serve as the main trap for all sediment derived from the highlands, with an average sediment deposition of about 400×10^6 tons yr^{-1} (Hurst and Chao, 1975).

The Rupal valley on the south side of Nanga Parbat has a number of large glaciers (Figure 2), averaging ~10 km in length, where ice accumulates on high cliffs and cirques and descends through ice falls or massive ice avalanches to the ablation areas below. The relief and high energy of the descent contribute considerable rock rubble that is the source of the extensive debris covers at lower elevations. Similarly, the 14 km-long Raikot glacier on the north side of the mountain descends steeply with fairly high velocities (Finsterwalder, 1937; Pillewizer, 1956) and has an extensive bouldery mantle (Gardner and Jones, 1993). Mattson and Gardner (1989) and Mattson et al. (1993) have noted that Himalayan glacier sites are subject to greater amounts of incident solar radiation than comparable glaciers elsewhere at higher latitudes and lower latitudes. The result is concentration of thick supraglacial debris covers that are easily remobilized in catastrophic floods.

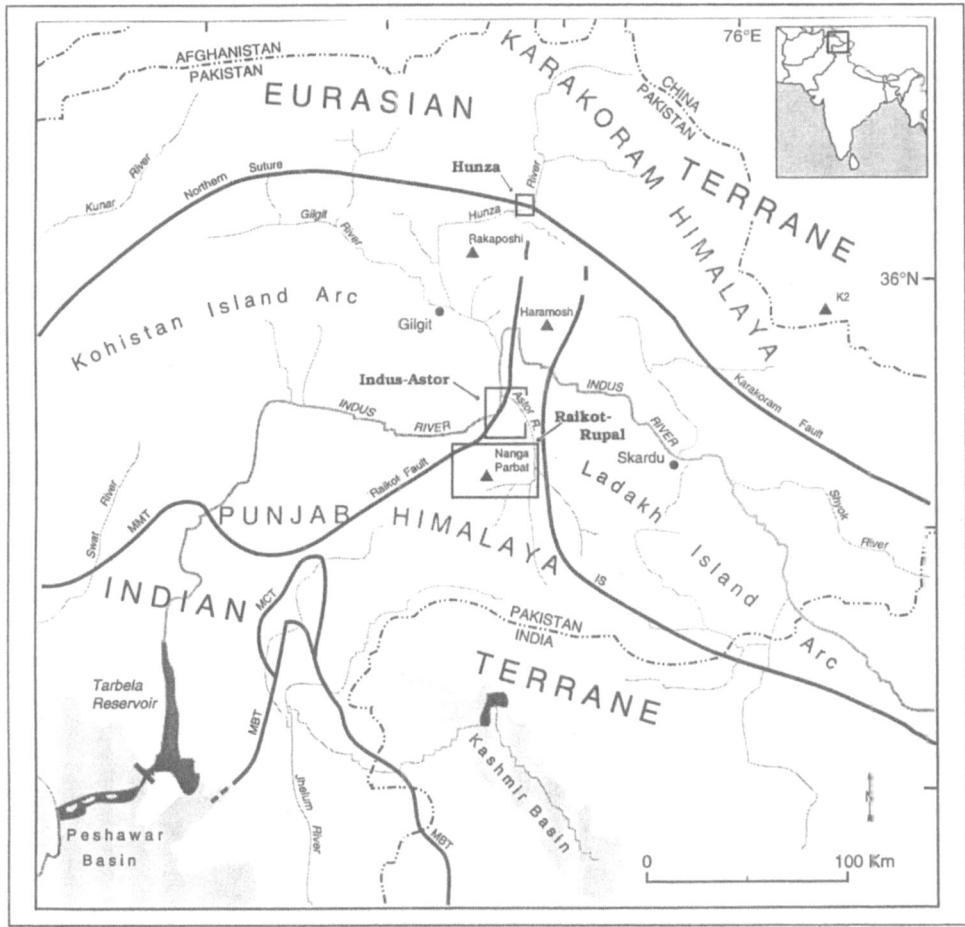

Figure 1. Index map of glacier and slope failure areas studied for this report. The Raikot-Rupal area is shown in Figure 2, the Indus-Astor area in Figure 8 , and the Hunza area in Figure 10.

Slope failures, another major hazard in the Himalaya (Shroder, 1989a & b) are not only hazardous in the initial event, but also later if they have chanced to impound a body of water behind them. Costa and Schuster (1988) classified all kinds of natural dams with those from landslides and glaciers being the greatest threat to people and property. Their classification of landslide dams was based upon the relation of such dams to the valley floor. Type I dams do not reach from one valley side to the other, whereas type II do extend all the way across. Type III landslide dams move both upstream and downstream and type IV involve contemporaneous failures from both sides of the valley. Type V forms multiple dams, whereas type VI extends under the valley bottom to emerge on the opposite valley side. To this classification we add the slight modification that type I dams may be (Ia) full, or (Ib) partial, in which flow is restricted but not stopped. Of the partial types, the dam may also be a single, one-time event,

Figure 2. Index map of the Raikot-Rupal and upper Astor parts of the Nanga Parbat area.

or it may be recurrent. In this latter case, continued undercutting by the river will remove shear resistance such that the landslide is repeatedly reactivated. In all of the type Ib partial dams, both upstream alluviation and downstream river gradients are increased with attendant hazard, while engineering or farming structures in the failure zone and along shore will be affected variously by ongoing rupture, undercutting, and alluviation. In our study area, all of the major river systems, the Hunza, Rupal, Astor, Raikot, and Indus have had abrupt landslide dams of types Ia , II, III, and IV, as well as type Ib partial and recurrent in the Astor river valley.

1.1. METHODOLOGY

In the course of ten excursions to the western Himalaya since 1984, we have collected field information on slope failures, glacier advances, glacier surge outbreaks and dams, catastrophic floods from a variety of causes, and other natural geomorphic hazards. We augmented our research with SPOT satellite imagery and high quality topographic maps at 1:50,000 scale. Our field mapping program was accomplished with a GPS locational system, laser theodolite, Brunton compass, Abney level, and tape measure. Photography from high altitude locations down to lower altitude landforms enabled further accurate plotting of phenomena. Soil pits, dendrochronology, ^{14}C and cosmogenic radionuclide dating provided relative and quantitative chronological control. River discharges were estimated from channel width, depth, and velocity measures. Former lake levels and volumes were measured from abandoned beach lines, and catastrophic floods were assessed from flood scour berms, abandoned channels, and large boulder lobes.

2. Glacier-induced floods from within glaciers

Floods from within glaciers are common in the western Himalaya and may occur as a result of abrupt drainage of surfacial meltwater lakes, or from blocked englacial and subglacial drainage ways. Sometimes if the water body is large enough, extensive damage may result, as when an extensive stretch of the the Karakoram Highway and the large, Chinese-built Friendship Bridge in Hunza was completely buried under massive quantities of water-borne debris in 1974 (Cai et al., 1980; Wang et al., 1984). Similarly, the bridge over Batura river in Hunza was washed away in 1973 when the channel changed abruptly in a summer of abnormal melt (Li et al., 1980; Zhang et al., 1980). In our study we focus attention upon a number of glaciers responsible for significant floods on Nanga Parbat; the 1994 flood from Raikot glacier on the north slope of the mountain, and the 1995 flood from Shaigiri glacier on the south side. A smaller flood from Sachen glacier in the 1920's is discussed as well.

2.1. SHAIGIRI GLACIER FLOOD - 1995

Shaigiri Glacier in the upper Rupal valley descends from the Mazano Peaks ridge eminating from the west of the Nanga Parbat summit and compresses the Rupal river against the mountain wall opposite (Figure 2). The glacier terminus is piled up in a huge rampart ~200 m high of moraine, and is extensively debris covered, although

Figure 3. View down on the terminus of Shaigiri glacier showing two light-colored zones of fine clastics deposited in lakes that broke out over the front, the most recent one of which occurred in summer 1995 on the left (west) side of the terminus.

Figure 4. View of lower Tarshing glacier from the distal part of the water plane impounded by the ice in the 1850's. The entire valley in view was filled with water from the foreground to a level just above the dark trees on the right lateral moraine in the photo center. Light-colored sand in the forground represents sand deposited at the delta head of the impoundment.

Figure 5. View down onto the Bazhin glacier on the left center. The Rupal river flows along the right (west) side. Linear beach lines marked by vegetation changes from water impounded behind Bazhin glacier occur on the alpine debris fan on the left.

Figure 6. View west up the Rupal valley from a position on the south face of Nanga Parbat near the head of the Bazhin glacier. Rupal glacier occurs at the extreme head of the valley. Shaigiri glacier is on the upper right. Tap glacier, with its impounded lake in the terminus, occurs at the lower left. The alluvial plain directly up-valley from Tap glacier reflects prior impoundment of a water body behind the Tap moraine of probable Neoglacial age.

ice cliffs protrude through the cover in a few places. The dominant meltwater escape since the 1930s is along the left flank on the north side of the ice and a prominent fan of debris has been spread from there to the Rupal river. Changes in the configuration of the subglacial channel geometry have occurred from time to time in the past 60 years since the area was first mapped. Two different supraglacial water impoundments at the terminus have deposited thick fluvial and lacustrine sediment above the terminus and caused subsequent breakout floods over the frontal moraine; with the most recent being in the summer of 1995 (Figure 3). The two floods cut longitudinal channels that are ~100 m long, several tens of m wide and ~3 m deep, which equates to an estimated minimum of 6000 m^3 of sediment that has been transfered into the river at this site. A rough minimum figure would thus be about 100 m^3 yr^{-1}, which neglects the contribution made by the main meltwater stream from Shaigiri, but which serves to adequately characterize some of the denudation contribution made by the glacier. The hazard attendant to these events has been the destruction of the main shepherd migration path and small bridges at the terminus. An alternate route along the south side of the Rupal river has been constructed with difficult and dangerous high bridges built up arduously to avoid the hazard from Shaigiri glacier.

In addition to the recent events at the Shaigiri terminus, Neoglacial (NG) moraine remnants of a few thousand years age occur on the south side of the river as well. Fine-grained sediments associated with them and with later ice-blocking events of probable Little Ice Age (LIA) time during the last few centuries show clear evidence that ice dams of the main Rupal river have occurred there. The river has removed all other terminal moraines, which shows that the Shaigiri glacier is highly efficient over the long term in transferring sediment to the river, but is of only average hazard because only a few trails and rude bridges are affected in this sparsely inhabited, high mountain area.

2.2. RAIKOT GLACIER FLOOD - 1994

In late 1993 or early 1994, the subglacial drainage system of the Raikot glacier on the north side of Nanga Parbat became blocked about 1.5 km up-ice from the terminus and stopped the Raikot river (Figure 2). After a number of spring months of restricted water discharge, a catastrophic breakout flood burst from four new portals; three on the west side and one on top of the ice above the terminus. After several months when all the impounded water was drained from its reservoirs and additional melt adjustment occurred, the Raikot river returned to its pre-flood drainage portal from the west side of the terminus. This flood event mobilized considerable debris from along the west side of the glacier, the lateral ablation valley, the nearby NG and LIA lateral moraines, and a small portion of the most distal of the Biale debris-flow sediment discussed below. All this sediment was transported into and flushed from the mountain in the main Raikot river that reestablished itself close to its original position. Raikot glacier has almost no terminal moraine left in its narrow valley, which suggests that other breakout flood events in the past have efficiently removed sediment from the terminal area.

A number of trails and a small bridge were washed away in this flood but in general the hazard was limited in spite of the apparently impressive breakout. An old man reported that the water burst sounded like cannon fire and blew ice blocks over the trees. We thought that statement to be rather exaggerated, but there was no question the event startled the villagers.

2.3. SACHEN GLACIER FLOOD - 1920'S

Sachen glacier on the east side of Nanga Parbat (Figure 2) has almost no firn accumulation zone in direct contact with the terminus, unlike the huge Raikot system, and instead is fed almost entirely by ice avalanches from high up the mountain. The lower efficiency of Sachen glacier is exemplified by the huge lateral and recessional NG and LIA moraine storages and the last glacial maximum (LGM) moraine at its terminus. Dendrogeomorphological event-response chronology (Shroder, 1978, 1980) shows that a minor ice advance and catastrophic breakout flood are known to have occurred on the north frontal lobe in the late 1920's but the volume of sediment moved was small and only the trees were disturbed (Spencer, 1985). No reports from the British agents at Astor were made that we know of so the assumption is that the hazard was limited to local effects only. The data suggest that significant denudation from the Sachen area could only be accomplished during times of major glaciation when greater volumes of meltwater and other river daming events could occur. Sachen glacier is thus an inefficient ice mass of low hazard.

3. Glacier-induced floods from behind ice dams

Floods caused by glacier ice dams across valleys are a common phenomenon in the Himalaya and numerous examples have been cited over the years (Mason, 1929, 1930, 1935; Hewitt, 1982; Goudie et al., 1984). Several valleys are notorious for such events, with the Shimshal valley in Hunza being among the most problematic. Around Nanga Parbat we now know of a number of glaciers with histories of damming and flood events.

3.1. TARSHING GLACIER FLOODS OF 1850-51 AND 1856

Tarshing (also Toshain or Tasching) glacier is formed by the confluence of Chungpar and Chongra glaciers that descend into the Rupal Valley from Raikot and Chongra Peaks on the southeast ridge of Nanga Parbat (Figure 2). In 1850-51, toward the end of the LIA, the glacier was against Gonnott rock and close to or at the Chichi (Zize) river, which was tributary to the Rupal river about 4 km further downstream (Drew, 1875; Kick, 1989). The glacier dammed the Rupal river and produced a lake which filled up to the top of the glacier, flowed over it and burst out in the summer of that year. The disastrous flood lasted for three days and washed away irrigation canals, houses and agricultural land at Tarshing, Gorikot, and Dashkin, although no lives seem to have been lost. In addition the lower course of the Chichi river was thereby diverted directly into the Rupal river close to the Tarshing terminus, away from its previous tributary junction downstream. This cut it off from the Zaipur alluvial plateau on which the nineteenth-century village of Choi was situated and reduced the prosperity of that place (Drew, 1875), although by the late twentieth century irrigation had been restored.

In 1856 Adolf Schlagintweit made sketch maps and painting of the Rupal area that also showed a new 'Rupal See' lake dammed up behind the Tarshing glacier (Kick, 1989). This is yet another lake that is presumed to have burst out in that year as well, although Drew (1975) only reported the one flood of 1850-51. Kick (1989) also thought that other lake impoundments had occurred before 1850. In any case it is clear that at least one major flood, and possibly others, occurred at this site. We have identified the most distal, upstream sandy end of the delta in one of the impounded lakes and calculated the length of the lake at ~2.2 km and a maximum depth of ~65m (Figure 4). Drew's (1875) figures were slightly larger at 2.4 km in length and 91 m in maximum depth. Our slightly more conservative figure would give a total water volume of about 16×10^6 m^3 that was involved in the three-day flood. Thus it can be seen that Tarshing glacier was highly efficient in transfering its debris cover into the Rupal river where it was carried off in at least one major breakout flood. Furthermore, in non-flood years all glacier sediment is also transferred directly into the river. The result is a highly efficient glacier-to-river transport system that accelerates denudation in the area.

In the winter of 1995-96 Tarshing glacier advanced across the Rupal river to the Gonnot rock wall on the south side once again. The river was blocked for a time and caused minor alluviation for about 100 m upstream before breaking through once more in the spring season. No damage occurred at this time.

3.2. BAZHIN GLACIER FLOODS OF ANTIQUITY

Bazhin Glacier descends from the summits of Nanga Parbat and Raikot Peak and crosses the entire Rupal valley from north to south, where it blocks the Rupal river. Sometime after the retreat from the LGM in late Pleistocene time the Bazhin glacier established itself athwart the Rupal river where it remained through the subsequent NG and LIA ice increases. In times of decreased ice movement the Rupal river has maintained a subglacial drainage way beneath the ice, as it does today. During greater ice buildup and movement, the subglacial tunnel has collapsed and blocked the Rupal river. Multiple lakes up to 3 km long formed behind the Bazhin ice dam and produced characteristic beach lines in topography, sediment, and vegetation (Figure 5). The presence of the Rupal river cutting through the lateral and terminal moraines of Bazhin glacier has enabled fairly efficient transfer of sediment from the ice to the river waters. Several prominent NG and LIA lateral moraines near the terminus have not been removed, however.

No historical records exist of past breakout floods at Bazhin but plentiful flood-scoured berms, terraces, and flood boulder lobes are spread out in the deep valley channel from Bazhin glacier for 5 km downstream to Tarshing glacier. Inasmuch as the flood channel is ~300 m wide and 50-100 m below the main valley floor, it was most probably eroded out by catastrophic flood waters and is sufficiently voluminous to accomodate any future flood waters without major hazard, except to a few low-lying houses and bridges.

3.3. TAP GLACIER BLOCK AGES

Tap glacier descends directly south from the summit of Nanga Parbat in a steep icefall to the Rupal valley below. During NG time when the glacier had expanded considerably, a series of moraines partly blocked the Rupal river in front and down valley from Tap glacier and formed a lake about 0.75-1 km long (Figure 6). Subsequent flood waters from upstream have greatly modified the blocking moraines and eroded a series of planar channels through the mass (Figure 7). At the present time the Tap glacier has melted back ~ 300 m from its LIA terminal moraine and has formed a lake which drains over the front. Since the lake was first mapped in 1934 it has increased in size by about 25%, but not sufficiently as yet to erode through the front and drain catastrophically. Still the potential for such hazard exists wherever large bodies of water are impounded behind unstable moraines (Watanabe, et al., 1994; 1995; Watanabe and Rothacher, 1996). Should Tap lake break out catastrophically in future, its floodwaters would be channeled directly into the Rupal river and thence to the Astor river where some roads, bridges, hydroelectric facilities, and a few buildings would be adversely impacted. In general, however, the hazard would not be too great, because of the remoteness of the upper Rupal valley where Tap glacier occurs.

Figure 7. View of Neoglacial Tap moraine washed over by floods from the impounded Rupal river upstream. Prior impoundment water levels by Bazhin glacier to the left (east) occur as zones of dark grass on upper left and lower center left.

4. Slope-failure-induced floods

Catastrophic floods from lakes impounded behind dams produced by massive failure of slopes in the Himalaya are well known (Hewitt, 1982; Shroder, 1993); less well understood is the fact that the landslide-generated flood of 1841 at the base of Nanga Parbat is the largest such dam and flood known in the world (Code and Sirhindi, 1986; Shroder, 1993; in press). Mass movement generation of the largest catastrophic floods at Nanga Parbat and Hunza are probably not high frequency events but do involve multiple occurrences in the same general area through time. Large, deep-seated failures in the nearby Hunza valley as a whole occur with a recurrence interval of about 1 in 50 years (Goudie et al., 1984) .

The western Himalaya contain some of the highest and steepest slopes on Earth. Goudie et al. (1984) detailed the pervasive effects of mass movement there because failure events occur more frequently than the landscape can recover, with the result that slopes are in a state of transience typical of frequent disturbance and continuous adjustment. The result is a vast quantity of debris delivered to the river systems for transport away, which produces rapid denudation. Davis (1984) assessed all of the natural hazards in the Hunza area and noted the nearly continuous to decadal occurrences of damaging mass movement events. He was impressed with the ability of the local 'survival artists' to cope with the hazards that continually threaten.

4.1. NANGA PARBAT SLOPE FAILURE DAMS AND FLOODS

A number of slope failures in the Indus and Astor river gorges at the base of Nanga Parbat (8,125 m) have dammed or restricted river flow and caused a variety of severe hazard, as well as moving copious quantities of sediment downstream (Figure 8). The area is famous for the seismic event in the winter of 1840-41 that triggered a massive fall of rock and debris from a spur of Nanga Parbat into the Indus. Water was impounded in a lake over 150 m deep that backed up some 30 km upstream until the dam broke in June, 1841, and wreaked havoc for over 400 km downstream. As the largest such event ever reported in historical times, elucidation of its morphology has resulted in a number of slightly controversial studies in terms of location and character , but each one of which has introduced incrementaly greater precision through time (Drew, 1875; Code and Sirhindi, 1986; Owen, 1989; Shroder, 1989a & b; Shroder et al., 1989; Shroder, 1993; Shroder, in press). Nine major slope-failure complexes exist in the area, seven close together on the Indus river and two at the mouth of the Astor confluence with the Indus. From upriver to downriver, the first seven are: (1) the 1841 Bunji rockslide triggered by the rising lake waters from the dam further downstream; (2) the prehistoric Thelichi debris slide with an unknown trigger that may have included catastophic flood undercutting or dewatering; (3) the prehistoric Hattu Pir failure caused by steep bedrock cliffs overlying the Raikot fault zone below; (4) a prehistoric Lichar failure caused by bedrock faulted over unconsolidated glacial gravel; (5) the main Lichar failure of 1841 caused by the same substrate condition and strong seismicity; (6) the Gor Gali failure of 1841 in which the slope opposite the main Lichar failure was also mobilized in the 1841 event by either the seismic shaking, or the later flood undercutting, or both; and (7) the Tatta Pani slides of 1841 and modern times, caused by unconsolidated glacial gravels cut through by the Raikot fault and being actively undercut by the Indus river.

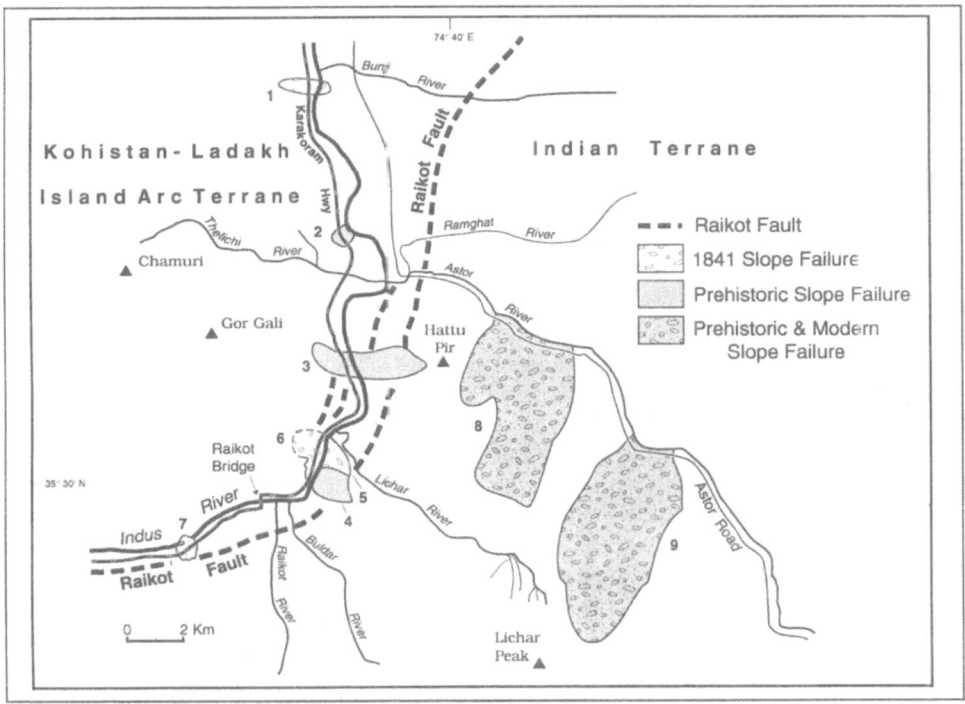

Figure 8. Index map of slope failures in Indus and Astor valleys that have caused considerable hazard in the past. 1. Bunji rockslide of 1841; 2. prehistoric Thelichi debris slide; 3. prehistoric Hattu Pir rockslide; 4. prehistoric Lichar rockslide; 5. Lichar rockslide of 1841; 6. Gor Gali debris slide of 1841; 7. Tata Pani debris falls and slides (probably 1841 and recently); 8. Doian slope failure; 9. Mushkin slope failure.

Collectively, these historic slope failures have killed tens of people and destroyed much property; when adding in the effects of the one known catastrophic breakout flood of 1841, the cost in lives went up by an order of magnitude and the property destruction is likely to have increased commensurately as well. The tremendous relief, steep slopes, seismically active fault zone, weak sediment and rocks, major river system, and presence of the high- traffic, strategic Karakoram Highway set up this area of the Indus valley for considerable future hazard.

4.2. ASTOR SLOPE-FAILURE DAMS AND CONSTRICTED FLOWS

The lower Astor valley has foliation planes dipping 25°-30° into the valley, which has facilitated massive failure into the river over at least the past two centuries, and undoubtedly for a much longer time as well. Conway (1894) noted the instability of the Doian slope close to the mouth of the canyon, and the German expedition of 1934 mapped the slope as 'bergsturz,' or mountain fall, as well. Directly ajacent, with only a narrow ridge intervening, the Mushkin failure has also clearly had a long history of motion, although we have no historical accounts to confirm this supposition based upon landform analysis. The Mushkin slope failure, itself, was overridden by glacier ice that came down the Astor valley in the last glacial maximum and deposited a light-colored, micaceous till across the slope to the ridge line between Mushkin and Doian, where it left a terminal moraine. Meltwater from this terminus in Mushkin may have facilitated failure at Doian. Following withdrawal of the ice, most of the till overlying the original rock and debris of the pre-glacial Mushkin failure area was carried down into the river in a series of slow debris slides, rockslides, and rapid, wet debris flows. Upstream terrace alluviation directly ajacent to the two failures indicates significant blockage at some time in the past. It is also likely that the river was entirely impounded by such motions in the past and then broke out in catastrophic floods downstream, but no cases are known in the historical record.

The long-term movement of rock and debris down the Doian and Mushkin slopes has choked the Astor river, increased its gradient, caused alluviation upstream, and forced profound erosion of the toe of the landslides. Throughout the 19th and most of the 20th centuires only the rough trails and an occasionally open jeep road were maintained across the unstable failure zone. Beginning in the late 1980's the Pakistani army began to blast and grade a wide road across the toes of the two landslides. The new highway lasted a few years until high snow melt and torrential summer monsoon rains combined to undercut both the landslide toes and the new roads, which caused renewed movement of the toes, as well as bringing massive quantities of debris down the slopes above. Several kilometers of the highway were alternately dropped into the river and swept away or buried under tons of debris and thereby destoyed (Figure 9).

The hundreds of homes, barns, people and livestock who inhabit the slower moving parts of the Doian and Mushkin slope failures are carried gradually downslope without significant disruption. The channelized and more rapidly moving parts are avoided as building locations, however. With the main lower road out of use since the early 1990's because of the disruption, a crude old high-road track is maintained across the more stable upper parts of both slope-failure zones. In summer 1996 the Pakistani army was once again blasting and bulldozing its way across the toes of both the Doian and Mushkin failure zones in order to reopen the lower road. The continued undercutting of the toes of the two zones, by both river and man, is likely to continue the hazard unabated.

4.3. HUNZA SLOPE FAILURE DAMS AND FLOODS

In the Atabad area the Hunza river cuts through the Karakoram batholith and along the Karakoram fault to expose metasedimentary rocks that dip steeply 40°-50° into the river canyon (Figure 1). This combination of dipping planes of weakness parallel with the steep south slope, with as much as 2100 m of relief along a major fault, set up a condition of extreme instability that culminated in the massive rockslide of 1858, and like the 1841 event at Nanga Parbat, produced a massive breakout flood (Figure 10). Seven major slope failures (Shroder, in press) have occurred here (from up-valley downstream): (1) the 1992 debris slides and flows mobilized by torrential monsoon rains in the late summer; (2) the older, and (3) the younger Serat rockslides that were prehistoric failures that impacted glacial lake beds; (4) the main 1858 Ghammessar slope failure which impounded a lake that backed up 28 km and finally broke out;

Figure 9. View southeast down onto the toe of the Mushkin slope failure, showing the multiple disruptions of the main Astor road in midsummer 1996. The largest light-colored rockslide disruption of the road visible in this photograph occurred between late summer 1995 and the time of this photograph.

(5) the Ghammassar retrogressive slump of the toe back into the gorge after the breakout flood; (6) the 1962 Ghammessar rockslide that came down the original 1858 slip surface to block the Hunza river again for several months; (7) the 1991 Sulmanabad rockfall from a cliff on the north wall of the canyon.

The first six of these slope failures all came from the south wall of the gorge and disrupted the transportation route on both sides of the canyon; the old silk road and the new Karakoram Highway. The 1858 event also produced a huge flood that caused extensive damage and probable loss of life for at least 500 km downstream where it was still 20 m deep. The rockslide of 1962 killed six people and destroyed a number of houses, but that of 1991 only killed farm animals.

5. Torrential rain-induced debris flows and floods

Debris flows in the Himalaya are probably the most common form of sizeable mass movement (Rickmers, 1913; Goudie et al., 1984). Surface debris is mobilized by rain or snowmelt, and successive waves of this slurry rush down gullies, cross fans, and devastate bridges and fields. Storms on the mountain peaks can generate small flows far down in the valleys made arid by the strong rain shadows. Large debris flows have temporarily blocked rivers and overwhelmed engineering structures, particularly bridges, highways, and buildings. Large debris flows occur in the same places at recurrance intervals of about 1 : 5-10 years in the Hunza valley, but many smaller debris flows are nearly annual events (Goudie et al., 1984).

5.1. BIALE DEBRIS FLOWS

The Raikot basin on the north face of Nanga Parbat has numerous scars on the rock faces and in the vegetation that are evidence of past torrential rain and debris flow events (Figure 2). One building near Biale was partially buried in 1990 by a debris flow (Figure 11). Two basins on the west side of the basin between Fairy Meadow and the Biale shepherd's hamlet were studied by us for this report (Figure 12). High re-solution SPOT satellite imagery and field traverses were used to plot debris-flow fan sediments and catchment perimeters at 1:50,000 scale. The north Biale catchment area is ~3.02 x 10^6 m^2 and the volume is ~5.9 x 10^7 m^3; the south basin area is ~4.56 x 10^6 m^2 and the volume is ~9.6 x 10^7 m^3. Assuming an initial smooth glacia-ted slope prior to catchment incision, basin volume and rock density allow calculation of denuded sediment tonnage from the north catchment at ~1.57 x 10^8 Mg and from the south at 2.52 x 10^8 Mg. The high lateral moraine that crosses the top of the catchments has a cosmogenic ^{3}He age of ~55,000 years. This allows calculation of a sediment production rate of ~1075 m^3 yr^{-1} for the north and ~1750 m^3 yr^{-1} for the south and denudation rates of 0.36 mm yr^{-1} for the north catchment and 0.38 mm yr^{-1} for the south. These figures are closely coincident to an independant cosmogenic de-nudation rate from garnets in the mouth of the south Biale catchment of ~0.4 mm yr^{-1} (Phillips et al., 1995). Dendrogeomorphological magnitude/frequency assessment of catastrophic debris-flow events in 1923 and 1990 averaging 1.4 x 10^5 m^3 equivalent bedrock volume, gave a 67 year return interval; reasonably close to the ~100 similar such events required in the past 55,000 yr to evacuate the basins and equating to an average bedrock denudation equivalent of ~1 cm yr^{-1} for each basin. The variations between the calculated denudation rates are caused by changing process types and rates through time, unaccounted for losses and gains of sediment, and uncertainties in field and laboratory measurements, but the rates are viewed as lower end members of gene-rally higher geomorphic process rates for the Nanga Parbat Himalaya.

These calculations show that debris flow frequency on the north face of Nanga Par-bat will likely continue to be one or two events per century, which mainly are only hazardous to trails, roads, and a few minor buildings. In general, local villagers are aware of such hazards and build their structures away from basins that are potential sources of debris flows, as well as off of debris flow fans. As the population of vil-lagers and foreign trekking hordes increases in this and other popular valleys, however, greater hazard of this kind is likely to occur.

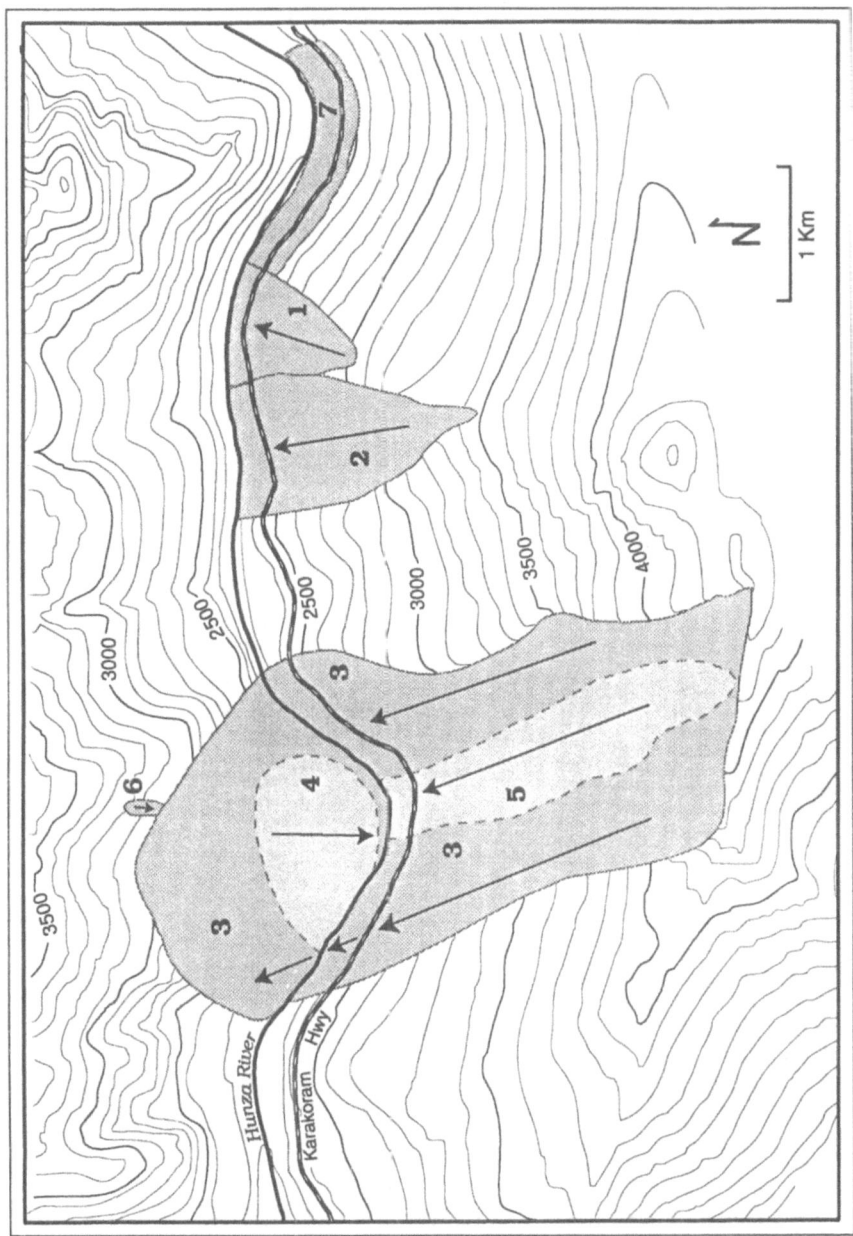

Figure 10. Index map of the Hunza river gorge. 1. older Serat rockslide; 2. younger Serat rockslide; 3. 1858 Ghammessar rockslide; 4. Ghammessar rockslide toe retrogressive failure following breakout flood of 1858; 5. 1962 Ghammessar rock slide; 6. 1991 Sulmanabad rockfall; 7. 1992 Serat debris fall, slides, and flows. Base map adapted from Finsterwalder (1995).

Figure 11. Remaining building at edge of Biale village partly buried by the south Biale debris flow of 1990 in the Raikot valley of Nanga Parbat. This building was not overwhelmed but boulders were piled up around it.

Figure 12. View of south Biale catchment that generated debris flows from torrential rains in 1923 and 1990. Light-colored scars from the 1990 event are visible. The mouth of the south Biale catachment in the photocenter provided a cosmogenic 3He denudation rate of 0.4 mm yr-1 (Phillips, et al., 1995).

Figure 13. View down on the Tato road from the KKH up to Raikot basin in summer 1996 showing slope failure being crossed by villagers on foot. The road was out of use to wheeled traffic all summer during the high traffic tourist season upon which the villagers depend as their only source of income in a subsistance economy. Such natural hazards are extremely common on the steep slopes that are characteristic of the Himalaya.

During the spring meltwater flows of 1996, the jeep road from the KKH up the Raikot gorge and into the upper basin was cut off by a large rockfall and debris slide (Figure 13). During a stong seismic event in 1984, this same road was cut in 50-100 places by rockfall, but was repaired at a fairly slow pace becausethe road was so new, people had not yet come to rely upon it. In 1996, however, the destruction of the road severely curtailed income from the many tourists who frequent the area during the summer season. Many Pakistani mountain villagers have come to depend upon summer revenue from the trekkers and tourists, as their chief, or only, source of real currency. Natural hazards, so common in the western Himalaya, are not only a source of direct loss of life and property, but also cause indirect loss of income to a poverty-stricken people.

6. Conclusion

Glaciers and slope failures in the western Himalaya have formed a variety of dams across rivers that have resulted in large water impoundments and catastrophic breakout floods. Torrential rains also produce major floods. Hazards include destruction by the inital slope failures or ice advance, quiet water inundation of property, erosion by breakout floods, burial of structures and property in sediment, subsidence and collapse of highways and bridges, and loss of life. We have studied these phenomena primarily to measure net denudation, but also to better understand the high magnitude - moderate to low frequency of these processes in the Himalayan landscape. The rapidly growing mountain industries of trucking, bus service, tourism, trekking, and mountain climbing depend wholly upon reliable and open roads and highways. Interdiction of these transport routes by the many slope failures, glacier variations, and catastrophic floods can be a serious problem.

The strategic nature of Himalayan highways and the Tarbela dam across the Indus are of major concern to the government of Pakistan. Most of the power and irrigation waters of the country come from the Indus river, so any disruption of this source would be a major catastrophe. Breakout floods from glaciers and landslides upstream could cause catastrophic failure of the earth-filled and easily erodable Tarbela dam. The potential for such hazard continues unabated in the western Himalaya. Several major floods occurred on the Indus and its tributaries in the 19th century, and the last major flood of any consequence was from breakout through a glacier dam in 1926. Many other smaller floods have occurred from a variety of causes since then. The long-term hazard of most concern to the Water Power and Development Authority (WAPDA) of Pakistan, however, is the vast quantities of denuded sediment pouring into the Tarbela reservoir and filling it prematurely. Upstream sediment check dams are being contemplated and preliminary studies have been completed, but right now, the attention of WAPDA is drawn elsewhere to more pressing problems of electric delivery so that long-term hazard mitigation is not much implemented. The sheer size of the problem, at the scale of the Himalaya, is daunting and requires a dedicated focus to the problems and probabilities. Education of students in the universities and personnel in the ministries and military about the potentials for continuing problems is a necessity. An additional benefit would be the importation of technologies and techniques from developed mountainous countries elsewhere in the world who have already successfully met the great challenges of living and working in such terrain. Finally, the maintenance by a government agency of accessible records on natural hazard events in Pakistan would be a valuable addition to development in the country. Such data are essential for predictive, protective, and remedial measures to mitigate hazards, a vital necessity cited by the U.S. National Research Council's recommendation for an International Decade of Hazard Reduction.

Acknowledgments

Valerie Sloan, Andrew Jacobson, and Douglas Norsby served most ably as field assistants for part of the project. We are grateful to Marvin Barton for the graphics. This work was sponsored by the National Geographic Society and U.S. National Science Foundation grant EHR-9418839.

References

1. Cai Xiangxing, Li Nienjie, and Li Jian (1980) The mud-rock flows in the vicinity of the Batura glacier (in Chinese with English abstract). In, Batura Glacier Investigation Group (editors) Karakoram Batura glacier exploration and research. (in Chinese) Academica Sinica, Lanzhou, Science Press, Beijing.

2. Code, J. A. and Sirhindi, S. (1986) Engineering implications of the Indus river by an earthquake-induced landslide. In R. L. Schuster (editor), Landslide dams: processes, risk, and mitigation. Geotechnical Special Publication 3: 97-110.

3. Conway, W. M. (1894) Climbing and exploration in the karakoram Himalayas. London.

4. Costa, J. E. and Schuster, R. L. (1988) The formation and failure of natural dams. Geological Society America Bulletin 100, 1054-1068.

5. Davis, I. (1984) A critical review of the work method and findings of the Housing and Natural Hazards Group. In K. J. Miller (editor), The International Karakoram Project, 2: 200-227.

6. Drew, F. (1875) The Jummoo and Kashmir territories: a geographical account. Reprint 1980 by Indus Publications, Karachi. 568 pages.

7. Ferguson, R. I. (1984) Sediment load of the Hunza river. In K. J. Miller (editor), The International Karakoram Project, 2: 581-598.

8. Gardner, J. S. and Jones, N. K. (1993) Sediment transport and yield at the Raikot glacier, Nanga Parbat, Punjab Himalaya. In J. F. Shroder, Jr. (editor), Himalaya to the Sea, Routledge, London, 181-197.

9. Goudie, A. S., Brundsen, D., Collins, D. N., Derbyshire, E., Ferguson, R. I., Hashmet, Z., Jones, D. K. C., Perrott, F. A., Said, M., Waters, R. S., and Whalley, W. B. (1984) The geomorphology of the Hunza valley, Karakoram Mountains, Pakistan. In K. J. Miller (editor), The International Karakoram Project, 2: 359-410.

10. Finsterwalder, R. (1995) Hunza - Karakorum 1:100.000. Topographic map produced from German-Austrian expedition 1954 and German expedition 1959.

11. Finsterwalder, R. (1937) Die Gletscher des Nanga Parbat: Glaziologische Arbeiten der Deutsche Himalaya Expedition 1934 und ihr Ergebnisse. Zeitschrift für Gletscherkunde 25, 57-108.

12. Hewitt, K. (1968) Records of natural damming and related events in the upper Indus basin. Indus, 10, 11-19.

13. Hewitt, K. (1972) The mountain environment and geomorphic processes. In H. O. Slaymaker & H. J. McPherson (editors), Mountain Geomorphology: Geomorphological Processes in the Canadian Cordillera. Tantalus, Vancouver, B. C. Geographical Series, No. 14, 17-34.

14. Hewitt, K. (1982) Natural dams and outburst floods of the Karakoram Himalaya. In J. Glen (editor) Hydrological aspects of alpine and high mountains areas. International Hydrological Association (IAHS) 138: 259-269.

15. Hewitt, K. (1992) Mountain hazards. Geojournal 27: 47-60.

16. Hurst, A. J., and Chao, P. C. (1975) Sediment deposition model for Tarbela reservoir. Symposium on Modeling Techniques, American Society Civil Engineers, 1, 501-520.

17. Kick, W. (1989) The decline of the last Little Ice Age in high Asia compared with that in the Alps. In J. Oerlemans (editor), Glacier Fluctuations and Climatic Change. 129-142.

18. Li Jian, Cai Xiangxing, and Li Nianjie (1980) Basic features of the meltwater of the Batura Glacier. Karakoram Batura Glacier, Exploration and Research (in Chinese with English abstracts) Lanzhou Institute of Glaciology and Cryopedology, 111-132.

19. Mason, K. (1929) Indus floods and Shyok glaciers. Himalayan Journal 1:10-29.

20. Mason, K. (1930) The Shyok flood: a commentary. Himalayan Journal 2:40-47.

21. Mason, K. (1935) The study of threatening glaciers. Geographical Journal 85:24-41.

22. Mattson, L. E. and Gardner, J. S. (1989) Energy exchanges and ablation rates on the debris-covered Rakhiot glacier, Pakistan. Yeitschrift für Gletscherkunde und Glaziologie 25, 17-32.

23. Mattson, L. E., Gardner, J. S., and Young, G. J. (1993) Ablation on debris covered glaciers: an example from the Raikot glacier, Punjab, Himalaya. IAHS Publ. 218,289-296.

24. Owen, L. (1989) Neotectonics and glacial deformation in the Karakoram mountains, and Nanga Parbat Himalaya. Tectonophysics 163:227-265.

25. Phillips, W. M., J. Quade, J. F. Shroder, Jr.. and J. Poths, 1996, Cosmogenic 3He in garnet, Raikot Valley, Nanga Parbat, northwestern Himalaya, Pakistan. 11th Himalaya-Karakoram-Tibet Workshop, Flagstaff, AZ, 116.

26. Pillewizer, W. (1956) Der Rakhiot-Gletscher am Nanga Parbat im Jahre 1954. Zeitschrift für Gletscherkunde und Glaziologie 3, 181-194.

27. Rickmers, W. R. (1913) The Duab of Turkestan. Cambridge University Press, Cambridge, UK.

28. Shroder, J. F., Jr. (1978) Dendrogeomorphologic analysis of mass movement on Table Cliffs Plateau, Utah. Quaternary Research 9:168-185.

29. Shroder, J. F., Jr. (1970) Dendrogeomorphology: review and new techniques of tree-ring dating and geomorphology. Progress in Physical Geography 4:161-188.

30. Shroder, J. F., Jr. (1989a) Hazards of the Himalaya. American Scientist 77, 564-573.

31. Shroder, J. F., Jr. (1989b) Slope failure: extent and economic significance in Afghanistan and Pakistan. In E. E. Brabb & B. L. Harrod (editors), Landslides: Extent and Economic Significance. A. A. Balkema, Rotterdam, 325-341.

32. Shroder, J. F., Jr. (1993) Himalaya to the sea: Geomorphology and the Quaternary of Pakistan in the regional context. In J. F. Shroder, Jr. (editor) Himalaya to the Sea. Routledge Press, London, UK.

33. Shroder, J. F., Jr. (in press) Slope failure in the western Himalaya. Geomorphology.

34. Shroder, J. F., Jr., Khan, M. S., Lawrence, R. D., Madin, I., and Higgins, S. M. (1989) Quaternary glacial chronology and neotectonics in the Himalaya of northern Pakistan. In Malinconico, L. L., Jr., and Lillie, R. J., (Editors) Tectonics and geophysics of the western Himalaya, Special Paper 232,275-293.

35. Shroder, J. F., Jr., M. P. Bishop, J. Quade, W. Phillips, P. H. Nieland, and A. M. Schmidt, 1996, Dendrogeomorphology and denudation efficiency, Nanga Parbat Himalaya. 11th Himalaya-Karakoram-Tibet Workshop, Flagstaff, AZ, 135-136.

36. Spencer, M. C. (1985) Geomorphologic analysis of the Sachen glacier, Nanga Parbat, Pakistan. Unpublished senior thesis, University of Nebraska at Omaha, Omaha, NE.

37. Wang Wenying, Huang Maohuan, and Chen Jinming. (1984) A surging advance of Balt Bare glacier, Karakoram Mountains. In K. J. Miller (editor), The International Karakoram Project, 1,76-83.

38. Watanabe, T., Ives, J. D., and Hammond, J. E. (1994) Rapid growth of a glacial lake in Khumbu Himal, Himalaza: Prospects for a catastrofic flood. Mountain Research and Development, 14,329-340.

39. Watanabe, T., Kameyama, S., and Sato, T. (1995) Imja glacier dead-ice melt rates and changes in a supra-glacial lake, 1989-1994, Khumbu Himal, Nepal: Danger of lake drainage. Mountain Research and Development, 15, 293-300.

40. Watanabe, T. and Rothachter D. (1996) The 1994 Lugge Tsho glacial lake outburst flood, Bhutan Himalaya. Mountain Research and Development, 16, 77-81.

41. Zhang Xiangsong, Shi Yafeng, and Cai Xiangxing (1980) The migrating subglacial channel of the Batura Glacier and the tendency of a new channel. Karakoram Batura Glacier, Explorating and research (in Chinese with English abstracts) Lanzhou Institute of Glaciology and Cryopedology, 153-165.

Authors

J F. Shroder jr., M. P. Bishop, R. Scheppy
University of Nebraska at Omaha
Department of Geography and Geology
Omaha, NE 68182
U.S.A.

LANDSLIDES AND DESERTED PLACES IN THE SEMI-ARID ENVIRONMENT OF THE INNER HIMALAYA

JUSSI BAADE, ROLAND MÄUSBACHER,

GÜNTHER A. WAGNER,

ERWIN HEINE, ROBERT KOSTKA

1. Introduction

Among other geomorphologic hazards, landslides are a common feature in high mountain areas like the Himalayan Range (cf. Shroder 1989). On the southern flanks of the Himalayas, specially the Low and Middle Himalaya, landslides, representing the dominant hillslope process (Whitehouse 1990), are of special concern as they cause permanent loss of agricultural land (Ives & Messerli 1981), block important roads and even destroy housing (cf. Schelling 1988). The dominance of landslides in this region can be attributed to lithology, the tectonic stress applied to the rocks and steep relief. In addition the climatic conditions promote deep weathering and provide landslide triggering rainfall events, generally mainly during the monsoon (cf. Kienholz et al. 1982, Bartarya & Valdiya 1989).

Mainly because of the much lower precipitation on the northern flanks of the Himalayas, landslides are believed to be of minor importance in that region (cf. Whitehouse 1990). Indeed, there are only a few reports of landslides or related landforms in the semiarid to arid environment of the Inner Himalaya (Fort et al. 1984, Fort 1987). On the other hand, there are reports in this region of a relatively large number of abandoned settlements and fields; including local tales about demons destroying these sites (Pohle 1994). The demons in the local tales are interpreted as reflecting natural hazards like floods, possibly caused by glacial lake outburst floods or mud flows. For these reasons, investigations into the relationship between natural hazards and the history of settlements were launched in 1993 within the framework of the Nepal-German Project on High Mountain Archeology (Haffner & Pohle 1993). The aim of this paper is to report on the first results of our field work in South Mustang and the findings of our reconnaissance survey to North Mustang.

2. Investigation area

The investigation area is located north of the High Himalayas in the Mustang District of Nepal. It extends roughly over 60 km from Tukce (28° 43' N, 83° 39' E) in the south to Mantang (29° 11' N, 83° 58' E) in the north (cf. Fig. 1).

49

J. Kalvoda and C.L. Rosenfeld (eds.), Geomorphological Hazards in High Mountain Areas, 49-62.
© 1998 *Kluwer Academic Publishers.*

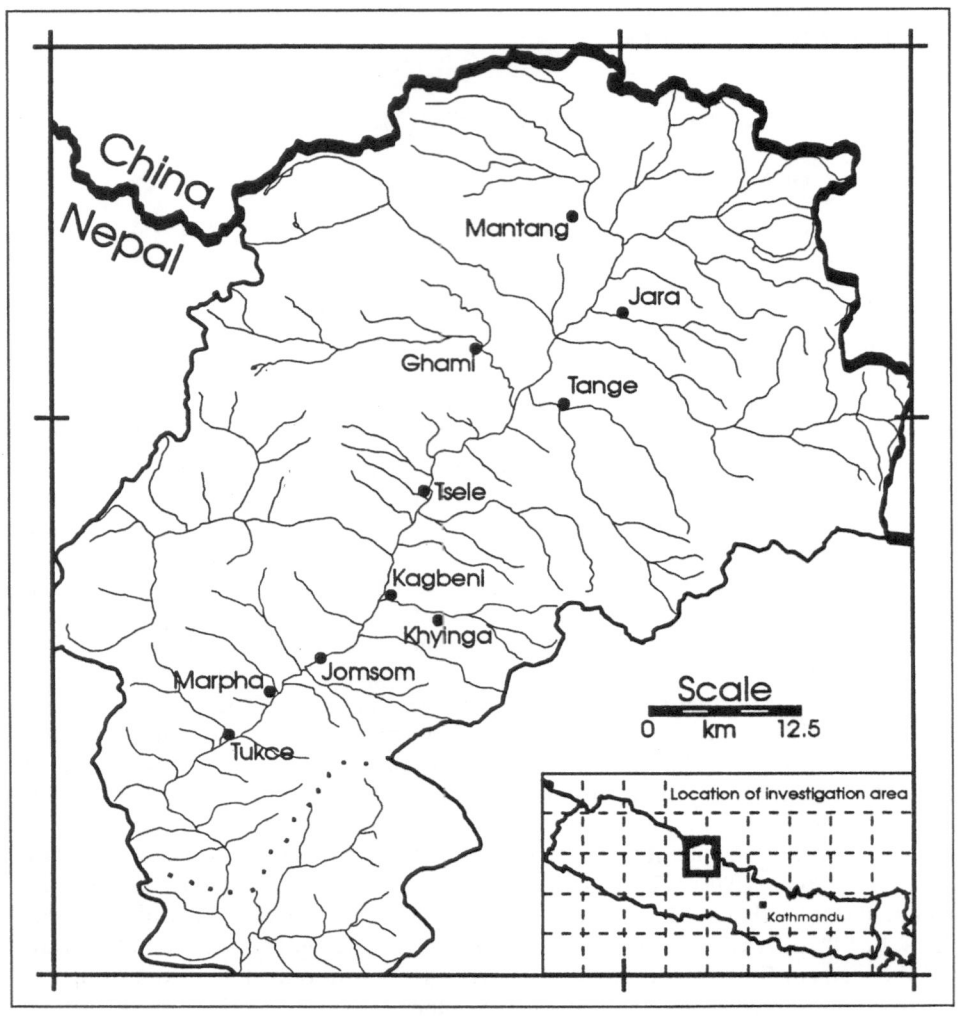

Figure 1. General map of the investigation area

However, because of easier accessibility, most of the research was conducted in the southern part of the Mustang District, i.e. south of Kagbeni (28° 50' N, 83° 47' E). Towards the south, the investigation area is delineated by the crest of the High Himalaya with peaks rising over 8,000 m a.s.l. Towards the west and east, several mountain chains like the Dhaulagiri Himal and the Mustang Himal in the west and the Muktinath Himal and the Damodar Himal in the east, rising to over 6,000 m a.s.l., form climatologicaly important boundaries. Towards the north the boundary is less pronounced (cf. Hagen 1968). The whole area, named Thakkhola, is drained towards the south by the Kali Gandaki river, which - north of Tsele - is locally called Mustang Chu (Fig. 1).

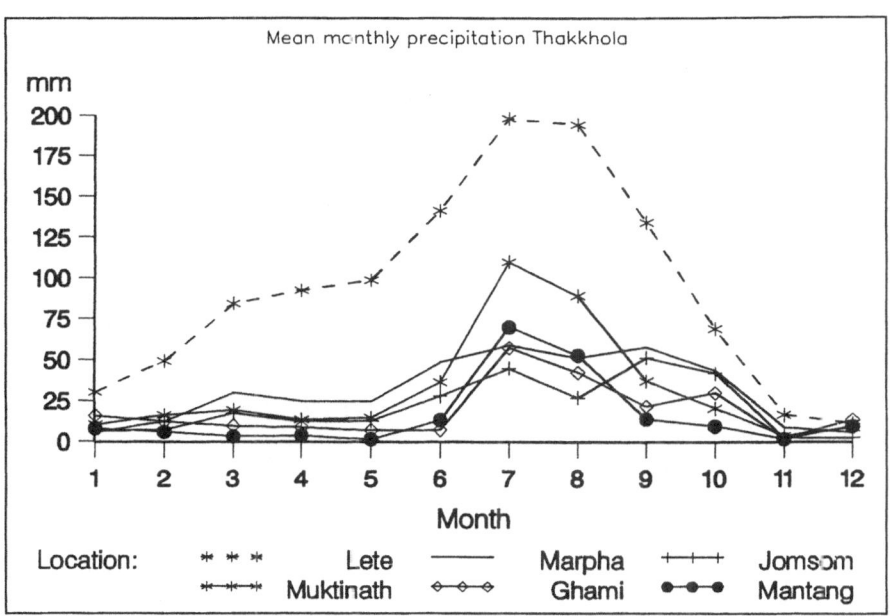

Figure 2. Mean monthly precipitation in the Thakkhola region, 1971-1986 (Data source: Climatological records of Nepal 1977ff.)

2.1. PRECIPITATION

With annual rainfall amounts from about 370 mm in the south (Marpha) to less than 200 mm in the north (Mantang) (cf. Tab. 1) this semi-arid area represents the driest region in Nepal (cf. Shresta 1989). The data given in Table 1 show a sharp decline in annual rainfall at the southern edge of the investigation area (between Lete south of the investigation area and Marpha) and a further decrease in annual rainfall with increasing latitude; i.e., with increasing distance from the crest of the High Himalayas. The higher precipitation at Muktinath may be attributed to its location well east of the main valley and the 'Troll effect'; the 'Troll effect' being a local climate phenomena which explains rainfall distribution within a valley with reference to diurnal air circulation (cf. Flohn 1970). Associated with the decrease of annual rainfall amount, an increase of the interannual variability of rainfall can be observed. An analysis of the mean monthly rainfall distribution (Fig. 2), shows an interesting pattern. For all stations rainfall for the year is strongly concentrated in the months of July and August. However, those stations with low annual rainfalls (Mantang and Ghami), during July and August have monthly rainfall totals equivalent to those stations with higher annual rainfalls (Marpha, Jomsom).

2.2. GEOLOGY

Lithologically the investigation area is made up mostly of Triassic limestones, sandstones, quartzites, schists and slates (von Rad et al. 1994) as well as Tertiary conglomerates (Fort et al. 1982). In the northwestern part, granites crop out (Hagen 1959, Le Fort & France-Lanord 1995). Because of their content of ammonites, which are worshipped widely in South Asia as Vishnu (cf. Messerschmidt 1989), the most well known rocks of this area are the dark yurassic shales (Spiti formation according to Fort et al. 1982). Hagen (1968) formerly used the name 'Saligram series' for this formation. An important geological (tectonic) feature of the investigation area is the Thakkhola graben, an asymmetric graben with a well defined line of fault steps in the west and a more flexure-like transition in the east (Hagen 1959, Fort et al. 1982). The displacement of rocks along the western margin of the graben is about 3,000 m (Hagen 1968, 155). In the south, i.e. around Tukche, the width of the graben is about 15 km. The maximum width (55 km) is reached around Tsele. Towards north the graben is narrowing again and around Mantang it reaches a width of 38 km (cf. Hagen 1968, 15).

Table 1. Annual precipitation (mm) in the Thakkhola region during the period 1971-1986
(Data source: Climatological records of Nepal 1977ff.)

Location	Lete	Marpha	Jomsom	Muktinath	Ghami	Mantang
Latitude	28°38' N	28°45' N	28°47' N	28°49' N	29°03' N	29°11' N
Longitude	83°36' E	83°42' E	83°43' E	83°53' E	83°53' E	83°58' E
Elevation	2384 m	2566 m	2744 m	3609 m	3465 m	3705 m
Length of recording	15 a	15 a	11 a	12 a	9 a	8 a
Precipitation mean annual	1114.7	368.4	251.8	377.3	225.7	191.6
min. annual	961.0	249.0	89.0	178.0	104.0	122.0
max. annual	1388.0	487.0	451.0	485.0	534.0	296.0
CV (%)	11.4	20.3	43.7	22.5	60.9	33.0

Note: the values given in this table are not fully comparable as the number of complete annual data sets and missing years, respectively, differs.

3. Landslides of the Thakkhola

As mentioned above, there are only few reports of landslides in the semi-arid environment of the Thakkhola region. This might partly be explained by the restricted access to the northern part of the Thakkhola as it was opened to a limited number of foreigners only five years ago. In 1995 we were able to conduct a reconnaissance survey in this region. In conjunction with the results of the ground survey, the areal extent of landslides in this region was estimated using a SPOT2 Image taken in 1990. The analysis of the area affected by landslides, seems to support the contention that landslides play a minor role in shaping the semi arid landscape and therefore might be neglected as geomorphic hazards. But, the importance of a geomorphic hazard is not only judged by the magnitude and frequency of the process and its areal extent. It also depends on the values at risk (Crozier 1986, 207f). It is therefore significant that, in this region most landslides are located close to recent settlements or abandoned sites.

3.1. LANDSLIDE AT JARA

A good example of the association of landslides and settlement is provided by the village of Jara (29° 06' N, 84° 00' E). In the vicinity of Jara an area of about 2.75 km^2 is affected by landslides. The village itself is located on currently stable schists, but most fields around the village already show signs of ground movement. West of the village, some fields have been abandoned because of the disturbance of the land as is evident in Figure 3. Two types of landslides can be identified in the photo presented: in the background, south of the river, a complex landslide with multiple rotational blocks, indicating the possibility of a curved shear surface, is present. This landslide seems to have been stable for a long period of time as fields have been established up its toe In the foreground, and north of the river, a planar slide is present. The former land surface, clearly showing terraced fields, is still preserved. The fields on the landslides suggest two periods of landslide activity. The landslide north of the river must have been active before the fields were established; contrary the landslide south of the river must have started recently. It is assumed that the agricultural use of the fields reflect the period of stability of the landslides. Due to the fact that the sediments of the fields have been build up by men (Baade et al. in press) the period of agricultural use might be dated using luminescence-dating methods (Wagner 1995, Lang 1995).

The field survey strongly indicated that the landslides around Jara were caused by the fluvial down cutting of the Puyan Khola river, flowing from the glaciated mountains in the east towards the Kali Gandaki in the West. In Figure 3 the river is not visible but runs laterally from left to right. This raises the question whether the periodicity of the landslide activity reflects a similar periodicity in fluvial down cutting; the changes in the fluvial activity being controlled by climatic change (Röthlisberger 1986) or tectonically induced down cutting of the Kali Gandaki (Iwata et al. 1982).

3.2. LANDSLIDE AT TANGE

Another example of the spatial proximity of settlements and landslides is Tange (29° 01' N, 83° 57' E). According to the interpretation of the SPOT-Image an area of 6.25 km^2 is affected by landslides. Figure 4 shows the village of Tange on the northern banks of the Yak Khola right in the centre. North of the village, and about 150 m above the recent flood plain a westward dipping terrace, is evident, extending further to the north.

This terrace covers the westward dipping Triassic and Tertiary sediments. Towards the east it extends up to the badlands developed in light coloured sediments. Further to the east, these sediments are removed and the underlying dark shales crop out in a large landslide (in the background of the picture). Another large landslide is visible in the left center of the picture. The toes of these landslides extend right down to the Tange Khola river joining the Yak Khola at Tangbe. The close up view (Figure 5) clearly documents the recent landslide activity. The toe of the landslide on the eastern banks of the Tange Khola is pushing the river towards west, causing undercutting and consequently slope failures along the edge of the terraced fields. An irrigation channel, delineated by trees growing at the edge of the terrace, is endagered as well.

Figure 4: The landslide at Tange (Photo: J. Baade)

Figure 3. Detail of the landslide at Jara (Photo: R. Mäusbacher)

Figure 5. Detail of the landslide at Tange (Photo: J. Baade)

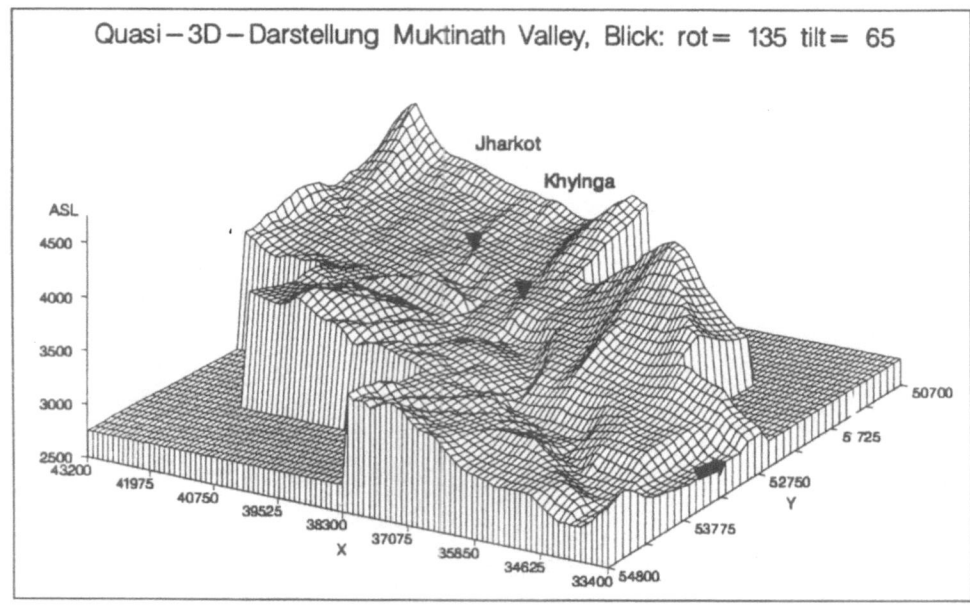

Figure 6. Digital elevation model of the Muktinath Valley

Figure 7. Detail of the toe of the landslide between Khyinga and Jharkot, Muktinath Valley (Photo: J. Baade)

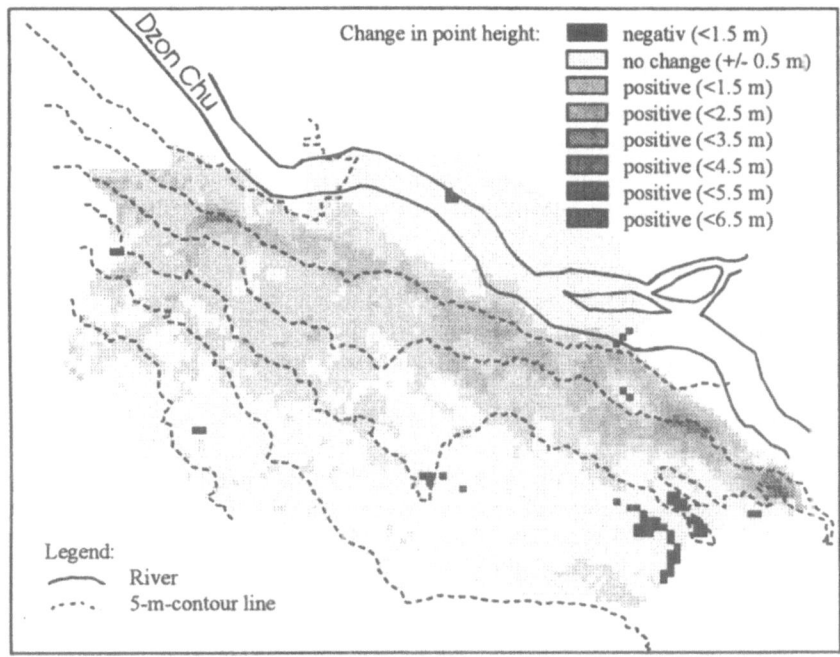

Figure 8. Changes in spot heights at the toe of the landslide between Khyinga and Jharkot

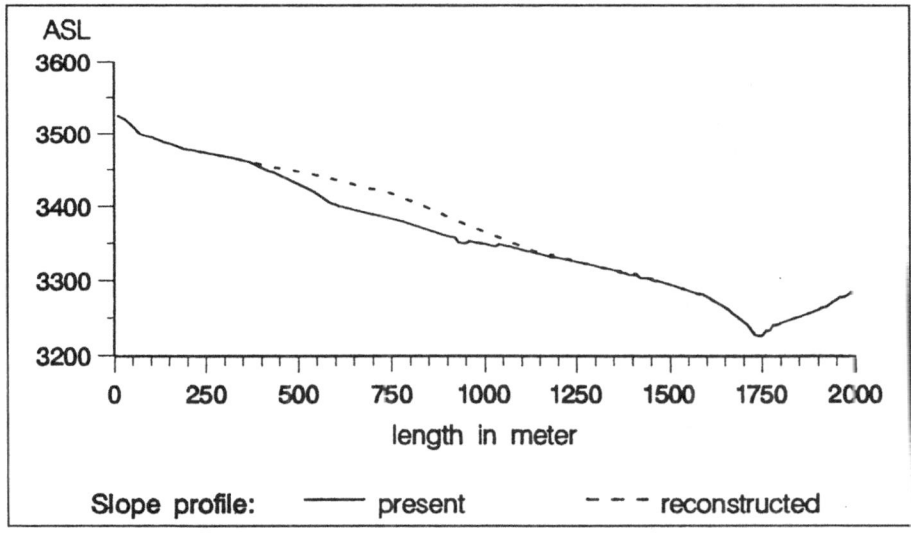

Figure 9. Selected slope profiles of the contemporary and reconstructed land surface of the Khyinga landslide

3.3. LANDSLIDES IN THE MUKTINATH VALLEY

The Muktinath Valley is located just south of the limit of the restricted area of North Mustang. Several landslides have been identified: near Kagbeni a relatively small rock slide occured on the northern bank of the Dzon Chu river. In the upper part of the valley several deep seated and very active landslides dominate the slope processes and disturb agricultural land in this area. One of these landslides, being very similar to the examples described from Northern Mustang, is found between Khyinga and Jharkot. The depression formed by this landslide, covering an area of about 340 ha is clearly recognisable in the digital elevation model of the Muktinath Valley (Fig. 6). The distance between the crown and the tip of the landslide is 1,500 m and the toe width, along the river, is about 1,000 m.

The activity of this landslide was monitored between 1994 and 1995 using terrestrial stereo photos, covering a 100 m wide sector (0.5 ha) of the toe (Fig. 7). From two pairs of stereo photos, taken in autumn 1994 and 1995, digital contour maps (scale: 1:200) with a 1-m-contour interval were produced using the photogrammetric software package CRISP (Fuchs & Leberl 1980) and the CAD-system AutoCAD (Autodesk Inc. 1995). From these maps two digital elevation models (DEM) were interpolated using the GIS IDRISI for Windows (Eastman 1995). Comparing these two DEM's the following results were obtained.

Figure 8 displays the spatial distributed differences in point heights in the area covered by the large scale digital contour maps. Positive changes up to 6 m occur all along the tip of the landslide forming the river bank. Negative changes are concentrated close to the areas of highest positive change. On average the height of the surface increased by 0.6 m, being equivalend to an increase of volume of this sector of the toe of 3,070 m^3 in one year. Relating this volume to the width of the upslope limit of the surveyed area gives a flux of 33 m^3 per meter in one year from the landslide's upslope reaches, outside the covered area. This is a minimum value for the period studied as the amount of material moving towards the river bed and being eroded during this period could not be accounted for.

A major aim of our work is to find out whether past and present geomorphologic hazards have an impact on the settlement history of this region. In order to get a first estimation of the time during which the landslide east of Khyinga might have been active, the original surface was graphically reconstructed based on a topographical map scaled 1:10,000.

Figure 9 shows a selected slope profile of the actual slope surface as well as a profile in the same position for the assumed original surface. Comparing the two profiles it is clearly seen that in reconstructing the former surface it was assumed that all the material missing above today's surface of rupture has been removed from the area and that no build up of the toe occurred. Again, two DEM's were interpolated and the difference in volume calculated. Assuming that the measured flux is equal to the mean flux for a long period, one ends up with a time span of approximately 500 years to remove the volume missing at the upslope part of the landslide. It should be pointed out that this is a first and very crude estimate as the measured value represents just a value for one year and takes no account of temporal variability in mass transport rates through time, being evident in the example of Jara. However, by coring and dating of organic matter deposited in hollows at the head of the landslide this estimate will be reviewed. Nonetheless, this date is in good agreement with the archeological findings for the abandonment of the near-by settlement of Khyinga-Khalun in the 15th century (Hüttel 1994) and the settlement of Phuzelin (Simons et al. 1994) further downstream. Allthough political changes as a reason for the abandonment of the settlement can not be ruled out, there is little doubt that disruption of the agricultural system by landslide activity would make it difficult to maintain human settlement within the area. A good contemporary example of the difficulties involved in maintaining the irrigation system is the water channel feeding the hydro power station just west of Jharkot. On several hundered meters plastic tubing is used to prevent blocking of the channel by a landslide.

The reconnaissance survey to the northern part of the Thakkhola clearly indicates that the Muktinath Valley is similar to the northern region on the basis of lithology and tectonic features as well as the spatial proximity of landslides and settlements and fields. On these grounds it is considered that the results obtained in the Muktinath Valley can be extended to the whole Mustang District or Thakkhola region. respectively.

4. Conclusion and discussion

The first results of the investigations into the relationship between natural hazards (landslides) and the history of settlements in the semi-arid environment of the Inner Himalaya can be summarised as follows:

The field work in Southern Mustang and the reconnaissance survey to Northern Mustang show morphological evidence of large-scale, deep seated landslides with major landsurface deformation. There is a clear concentration of large scale landslides in the eastern part of the Thakkhola where the strata are dipping westward. This can be attributed to the asymmetric structure of the Thakkhola graben. A comparison with the geological map published by Hagen (1968) shows a striking coincidence of the areas affected by landslides and the areas mapped as 'Saligram series'. In addition to geological conditions other factors contribute to the inherent instability of certain slopes. These include the removal of lateral support by active stream incision, ultimatively controlled by the incision of the Kali Gandaki, and possibly differences in ground water conditions. The analysis of the precipitation pattern in the investigation area suggests that ground water conditions enhance slope movement during the monsoon. The specific triggering agent is unknown, but both climatic and seismic (cf. Bilham et al. 1995) triggering factors are a possibility. On the Holocene time scale, possible climatic change has to be taken into account (Röthlisberger 1986).

The deformation of the land surface causes physical disruption to agricultural infrastructure such as field terraces and irrigation channels. As crops can be grown only on irrigated land in this area this must have a significant effect on food supply, especially in former times when transportation was more difficult than today. In this context the observed proximity of settlements to unstable areas raises the question as to whether the relatively high water content associated with landsliding was also a factor attracting settlements.

At all sites there is evidence of former and contemporary landslide activity, suggesting a considerable period of slope instability. The estimate of the onset of landsliding in the Muktinath valley suggests that slope instability was a possible cause for the abandonment of the settlements of Khyinga-Khalun and Phuzelin. However, further work is needed to investigate the timing of the onset and the temporal variation of landslide activities. As mentioned above, anthropogenic field terraces have a key position because they allow to date the beginning and ending of agricultural use on the surface of the landslides, representing the periods of stability.

Acknowledgements

The field work was supported by the German Science Foundation (DFG, Ma 1308/5-1 - 5-2) within the framework of the Schwerpunktprogram 'Siedlungsprozesse und Staatenbildung im Tibetischen Himalaya'. We are grateful to Mike Crozier for his constructive comments on an early draft of this paper.

References

1.   Autodesk Inc. (1995) AutoCAD Release 13 - „User's Guide", Autodesk-Trademark, N.N.

2.   Baade, J., Mäusbacher, R., Wagner, G.A. (in press): Der Einfluß von klimatischen Veränderungen und Natur-ereignissen (Katastrophen) auf die Siedlungsprozesse in Mustang, Tibetischer Himalaya, Jenaer Geographische Manuskripte.

3.   Bartarya, S.K., Validiya, K.S. (1989) Landslides and erosion in the catchment of the Gaula River, Kumaun Lesser Himalaya, India, Mountain Research and Development 9(4), 405-419.

4.   Bilham, R., Bodin, P., Jackson, M. (1995) Entertaining a great earthquake in Western Nepal: Historic inactivitiy and geodetic test for the development of strain, Journal of Nepal Geological Society 11, 73-88.

5.   Climatological records of Nepal 1977ff., published by Department of Irrigation, Hydrology and Meteorology, Ministry of Food, Agriculture and Irrigation, Kathmandu.

6.   Crozier, M.J. (1986) Landslides: causes, consequences and environment, Croom Helm, London.

7.   Eastman, J.R. (1995) Idrisi for Windows, User's Guide, Worcester.

8.   Flohn, H. (1970) Beiträge zur Meteorologie des Himalaya., Khumbu Himal 7(2), 25-45.

9.   Fort, M. (1987) Geomorphic and hazards mapping in the dry, continental Himalaya: 1:50,000 maps of Mustang District, Nepal, Mountain Research and Development 7(3), 222-238.

10. Fort, M., Freytet, P., Colchen, M. (1982) Structural and sedimentological evolution of the Thakkhola Mustang Graben (Nepal Himalayas), Zeitschrift für Geomorphologie, N.F., Suppl.Bd. 42, 75-98.

11. Fort, M., White, P.G., Shrestha, B.L. (1984) 1:50.000 geomorphic hazards mapping in Nepal., FLAGEOLLET, J.-C. (Ed.): Mouvement de terrains. Communications du colloque, Caen, 22-24 mars 1984 185-194.

12. Fuchs, H., Leberl, F. (1980) „CRISP", A software package for close range photogrammetry for the KERN DSR1, Aarau.

13. Haffner, W., Pohle, P. (1993) Settlement processes and the formation of state in the high Himalayas characterized by Tibetan culture and tradition., Ancient Nepal 134, 42-56.

14. Hagen, T. (1959) Geologie des Thakkhola (Nepal)., Eclogae Geologicae Helvetia 52, 708-720.

15. Hagen, T. (1968) Report on the geological survey of Nepal. Vol. 2: Geology of the Thakkhola including adjacent areas, Denkschrift der Schweizerischen Naturforschenden Gesellschaft 86/2..

16. Hüttel, H.-G. (1994) Archäologische Siedlungsforschung im Hohen Himalaja. Die Ausgrabungen der KAVA im Muktinath-Tal/Nepal 1991-1992, Beiträge zur allgemeinen und vergleichenden Archäologie 14, 47-147.

17. Ives, J.D., Messerli, B. (1981) Mountain hazards mapping in Nepal - Introduction to an applied mountain research project., Mountain Research and Development 1(3-4), 223-230.

18. Iwata, S., Yamanaka, H., Yoshida, M. (1982) Glacial landforms and river terraces in the Thakkhola Region, Central Nepal, Journal of Nepal Geological Society 2 (Special Issue), 81-94.

19. Kienholz, H., Hafner, H., Schneider, G. (1982) Zur Beurteilung von Naturgefahren und der Hanglabilität - Ein Beispiel aus dem nepalischen Hügelland, HAFFNER, W. (Ed.): Tropische Gebirge: Ökologie und Agrarwirtschaft, Gießener Beiträge zur Entwicklungsforschung, Reihe I, 8, 35-56.

20. Le Fort, P., France-Lanord, C. (1995) Granites from Mustang and surrounding regions (Central Nepal), Journal of Nepal Geological Society 11, 53-58.

21. Messerschmidt, D.A. (1989) The Hindu Pilgrimage to Muktinath, Nepal. Part 1. Natural and supranatural attributes of the sacred field, Mountain Research and Development 9(2), 89-104.

22. Pohle, P. (1993) Geographical research on the cultural landscape of Southern Mustang. The land use map of Kagbeni as a basis, Ancient Nepal 134, 57-81.

23. Pohle, P. (1994) Wüstungen als Zeugen von Siedlungsprozessen im Tibetischen Himalaya (Süd-Mustang)., Fehr., K. (Ed.): Siedlungsforschung. Archäologie - Geschichte - Geographie. 12, Verlag Siedlungsforschung, Bonn, 327-340.

24. Röthlisberger, F. (1986) 10 000 Jahre Gletschergeschichte der Erde, Verlag Sauerländer, Aarau.

25. Schelling, D. (1988) Flooding and road destruction in Eastern Nepal., Mountain Research and Development 8, 78-79.

26. Shrestha, S.H. (1988) Nepal in maps, White Orchid Books, Kathmandu.

27. Shroder, J.F.Jr. (1989) Hazards of the Himalaya, American Scientist 77, 564-574.

28. Simons, A., Schön, W., Shresta, S.S. (1994) Preliminary report on the 1992 campaign of the team of the Institute of Prehistory, University of Cologne, Ancient Nepal 136, 51-75.

29. von Rad, U., Dürr, S.B., Ogg, J.G., Wiedmann, J. (1994) The Triassic of the Thakkola (Nepal). I: stratigraphy and paleoenvironment of the north-east Gondwana rifted margin, Geologische Rundschau 83, 76-106.

30. Whitehouse, I.E. (1990) Geomorphology of the Himalaya: A climato-tectonic framework, New Zealand Geographer 46(2), 75-85.

Authors

Jussi Baade, Roland Mäusbacher
Department of Geography, Friedrich-Schiller-University, D-07740 Jena,
Germany

Günther A. Wagner
Forschungsstelle Archäometrie der Heidelberger Akademie der Wissenschaften am
Max-Planck-Institut für Kernphysik, P.O. Box 103980, D-69029 Heidelberg,
Germany

Erwin Heine, Robert Kostka
Department of Remote Sensing, image processing and cartography, Institute for Ap-
plied Geodesy and Photogrammetry, Graz University of Technology, Steyrergasse 30,
A-8010 Graz,
Austria

GLACIER-INDUCED HAZARDS AS A CONSEQUENCE OF GLACIGENIC MOUNTAIN LANDSCAPES, IN PARTICULAR GLACIER- AND MORAINE-DAMMED LAKE OUTBURSTS AND HOLOCENE DEBRIS PRODUCTION

MATTHIAS KUHLE, SIGRID MEINERS, LASAFAM ITURRIZAGA

1. Introduction (M. Kuhle)

With the help of representative examples this paper attempts to infer the damaging effects, induced by glaciers, not only - as is normally the case (cf. amongst others, Hewitt 1982, 1988, 1995) - from the geomorphodynamics, observed directly on the spot, but from the entire glacigenic character of the high mountain landscape. Accordingly, not only the current changes of the glacier, which produce damage - as for instance the shifting of the glacier termination (cf. 3.) - are on the focus of interest, but also the development of the glacigenic relief during the High- to Late Glacial (cf. 2.). From the mountain shaping through prehistorical glaciation, partly filling up the relief, a transformation of the valleys resulted, which was not stable during the Interglacials. Characteristics of the glacial relief, such as U-shaped valleys, subsequently crumbled away on their typically over-steepened valley flanks and collapsed as soon as, following deglaciation, the abutment of the ice filling was absent in the valley. This is a process which, in the Interglacials, the preglacial U-shaped valley relief - depending on the length of time passed since deglaciation and according to resistance of the valley flanks - more or less quickly reestablishes. Through the processes proceeding as a result, i.e. wet and dry mass self-movements such as rockfall, rock slide, landslide etc., debris bodies are built up on the valley floors. These directly or indirectly result in or induce in the form of fans, cones and screes numerous damaging effects (cf. 4.).

This approach of geomorphological landscape-genetic explanation of hazards is based on the Pleistocene climatic history and the connected shaping of the relief through glaciation during the Ice Age. For this reason, a chronological concept must be followed. Thus, in the first chapter (2.), the glacial- geomorphological characteristics of valley forms are systematized in terms of their importance for glacier-induced hazards. These are forms, which have been developed from the Main Ice Age to the Late Ice Age through glacier erosion of the last melted ice stream network and then, finally, through the ever smaller dendritic valley glacier systems in the Himalaya and Karakorum.

The related lowering of the ice levels during the Holocene came to an end in the surfaces of the historical to modern glacier positions.

For this reason, the following section (3.) is concerned with the risks and hazards that are connected with the current glacier oscillations.

The last section (4.) is devoted to the risks and hazards that develop from the increasing interglacial detrital formation of the Ice Age glacier valleys.

J. Kalvoda and C.L. Rosenfeld (eds.), Geomorphological Hazards in High Mountain Areas, 63-96.
© 1998 *Kluwer Academic Publishers.*

For all three sections, study areas with regional examples from four mountain ranges of High Asia, namely, Karakorum, Himalaya, Tien Shan and Tangula Shan
(Central Tibet) were chosen (Fig. 1).

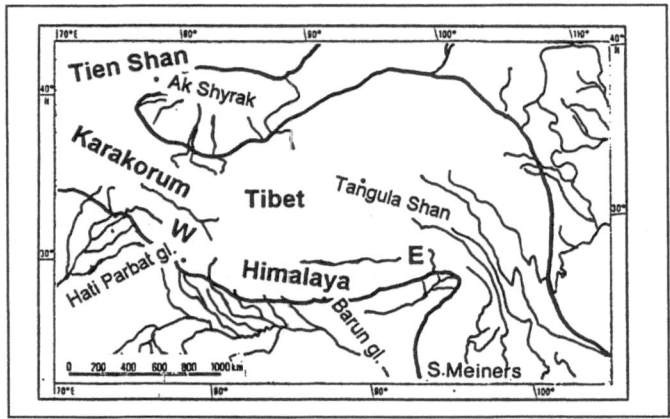

Fig. 1 Areas of investigations

2. Case-example of the significant features, which are characteristic for hazards in the glacigenic mountain landscapes of High Asia (M. Kuhle)

On the importance of glacigenic large-scale forms:
For this, the region of the at least 120 000 km² (Kuhle 1988a: 606/607) large Ice
Age Karakorum ice stream network is valid as a model. In its central parts the individual valley glacier streams reached ice thicknesses of at least 1300 m to approximately
2000 m.
Evidence of this is provided by lateral moraines, erratic boulders on high-lying flat
areas, such as valley shoulders and transfluence passes between adjacent valleys, as
well as by glacigenic flank polishings and abrasions, truncated spurs and highest glacial polish lines (Trinkler 1930, Wiche 1958, Haserodt 1982, 1989, Derbyshire et al.
1984, Schroder et al. 1988, Kuhle 1988-1994, Xu Daoming 1991, amongst others).
This was true for the section of the Hunza valley, selected as an example (Photos 3 and
4), and in both longitudinal valleys - the Hassanabad and Shimshal (Fig. 2) - which are
side valleys of the Hunza valley. The Hunza valley is, as a large Karakorum transverse
valley, connected to the Indus. The Shimshal valley runs northwards of the Hispar
Muztagh with the Distaghil Sar- (7885 m) and the Kanjut group (7760 m) (Photos 5-
8), the Hassanabad valley in the Batura Muztagh southwards from Shispare (7611 m)
and Sangemar Mar (7050 m) (Photo 1 and 2). The glacier levels, marked in the photos
by (----), indicate glacigenic flank abrasions as a confirmation of Ice Age to Late Glacial glacier thicknesses. This makes clear, that the trough profiles and partly very
steeply-flanked U-shaped profiles, which have an important part of the entire valley,
are of glacigenic origin (Photo 1-3 ----). The significant trough valley element is - in
the context of risk potential and the preparation of hazards - the very steep course of

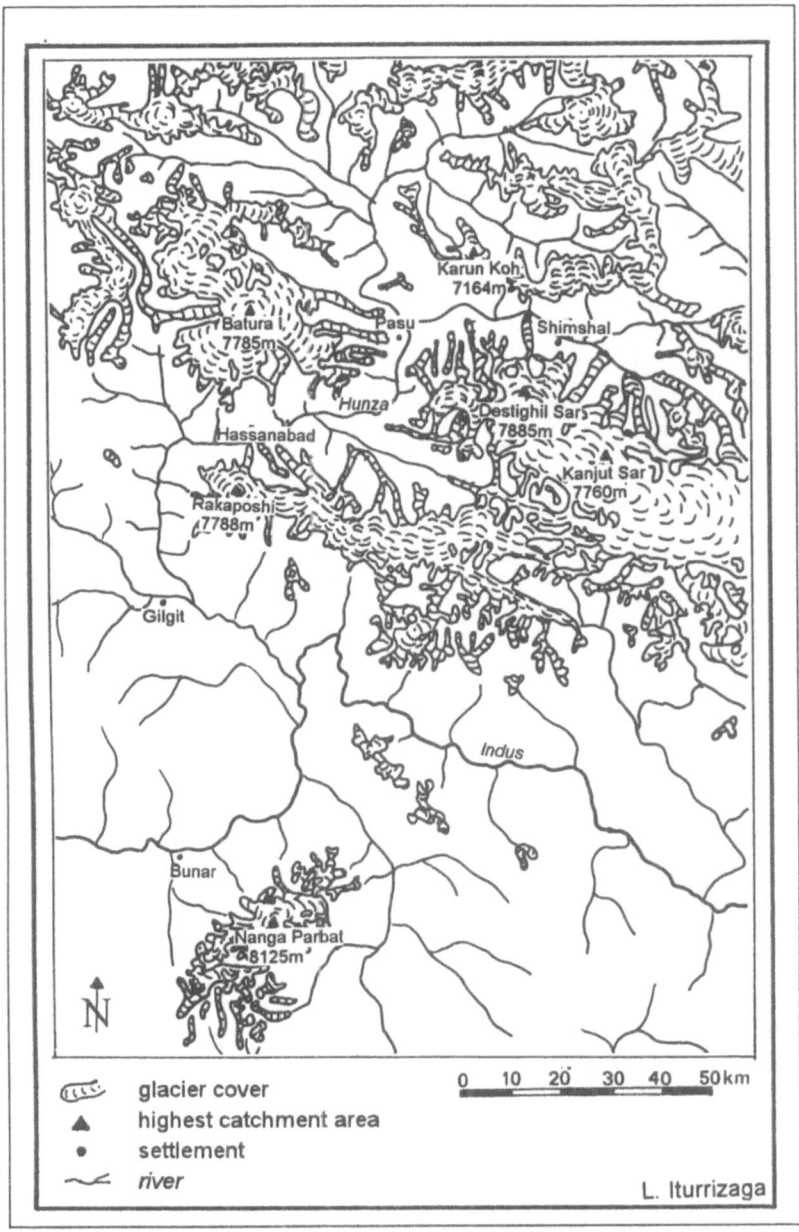

Fig. 2 The research area in the NW Karakorum and W Himalaya

Photo 1. View from 1970 m a.s.l. from the inflow of the Hassanabad valley into the Hunza valley, seen towards the N into the Hassanabad valley.(|)marks the c. 7050 m high Sagemar Mar (satellite of the Karakorum main ridge). (---- O) shows the minimally demonstrable valley glacier level, probably still reached in the Last Late Glacial, below of which the steep-flanked trough valley cross-profile was abraded by glacier erosion. (▶) indicates the steep flanks of the trough profile endangered by collapses; (■) the end moraine, which the Hassanabad glacier had still reached in 1925. (▽) shows debris cones and -screes, which obtain their supply from the deposits of Late Glacial ground moraines which adhere to the trough flanks (□). They threaten the houses with rockfall. Photo M. Kuhle 22.9.87.

the central part of the slope in comparison with a preglacial V-shaped valley profile (Photos 1-8 ▶ ◀). It normally extends over several hundred metres and lies in the middle of the sinusuidal slope course between the, above, convex and, below, - towards the valley floor - concave sloping curve. This slope- or wall-section is most affected by mass self-movement after the melting of the Main Ice Age glacier. Here, the force of gravity is able to attack most directly and thus takes away the rock-boulders and - debris, loosened by glacial frost weathering, nearly unhindered out of the rock bond. This section of the trough-flank, which is most exposed to the force of downward slope gravity, can therefore be described as over-steepened. That slope- or wall section was certainly stable subglacially, but tends since the melting of the valley glacier, i.e. sub-aerially, towards collapsing or crumbling away and hence to the formation of stretched V-shaped slopes.

This is confirmed by the debris cones and screes in the Shimshal valley, which start to develop precisely at these steep slopes of the U-shaped profile (Photos 5-8 ∇). The stratified bedrocks and phyllites (crystalline schists) in this valley produce debris bodies, which are composed of relatively homogeneous grain sizes. The falling down of these debris components takes place more or less successively; i.e. through rockfall. It endangers communication routes, such as roads and alpine paths (Photo 6 ↑).

2.1 A CONCRETE EXAMPLE OF A DANGEROUS, VERY FAST RESULTING CRUMBLING AWAY FROM THAT STEEP AREA OF THE TROUGH FLANK (SEE ABOVE) WILL BE PRESENTED FROM THE HASSANABD VALLEY.

This crumbling away is located at 36°20'N/74°35'E on the orographic left trough flank, 4.1 km inwards of the junction of this valley with the Hunza valley (Fig. 2), and approximately 2 km down-valley of the present (September 1992) termination of the Hassanabad glacier (2570 m a.s.l.). It lies in the valley cross section with an altitude of the talweg at 2270 m a.s.l. The fresh and active region of crumbling away itself is located 200-400 m above the talweg (Photo 2 □). Beneath this region, a debris cone of considerable dimension is heaped up from the rock slide material of gneiss (Photo 2 ∇). This young cone fills the valley floor up to two thirds of its width. It is crossed by two paths, which lead to the nine alpine pastures of the main settlements lying in the Hunza valley - Hassanabad, Aliabad, Murtazabad and others - up into the Shispar valley, the left branch of origin of the Hassanabad valley. The upper path (Photo 2 right above the ∇) is so endangered by rock fall and rock slide that it was nearly abandonned. In any case, it is no longer gone along by animal herds (goats and sheeps). Instead, the herds are driven up- and downwards along the less rock fall endangered, orographic right side of the Hassanabad Nala (Hassanabad river). On this right-hand side path they cross the tongue of the Hassanabad glacier, in order to avoid crossing the river. Only small groups of ibex hunters and collectors of precious stones can risk crossing the extremely active debris cone under the protection of its rock fall boulders - which are very large in the distal part (left from | in Photo 2) - on the path below on the orographic left side of the river. However, for reasons of safety this always only takes place following the shouted advice of a person, observing the debris cone and rock fall. On this route the authors have crossed the rock slide cone in the second half of September 1992 in the early evening hours. They had good weather during the ascent and descent. In both cases rock fall was observed: on the 20.09.92, indeed, a rock slide during which a several metres long, sharp-edged, freshly broken boulder fell on the cone. Its kinetic energy allowed to plough up the proximal cone-sector, composed of fine-grained debris, and to tear metre-deep holes in its debris coat before it came to rest (Photo 2 right above ∆). Normally such gneiss boulders bounce and roll right into the lower marginal regions of the debris cone. Here, they build up a bulgy, very wide peripheral rampart, rising over the further above stretched profile line of the cone (Photo 2 |). In the context of this study it is not possible to go in detail into the eventful history of the oscillations of the Hassanabad glacier tongue termination (cf. Mason 1935, Visser 1938, Goudie 1984, Pillewizer 1986, Iturrizaga 1994, Meiners 1996).

However, the following is pointed out with regard to the formation of this debris accumulation: still in 1925 the Hassanabad glacier terminated at only 2150 m a.s.l. (Photo 1 ■), approximately 400-500 m below its final height in 1992. This means, that the end of the glacier, which at that time lay 5 km further down-valley, has released the valley cross-section of the investigated collapse only after 1925. Consequently, that debris- and boulder cone was formed within a time period of less than 67 years from this rockfall, following deglaciation. The extremely active morphodynamics, induced by deglaciation, in this case did not affect the risks of an alpine path and its users alone, but also the construction and maintenance of an irrigation canal. About 7 years ago, the canal was led through this rock-flank from the terminus of the Hassanabad glacier, near the confluence of the Mutchual glacier and the Shispar glacier, up to the Hassanabad settlement. The trough flank collapse discussed, experienced an artificial undercutting as a result of the canal guideway, i.e. damage, and with that additional instability. The crumbling-away-activities, amplified by the canal, which through the canal water was still further boosted by frost weathering, have damaged and interrupted the canal (verbal information from residents, September 1992). As the cubic-angular boulder forms and the partly overhanging stepped forms of the breaking off (Photo 2 □) demonstrate, this occurs under the control of the clefts. It is oriented on the AC and BC levels of the release joints.

2.2 TWO FURTHER CASE EXAMPLES OF ENDANGERING OF COMMUNICATION ROUTES FROM THE ·HUNZA VALLEY SHALL BE MENTIONED

Photographs 3 and 4 show the exemplary localities along the orthogonal course of the valley between Pasu and Baltit. They lie in the Hunza-bend (Fig. 2) on the orographic left valley flank south of the Saret settlement. During the heavy rainfalls from 7th-9th September 1992, the Karakorum highway was damaged, blocked and also completely destroyed over distances of tens of kilometres. The case in Photo 3 (36°18'20"N/74°50'E; 2300 m a.s.l.) relates to a collapse resulting during this event, for which the High Glacial trough valley form was decisive. It broke out in gneiss, permeated by dykes (concerning the geology, cf. Schneider 1956: 10, 1957: 442 pp). The up-valley visible trough-walls (Photo 3, background), in their overall form still extensively undamaged (▼), were substantially undercut at this locality (Photo 3, foreground) by the highway route. Since the absence of the Ice Age glacier filling, i.e. since 13,000-15,000 years until 1981, collapses occurred only naturally. However, during the until then (1992) more than ten years existence of this undercutting (the Karakorum Highway was completed in 1981) the collapses have increased many times over (cf. the analogue effect of the Hassanabad irrigation canal). As a result of the outward-pushing pressure and traction vectors of the bedrocks in the steep rock face areas, fissures develop. Precipitation water infiltrates them. The fissures are widened by frost pressure and congelifraction and are disturbed in terms of their coherence by the eroding effect of rain and meltwater. Finally, the bedrock formation breaks apart when heavy rainfalls occur. At the exemplary locality in Photo 3, the partly very large,

Photo 2. View from 3.5 km further up-valley than Photo 1 from 2220 m a.s.l., looking into the steep glacigenic trough of the Hassanabad valley (to the N) on to collapses in the bedrock gneiss (□) and the fresh debris cone (Δ). (---- O) indicates the minimal level of the Last Glacial Maximum (LGM). (▲) marks the very steep to, in places, even over-steepened (i.e. slightly overhanging) trough walls, ready to collapse. Photo M. Kuhle 20.9.92.

angular boulders have fallen directly out of the formation of the bedrocks on the asphalt surface of the highway. They could only be removed in the following weeks with heavy road construction machines and by blasting.

From this locality only a few hundred metres the Hunza valley   upwards (36°18'20"N/74°50'01"E; 2300 m a.s.l.), another type of process occurred during the same incidence of precipitation. It has, however, the same root, namely the shape of the Hunza valley flanks, created by glaciers during the Ice Age (Photo 4). At this position, the round and smooth trough flanks, left behind by the glacial polishing (Photo 4 ⌒), are still extensively undamaged. Because of this form, they offer the highway route and its partly artificial deposits only little bearing surface and friction, i.e. too little hold and stability. Because of that, at this locality the complete route has

Photo 3. View from 2300 m a.s.l. from the orographic left flank of the Hunza valley, opposite the Saret settlement, looking up-valley to the ENE. The round-polished rocks of the valley flank (⌒) demonstrate the erosive
work of the valley glacier up to its minimal level (----O). (▼) marks the smaller collapse niches on the steep
sections of the U-shaped valley flanks. In the foreground (behind the person), a fresh collapse is visible, which fell
on the recently repaired asphalt surface of the Karakorum highway. Here, the route undercuts the steep valley
flank and thereby additionally makes it unstable. Photo M. Kuhle 18.9.92.

slid down by approx. 20-30 m in height over a distance of approx. 300 m, only interrupted by a few still existing remains of the route (x), up to the level of the Hunza
river. This process was combined with the coming down of mudflows (□□). These are
wet sliding and mushy flowing movements of higher-deposited loose materials from
debris cones (O). The debris masses, slid down and partly flowed off as mudflow, in
which the material from the highway route is incorporated to the point of being not
recognizable, are surged as far as into the Hunza river (■). Their small-scaled, hilly
Toma landscape constricts the Hunza river and is, because of that, undercut and eroded
by its lateral erosion.

The gneiss walls, polished smooth by glaciers ($\subset$), consequently offered the route too little primary hold and did not let infiltrate the rainwater. The water gushed down the compact gneiss walls and concentrated at the foot of the rock face and on the highway route. At the same time, this was potholed and underwashed at certain points. The additional increase in weight, resulting from the thorough wetting of the route body and the reduction in the adhesive friction coefficient on the wet rock faces of the trough flanks, were the triggering mechanisms for this hazard phenomenon, for which the incidence of heavy rainfalls was the impulse.

2.2.1 Effective repair measures

Consequently, such slid-down routes can only be lastingly repaired by - 1. Their anchoring in the rock. For this, rust-free steel girders with T-profiles must be embedded in concrete in bore holes. - 2. Below those rock face sections, where the rainwater concentratedly runs off due to channel-related hollowed forms, the route must be drained through stable underpass pipes made of steel and concrete.

3. Historical to Holocene glacial expansion as the basis for glacial hazards (S. Meiners)

Moraine- and glacial dammed lake outbursts can lead to damaging effects in the boundary area between ecumene and sub-ecumene. In order to explain the causes, the current changes in the glacier, such as a shifting of the glacial termination, must be placed in the complete context of the glaciation history. The hazard potential results from the effect of a melting ice stream network during the Late Glacial up to the distribution into dendritic tributary streams with historical and postglacial oscillations, or the uplift or lowering of glacier surfaces. For this, case examples from the NW Karakorum (Shimshal valley), from high plateau areas (Central Tibet, Tien Shan) and from the W and E Himalaya south face (Barun glacier, Khumbu Himal; Hati Parbat glacier, Kumaon Himal) have been chosen (Fig. 1).

3.1 ON THE GLACICAL HAZARD IN THE NW KARAKORUM, SHIMSHAL VALLEY

For investigating the connection between Holocene valley formation, recent glaciation and the danger potential developing from them, the transverse valleys in the NW Karakorum offer themselves. In a distance of about 20-30 km they join longitudinal valleys, which, for their part, have connections to larger transverse valleys. For this the Shimshal valley on the northern slope of the Hispar Muztagh (Fig. 2) is valid as a model. From the on average over 7000 m high catchment areas glaciers come down, which to a great extent fill their valleys with ice up to the junction with the Shimshal longitudinal valley.

The longitudinal gradient of the Shimshal valley, which runs in a westerly direction, over a distance of approximately 57 km from the valley origin at 3100 m a.s.l. up to the valley outlet into the Hunza valley at 2450 m a.s.l., is small. In the longitudinal valley course broad valley chambers alternate with gorge-like incised ravine sections.

The potential glacial hazard risk in this valley derives from the glaciers, the tongues of which flow into the Shimshal valley and come close to the Shimshal river. Amongst these are the glaciers of the upper valley, such as the Khurdopin/Yukshin (Photo 6 ●), the Yazghil glacier (Photo 5 ■) and the Malangutti glacier (Photo 7 □). Their oscillations can cause the formation of dammed lakes as a result of the recent ice body, which is stabilised at the glacier termination by the cake-shaped, surrounding historical to Post-Glacial end- and lateral moraine.

Apart for the permanent settlement of Shimshal at 3000 m a.s.l. (Photo 8 □), which is situated safe from high water between the Malangutti and the Yazghil glacier on a glaciofluvial alluvial fan about 100 m above the valley floor, smaller farmsteads are built on flat alluvial fans, such as on the orographic right side of the valley opposite the Malangutti valley exit or on its orographical left lateral valley mouth. In addition, simple rest houses are erected along the orographic left side, which are used by the Shimhalis wandering to the valley and carrying loads.

The glaciated tributary valleys are used for pasturing. In the narrower sense, the extended, westerly exposed lateral valleys, situated 100-200 m above the recent glaciers, and the valley slopes are put to pasture. Potential dangers that exist here are of rather small dimensions, i.e. they have smaller damaging effects. In this way, in particular in the case of glaciers the surface levels of which increase as a result of positive mass balance, it can happen that the lateral moraine valley run-off is dammed up by glacier ice, which overthrusts the lateral moraine ridge. Such potential lake outbursts make walking on the glacier by herdsmen, and also rucksack tourists, a risk. A further problem are the lateral moraine valley run- offs which, at appropriate positions, incise right through the consistent moraine material down to the surface of the glacier. These grooves are frequently used as possibilities for ascent on the alpine pastures in the lateral valleys, such that man and animal go closely one behind the other along the stream- bed on the loosely lying, outwashed moraine boulders. In this case, the unforeseen sliding of loose material can come about, which causes a burying. These risks are known to those using them and considered.

3.1.1 Case example: the Yazgil and Malangutti glacier

Inferences from the behaviour of glaciers during the Holocene allow an estimation of the changes in landscape as they are also to be expected in future, with changed climatic conditions and with a larger glacial expansion, connected with a lowering of the snowline.

The 30 km long Yazghil (36°13'-23'N/75°14'-25'E) descends as a firn basin glacier with an average catchment area height of c. 7000 m a.s.l. to 3060 m a.s.l.. Only little surface moraine debris comes on the glacier which, below the snowline at 5100-5200 m a.s.l., increasingly is covered by firn pyramids. The glacier termination extends hammerhead-like on the Shimshal valley bottom, so that the eastern branch reaches up to the Shimshal river bank (Photo 6 □ , Fig. 3), while the distance of the western branch with approximately 30 m is larger. Oscillations of the end of the glacier have been described since 1892 in the literature, from which it follows that the front of the glacier reached up to the river in 1892, while in 1908 the eastern branch lay 44 m from the right-hand side of the valley flank and the western part lay only 9 m away (cit. from Visser, 1938: 160). In 1925, both ends of the glacier are supposed to have been

Photo 4. View from 2300 m a.s.l. from the orographic left flank of the Hunza valley, opposite the Saret settlement, looking up-valley to the ENE. Up to (---O) the High- (LGM) to Late Glacial valley glacier has polished a very steeply flanked U-shaped valley profile into the bedrock gneiss. (▲) marks steep wall areas which are predestinated to the breaking-off of the glacigenic round polished rocks (⌒). (x) indicates a remnant of the route of the Karakorum highway, which slid down on the polished trough walls (⌒). In the fore- and background the still-existing highway route is likewise visible. (■) marks the debris which slid down with the route. Photo M. Kuhle 18.9.92.

Photo 5. From 3100 m a.s.l. (36°25'30"N/75°23'E) from the edge of an orographic right mudflow fan (it comes
from the Zadgurbin valley; the Bandasar settlement lies on it), seen the upper Shimshal valley upwards to the
ESE. In the middle ground the gravel floor (□) cf the Shimshal river, which during the period of snow melting is
overflowed in an extended width. Below the collapses in the bedrock (▼), debris cones (∇) and scree, an alpine
way endangered by rockfall runs up-valley up to the tongue end of the Yazghil glacier (■). This glacier, coming
from the orographic left side valley from the Karakorum main ridge, blocks off the Shimshal main valley (see
Photo 6). It could, on a future advance, dam-up the Shimshal river to a glacial dammed lake (cf. Iturrizaga
1996a). With that its behaviour involves the danger of a glacier lake outburst and the destruction of down-valley
settlements as for example Shimshal (cf. Photo 8). (---- O) marks the High- to Late Glacial and (----) the Late
Glacial glacier level. Photo M. Kuhle 25.8.92

pushed over the river without reports of the existence of a glacial dammed lake (ibid.).
These newest advances force over an older, neoglacial lateral moraine border (Meiners,
1996: 161), which is only preserved in the glacier centre, in the flow shadow of the
divergent termination of the glacier (Fig. 3), as a remnant of the former border. The
neoglacial advance in a direction directly down- or up-valley is consequently no longer
continued. Dead ice fills the space between the neoglacial moraine border and the
recent ice body. The post-glacial to historical glacier levels are only represented in the

fluctuations in surface height, which are canalised by the outermost moraines (Photo 5 ■). Potential danger derived from this, comparable to that of today's. The positive mass balance causes an uplift of the glacial surfaces in the middle sections of the glacial tongue, which results in a side-expansion of the glacier body over its historical lateral moraines and the cutting off of the lateral valley run-off. An advance of the glacier end is to be expected in particular, when a kinematic wave crest on the glacial surface continues as the expression of a positive mass balance up to the end. At the same time, a temporal delay is to be taken into consideration.

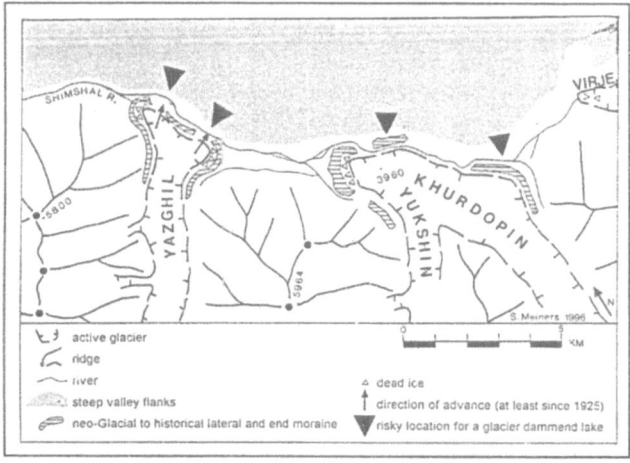

Fig. 3 Sketch-map of the Yazghil and Khurdopin/Yushkin glacier hazard potential

The endangering of the upper Shimshal valley increases through the position of the valley-damming, united Khurdopin/Yukshin glacier (Photo 6 ●). It allows the recent meltwater run- off of a side valley from the Virjerab, joining from the right, to pass without hindrance (Fig. 3). The damming up of the Virjerab run-off through the Khurdopin glacier led to the formation of a glacial dammed lake, for which a lake overflow was documented at the beginning of this century, in 1907. It took place within 11 days and left behind no serious damages. This dammed lake also existed in 1925 (Visser 1938: 162) and was still mapped in the map of the Survey of India from 1957. For 1959, a lake overflow was documented (Goudie et al. 1984:389), which had cut the top of the orographic left settlement terrace of Shimshal by about 300 m. According to eyewitness reports, in 1965 the lake is said to have once more run out, so that the tidal wave shifted the settlement terrace back this time by a further 30 m.

The Malangutti glacier (36°20'-29'N/75°14'E), which blocks off the wide valley chamber of the Shimshal village, ends directly at the orographic left Shimshal riverbank (Photo 7 ■). From the debris-covered end of the glacier, surface moraine boulders fall into the river, which are immediately transported away. Confined by the glacier, the river runs on the right side of the valley and undercuts the glaciofluvial ground moraine material which is attached to the flank (Photo 7 ■). In the current state (1992), the danger of a pushing forward and damming up of the main valley run-off is not very significant, because the glacier tongue surface, stretched forward like a cat-paw, in consequence of its concave cross-profile suggests a stagnation. Using the Malangutti glacier as a model, it can be shown, that even an advance up to the opposite-lying side of the valley must not inevitably cause a blocking up of the main valley, because the glacier can overthrust the smooth-polished bedrock valley floor, into which the Shimshal river has also cut. Moreover, a large basin-like gap at the upper edge of the post-glacial lateral moraine suggests that, in the course of historical phases, a fluvial surging of water or overflow over the glacier must have resulted (Meiners 1996:149). The historical to post-glacial glacier expansion of the Malangutti, which is established by the towering of the outer lateral moraine up to 200 m over the valley floor, determines the readiness to damming up. This arises from the relationship of the glacial surface increase to the behaviour of advancing. Within the last 100 years the position of the Malangutti glacier must have been rather stable. However, in 1908, the glacier stretched over the river, while in 1913 it already released the river again (cit. from Visser 1938: 158, 159). Similar as the Yazghil glacier, the Malangutti is a non debris-covered firn basin glacier type with a flat sheer-ice tongue, covered with firn pyramids.

In the widened lateral moraine valley exit an intensive utilization through the cultivation of cereals and field crops takes place. Because of the subsided glacial surface the meltwater is no longer tapped for the irrigation, but the discharge of the lateral moraine valley and the valley slope is drained off. Today, irrigation canals at two altitudes on the valley slope, which are no longer held in condition, trace the subsiding of the glacial surface. However, the giving up of irrigation by meltwater must not be a conclusive indication of a glacial retreat. In this case, it is rather an intentional giving up for economical and socio-culturally changed circumstances. Up to these days, the main path to the village of Shimshal leads over the termination of the glacier tongue. In the planned and already started construction of the jeep road, this glacier can only be circumvented, i.e. the way is bound to run along the opposite lying flank of the valley. In this connection, the construction should take place at a height above the level of the neoglacial lateral moraine i.e. about 200 m over the Shimshal river (Photo 7 ⇓).

3.2 ON THE GLACIAL HAZARD IN HIGH PLATEAU REGIONS OF THE TIEN SHAN (AK SCHIRAK) AND IN CENTRAL TIBET (TANGULA SHAN).

In the high continental plateau regions of Central and High Asia, the present glaciation is only insignificant. Typical in the case of small relief energies of 500-1000 m are short, cold firn basin and firn cap glaciers with temperatures in the snowline level of -6 to -8°C. The glaciers terminating in the periglacial altitude level, for the most part break off in a straight front to the end, which lies on a flat valley floor. A surface moraine cover is either only small or not present at all. Glacier tongues have few crevasses, so that the meltwater, coming down only in summer, to a great extent also flows off supraglacially.

3.2.1 Case example Ak Shyrak, Tien Shan

The Ak Shyrak plateau of the inner central Tien Shan (Fig.1) shows a total glaciation area of 411 km., which is spread out over 176 glaciers (Dyurgerov et al.1991:9,10). The average plateau height is about 3600 m a.s.l.. With 300 mm of annual precipitation and a mean temperature of -7°C, the Ak Shirak belongs to a dry, cold high continental climatic regime. The meltwater run-off takes place only in the summer months, in which the main amount of precipitation falls. The 9 km long Petroff glacier (41°50'N/78°09'E) on the west side of the plateau, with a minimum altitude of the catchment area of 4500 m a.s.l., breaks off recently with a straight front-line at 3750 m a.s.l. into a moraine dammed lake. In 1943 it was entered as somewhat wider than today in a map "Changes of glaciers in the Ak Shyrak-Range from 1943-1977". Ice-floes swim in the lake in the proximity of the break off, while farther parts are interspersed of kidney-shaped sand banks. In the today dried up bank regions exposed to the west, round kame-accumulations, containing an ice core, indicate the former ice margin. The lake is dammed-up by the moraine material, which was deposited by the Petroff glacier as a dumped end moraine during the Little Ice Age (Meiners 1996:15) and is underlaid by dead ice. To this an older, i.e. neoglacial dumped end moraine (3000 YBP) is adjoined. The position of the Petroff glacier is at present (1991) stable. The lake run-off takes place through an only 1-2 m wide opening in the historical end moraine, which can at any time be shifted or even obstructed by the unstable dead ice bridges in the moraine. This moraine dammed lake is no immediate danger, because firstly, permanent settlements are absent in these high valleys and secondly, a discharge would spread out in the wide valley area without causing high water damage. Large herds of sheep are driven over the plateau and also in the glacier forefields, so that here, at certain points, a danger exists. This example was chosen in order to gain information about the starting conditions of moraine dammed lakes and to carry out an assessment of risk in comparison to other regions.

3.2.2 Case example Geladaindong massif, Tangula Shan, Central Tibet

The Tangula Shan (32°05'N/ 91-92°02'E) stretches as a mountain range in a W/E direction through Central Tibet (Fig. 1). An annual precipitation of 400 mm per annum distinguishes the arid western part from the more humid eastern Tangula Shan with an annual precipitation of 400-700 mm per annum (Zheng 1983: 37-39). In the investigated Geladaindong massif at the central NE Tangula Shan an average temperature of 1.7°C and average highest daily temperatures of 4.7°C were measured (Meiners 1991:287) during a two-week long period at 5260 m a.s.l. in August 1989. The glaciation on the NE slope of the Geladaindong massif as the highest peak (6625 m or 6621 m) extends up to a height of 5400 m a.s.l., with which the average plateau height is approximately reached. A firn basin glacier south of the Geladaindong massif with a catchment area of 6361 m a.s.l. calves in a lake without run-off, dammed up by youngest end moraine (1850-1900 YBP), which is covered with ice-floes. The lake is bordered by the pertinent lateral moraines, which show no vegetation cover. The historical end-position of the glacier is fringed by an older historical lateral and end moraine (Little Ice Age ?), up to which the lake had once extended. The further glacier forefield - as a flat and over-deepened former tongue-basin - is bordered at 5400 m a.s.l. by a wall-like end moraine, which blocks the 1 km wide valley area. This very young lake formation could become a potential danger if, on the occasion of a glacial retreat - as it takes place at the adjacent Geladaindong glacier - large ice-floe-break-offs suddenly raise the lake level and bring it to a surge-like overflow.

The glacier forefield of the modern glacier, covered with thick ground moraine material, which is already saturated with water during the summer months, in this case would become additionally wet. In the peaty foreland, permeated with earth hummocks and hollows, which represents a temporarily used grazing area for several hundred numbers of yaks, sheep and goats, humans and animals could be endangered from a surge-like discharge of the lake. From the geomorphological point of view, the danger of an outflow of the lake exists in particular, when the glacier advances. Because of the small incline and the insignificant dissection, a flood wave might cause no considerable damage in the permament settlement region, situated 100 km away.

3.3 EXAMPLES FOR GLACIER HAZARD FROM THE SOUTH SLOPE OF THE E AND W HIMALAYA (BARUN GLACIER, KHUMBU HIMAL; HATI PARBAT, KUMAON HIMAL)

The glaciation on the southern slope of the high Himalayan bend is not so much pronounced as that of the Karakorum. The glaciers of the southern side, facing the summer monsoon, often reach the forest belt with their tongues.

3.3.1 Case-example Barun glacier, upper Arun valley, E Himalaya south slope.
The catchment area of the upper Barun valley (27°47'-58'N/ 86°58'-08'E) is formed by the SW to SE side of the Makalu main peak (8463m) up to a peak that forms the E boundary of the Everest massif. The lower Barun valley forms an impassable ravine section, which terminates in a narrow gorge towards the Arun valley at 1100 m a.s.l.. The upper Barun glacier is completely covered with surface moraine and ends flatly immediately at the foot of the SE steep flank of the Makalu at 4900 m a.s.l.. Spallings of ice balconies and flank ice or short hanging glaciers drop into the depth line from the Makalu wall, towering 4000 m. Here, a moraine lake is dammed up, in which the meltwater of the Barun glacier and the Makalu flank is collected. Fresh, probably historical, hill-like heaped up end- and lateral moraine material, which rises 80-120 m above the valley floor, forms the boundary of the lake, the level of which lies at least 50 m lower. A narrow outlet drains the lake. Kalvoda (1979: 23) describes the origin of three moraine dammed lakes in the forefield of the upper Barun glacier in a phase of glacier retreat between 100 to 1000 YBP, of which only the Barun lake still exists.

The lower Barun glacier (avalanche cauldron glacier) which comes from a less higher catchment area (6809 m) flows 5 km away from the upper Barun glacier termination in a longitudinal direction into the valley, such that the main valley run-off remains free in the south-exposed orographic left lateral valley. In contrast to the upper Barun glacier, the lower Barun glacier, likewise covered with surface moraine, forms a steep ice edge against a moraine dammed lake, which is fringed by dead ice blocks within a high lateral moraine border.

Similar to the comparable Imja Khola Lake in the Khumbu Himal (Watanabe, Ives, Hammond 1994), the stable lateral moraine border of the historical to neoglacial glacier expansion is responsible for the damming up. Downwards of the lower Barun glacier at 4500 m a.s.l., the main valley gradient curve descends into the middle Barun valley by 1000 m.

Access to the Barun valley can only take place over two 4100 m high passes, leading into the middle Barun valley. Because of its difficult accessibility, the Barun valley cannot be permanently settled. The upper to middle valley up into the glacial forefields is used as a pasture area. A fixed, temporarily used dairy settlement lies

Photo 6. View from 3230 m a.s.l. from a position on the tongue of the Yazghil glacier (see Photo 5), 180 m above the Shimshal river (36°24'N/75°23'30"E) facing ESE. In the background the tongue termination of the 47 km long Khurdopin-Yukshin glacier (●), which until 1959 and finally in 1965 dammed up a glacial lake in the upper Shimshal valley (cf. outburst of this lake, Photo 8). (----) marks the Late Glacial, (---- O) the minimal High Glacial valley glacier level. In the region of the orographic left valley flank, below the collapses in the bedrock (▲), rockfalls (◤) very frequently take place which build up the debris cones (Δ) and spread and endanger the alpine path leading above the outer slope of the river. The present (1992) end of the 31 km long Yazghil glacier (□) reaches the Shimshal river (⇩) (Meiners 1996: 161) and will dam it up on a further advance to a glacial lake in the region of the higher valley section (O), the outburst of which endangers the downwards-located valley settlements (e.g. Shimshal, Photo 8). The 250 m high Neoglacial moraines on the orographic right flank of the Shimshal valley (■) confirm such glacial lake cammings for the last millenia of the Holocene up to the historical time. The position of the glacier tongue is now similar to that mapped by Visser (1938: 160) in 1925. Photo M. Kuhle 23.8.92.

Photo 7. View from the orographic left lateral moraine landscape of the Malangutti glacier from 3100 m a.s.l. (36°29'30"N/ 75°13'E) on its tongue, covered by surface moraine debris (□), which reaches the Shimshal river at 2900 m a.s.l. (Meiners 1996: 149). The river even undercuts the present (1992) glacial tongue (◣). At that moment, when a kinematic wave crest causes a sudden further advance of the glacier tongue of this 24 km long glacier, the Shimshal river is also dammed up here (cf. Photo 6). With that, there exists the potential risk of a glacial dammed lake and its catastrophical-like outburst in the near future, i.e. in the coming years up to decades. This periodic blocking-off of the depth line and the damming up of the river to a glacial lake is proved in an inverse-chronological way by the 200-300 m thick kames- and moraine deposits of the Malangutti glacier (■) on the opposite flank of the valley. On these moraines, younger debris cones (∇) are deposited below collapse-exposed trough walls (▼). The position of the glacial tongue in 1992 is approximately the same as Visser (1938: 159) mapped in 1925. The photograph was taken in a NE direction. Photo M.Kuhle 16.8.92.

Photo 8. View from 3060 m a.s.l. over the edge of the settlement terrace of Shimshal (36°26'N/75°18'E), seen up-valley in an NE direction. (---- O) marks the minimal height of the Ice Age to Late Glacial glacier level, which is confirmed by the form of the valley flanks, (▼) collapses in the rock and (∇) the development of debris cones and screes during the Holocene. The settlement terrace of Shimshal (□) was swept away by the high water wave in a width of 300 m by way of the glacier lake outburst of the Khurdopin glacier lake (cf. Photo 6) in 1959. In 1965, another outburst of this lake followed, which shifted back the edge of the terrace by a further 30 m and, as a result, destroyed a few farm houses and their terraces with working areas (↘). (More detailed information about this in Iturrizaga 1994: 64-80; 1996a:4ff). Photo M. Kuhle 3.9.92.

Photo 9. A small hanging glacier (■) in the Hati Parbat tributary valley, upper Alaknanda river, Kumaon Himalaya   (30°30'N/79°30'E) narrows the cross profile. Pressed hard by the completely debris-covered glacier tongue
end (■), the river undercuts (↓) the opposite-lying side covered by moraine material. Even a small advance can
cause a blocking up of the valley run-off, which leads to the development of a glacier dammed lake. Photo: S.
Meiners 3.10.1993.

protected from high water on a valley-blocking rock bar below the lower Barun glacier, while further resting places are spread out in the middle Barun valley. An outburst of the lake would significantly raise the run-off-peak of the Arun river.

3.3.2 Case-example Hati Parbat West glacier, upper Alaknanda valley, Kumaon Himalaya.

The Hati Parbat massif (6727 m) of the Kumaon Himalaya (30°30'N/ 79°30'E) i.e. W Himalaya, is cited as an example of a strongly monsoon-influenced Himalayan south side valley (Fig. 1). The 8 km long Hati Parbat W valley joins - as a tributary valley of the third order - the upper Alaknanda valley at the settlement of Govindghat at 1700 m a.s.l..

Between the settlements of Badarinath and Joshimath the upper Alaknanda-valley has for approximately 32 km a pronounced ravine-like character, such that a difference in altitude exists of 1780 m. On both sides of the ravine are high terraces leaning on the valley slopes.

The Hati Parbat W glacier fills the upper valley chamber up to 4050 m a.s.l. Historical and neoglacial lateral moraines frame the surface moraine-covered firn cauldron glacier. Beyond the glacial termination, the valley chamber widens over a length of 3-4 km to a large broad valley which, at a topographically determined narrowing of the valley at 3450 m a.s.l., is almost obstructed by a hanging glacier, joining from the orographic right side. The main valley run-off already undercuts the high-rising glacier front and the outcropping rock of the opposite lying valley flank (Photo 9 ↙). Especially on its right-hand side this glacier is flanked by a 150 m high lateral moraine, on the outer slope of which crooked birch trees grow. If the glacier advances only a few metres or if an entire piece of the tongue breaks off, as in this case appears absolutely possible, then the possibility exists of the formation of a glacier-dammed lake.

3.4 SUMMARY

While for the NW Karakorum glacier-dammed lake formations occur most frequently, the debris-covered glaciers in the valleys of the Himalayan south side form predominantly moraine-dammed lakes. They are surrounded by historical or Holocene lateral- and end moraines, such that the proportion of dead ice in the moraine plays an important role with regard to the stability. In the high plateau areas, sheer-ice firn basin glaciers often break off with their glacier fronts into moraine dammed lakes. The latter are dammed up by historical, partly hilly heaped up dumped end moraines. It seems, that the endangering by glacial hazards in the Karakorum is higher than in the Himalayas, where the pressure of population does not have an effect in the closer surroundings of the recent glaciers.

4. The geomorphological transformation of debris accumulations in the NW Karakorum (L. Iturrizaga)

The Late Glacial glaciation in the NW Karakorum (cf. Kuhle 1988: 271-273) left behind - apart from sedimented kames formations against the glacier ice - a mountain landscape, largely cleared of slope debris bodies. The valley basins were, following deglaciation, dressed with ground moraine material, on which the postglacial debris bodies successively were accumulated. Thus the debris bodies, such as alluvial and mudflow cones, which today serve as settlement ground, could only completely develop after the retreat of the ice 10,000 to 5000 years ago. The debris production, induced in connection with the history of glaciation and the making available of loose material deposits in the semi-arid high mountain region is, through its resedimentation, e.g. in the form of rockfall or mudflow, a source of danger for the local inhabitants.

4.1 THE DEBRIS ACCUMULATIONS AS A SUPPORTIVE MEDIUM OF SETTLEMENT IN THE NW KARAKORUM

The vertical relief distance of up to not quite 6000 m (e.g. between Rakaposhi (7785 m) and the Hunza river (1850 m, Fig. 2)) over only a few kilometres in the horizontal direction, illustrates the steepness of the relief supply and implies at the same time its unsuitability for the purpose of settlement over wide areas of the valley flanks. Extended areas of debris-free zones caused by valley flank gradients of more than 40° are facing the lush sedimentary valley floor fillings (Photo 1, 8 □ and 10 ▼). The high vertical relief distances favour not only the occurence of high kinetic mass-movements, but connect the settlement areas lying in arid valley locations with the nival and glacial catchment areas. They deliver the meltwater, which is necessary for the mobilization of the loose material to be found in the valley sites, deficient in transportation. An upward settlement expansion is prevented, in addition to the steepness of the relief, by the series of debris scree, which border the valley flanks up to several kilometres (Photo 10 ●). In this way, the locations of the permanent settlements in the NW Karakorum are confined chiefly to the debris bodies of the valley floor, as to the sediment fans, i.e. alluvial and mudflow cones (e.g. Bandasar (3100 m)), the morainic depositions, such as ground and end-moraine material (e.g. Shimshal 3080 m), the glacio-fluvial terraces (the main concentration of settlements of the NW Karakorum, e.g. Karimabad 2300 m), as well as glacial and fluvial alluvial fans (e.g. Pasu 2650 m) and sediments of dammed lakes, caused by glaciers and landslides (likewise Pasu). The alpine settlements are localised, for the most part, in the lateral moraine valleys (e.g. on the orographic right and left sides of the Batura valley) or also in the pass sites (e.g. Shurt 4450 m).

The bringing of the soil into an agriculturally viable condition is - because of the aridity in the NW Karakorum with an average of 130 mm/year precipitation (Kreutzmann 1989: 69-71) - only possible with the help of irrigation measures which substantially dictate the choice of location of settlements. As the conception of the irrigation diversion is purely gravitional, the locations of settlements are tied to a low-lying settlement locality, which cuts the active slope foot zone.

The type of agricultural irrigation places the subsistence-economy oasis settlements in close dependence on the functioning of the canal supplies. The conditions of the water supply determine the possibilities of existence and expansion of the settlements. Because of the exposed sloping location, the canal paths, which are frequently laid in debris screes are, however, exposed every year to a high probability of destruction by mass movement processes (Photo 10 ↘).

Photo 10. View from the Shimshal valley gravel floor from 3100 m facing W to the orographic left, glacially polished valley flank, covered with moraine material. Down-slope of the high-lying, strongly dissected moraine deposits (↙) debris cones are joining (•), furrowed by mudflow processes (↘). The main settlement area of Shimshal (▼) is situated on a glaciofluvial alluvial fan close to the distal parts of the debris cones. Numerous canal tracks (↓), destroyed by the debris movements, end abruptly in the debris cones. Sallow thorn plantations are to consolidate the slope. The settlement unit of Barkut is located in the transition of the side moraine of the Hodber Valley (⋎) to the adjacent slope. Photo: L. Iturrizaga 10.8.1992.

4.2 IMPEDING OF SETTLEMENT THROUGH GLACIAL-GENETICALLY INDUCED MASS MOVEMENTS DEMONSTRATED BY THE CASE EXAMPLE OF THE HIGH MOUNTAIN SETTLEMENT SHIMSHAL (3080-3200 M)

Debris accumulations in the form of moraine accumulations, alluvial and mudflow cones, as well as the formation of kames serve in the upper Shimshal valley section as supportive mediums for the permanent settlement of Shimshal (3080-3200 m, 36°26'N/75°17'E, Fig. 2). They are more or less consolidated debris accumulations, which were originally sparsely occupied by vegetation and are now taken into culture. The following observations are based on an investigation carried out in 1992 on potential natural dangers of the Shimshal valley (Iturrizaga 1994, 1996 a). The settled sediment strips, principally located on the orographical left side of the Shimshal valley (Photo 8 □), are flanked in the sloping direction by screes, which upwards are fed by high-situated moraine deposits (Photo 10 ↙). On the orographic right side of the Shimshal valley only two mudflow cones are settled by ten households.

4.2.1 The mudflow cone as a product of the resedimentation of glacigenic deposition

The mudflow cones offer, on the one hand, high water and mudflow protection with regard to the main valley, on the other hand they are exposed in principle to the identical types of danger from the tributary valley, their catchment area and site of origin. This constellation is interesting, because the partly catastrophically occurring mudflow processes, which in the past contributed to the building up of the cones and, with that, to the location of the settlement, can today be, in the same way, cause of the destruction of the cones and, with that, of land for settlement. The formation of the extensive sediment fans lies in the altitude zone between approx. 1000 and 3000 m in the NW Karakorum (Iturrizaga 1996b: 2). The Zadgurbin valley exit on the orographic right-hand side of the Shimshal valley (Photo 5), for example, is bordered by a mudflow cone with a diameter of about 1.5 km and with a bank tilt height of 60 m. Only a central run-off canyon divides the cone into two halves. The up to more than 6000 m towering catchment area of the Boesam Pir holds several glaciers of 1-4 km in length with glacial terminal tongue sites around 5000 m. In the lower valley area, the extremely steep Zadgurbin valley is completely cloaked with Late Glacial moraine depositions, which form the starting material for large-scale mudflow courses. This strongly inclined mudflow cone with a bank edge, that ends almost vertically towards to the gravel floor, is occupied by two branch settlements of Shimshal with only five to six households. During the Late Glacial period, as the Shimshal valley was filled with glaciers, the debris load from the side-valleys and ravines was deposited against the glacier in the form of kames (Photo 7 ■). Only after deglaciation this mudflow cone was able to spread out in the form of a semicircle. The distal cone sections are, however, recently undercut, in particular in the summer months and maintained successively steep by fluvial processes. Debris cone formation out of the mudflow material at the base of the steep bank edges provides evidence of the disintegration of the mudflow cone. Today, mudflow movements clear.out the central mudflow cone run-off every year and form a secondary cone deposit. The building up and degradation of the mudflow cone lie closely together. The genesis of the mudflow cone resulted from a few catastrophic events through the resedimentation of the moraine material, made available in the catchment area. The steep tributary valley inclines in combination with the, at times, high rates of meltwater, delivered favourable preconditions for the removal of the moraine material. The nowadays up to an altitude of 2700 m advancing glaciers in the NW Karakorum show current formation of kames at lower sites (e.g. on the orographic left Batura valley side). They represent the potential permanent settlement sites if the glacier retreats.

In Shimshal, the sequence of colonization took place from the lowest accumulations, i.e. from the glacio-fluvial sediments, deposited between two end moraines (Photo 8 □ and 10 ▼), via the alluvial fans of Chukurdas and Aminabad and finally to the high-lying settlement locations on the mudflow cones on the orographic right side of the Shimshal valley. Here, a settlement policy can be observed that shows a reverse direction to the argument of high water and mudflow protection. The advantage of irrigation and cultivation appears to dominate over the criterion of settlement security.

4.2.2 High-lying moraine deposits as the basic material for mudflow movements

Particularly in the extended valley chambers (such as at Sost 2850 m, Garkuch 1800 m or Gilgit 1450 m), the flanks are covered up to several hundred metres above the valley floor with ground and lateral moraine material, into which also some lake sediments are incorporated. In Shimshal as well morainic, coat-like complete covers occur on the glacially polished valley flanks on the orographic left side of the Shimshal valley (Photo 10). These deposits are the remnants from the Late Glacial glaciation at a lowering of the snowline by 1200 m. The valley flanks reach up to almost 5000 m and show, in the month of August, snow deposits at high altitudes. The high moraine deposits are at many locations already strongly furrowed. On the one hand, the moraine material preserves the slope, in that it protects the outcropping rock from atmospheric and fluvial erosive processes; on the other hand, it provides to a large extent the basic material for dry and humid mass movements which, in particular, is mobilised in the time of snow melting in spring, as well as with the occurence of precipitation. The behaviour of moraine material could be observed during an exceptional event of precipitation in September 1992 (see below) on the orographic right side of the Batura valley, near to the alpine pasture Yunzbin (2750 m, 36°30'N/74°50'E). The precipitation, which at the beginning fell as rain and later passed over into sleet or snow, loosed over two days without a break rock material, which was attached to the slope, out of a moraine wall. First the fine material was washed away, so that larger boulders lost their fixture and slid down. The stones were slowed down in a sand basin, lying at the foot of the moraine slope, so that the alpine hut, situated approximately 40 m away, was no longer reached by the rockfall processes. As soon as the precipitation ceased, the moraine material was "baked" as firmly as cement. The young orographic right-hand lateral moraine of the Batura glacier at 2700 m is composed of significantly larger rock boulders than those of the moraine wall described. If a larger rock block is set in motion, then its absence from the rock face can take away the abutment from the moraine material lying above and cause large-scaled collapses. In consequence of the precipitations, this c. 100 m high moraine slope slid down to a great extent in its lower section, i.e. over a horizontal distance of 75 m, and then completely tore away the constructed alpine path. After this event of falling down and sliding, the material of the fresh, shifted moraine deposits, which showed very little fine material on the surface, was very firmly interlocked.

4.2.3 The pure and moraine-marked debris cones: their mudflow and rockfall susceptibility

In addition to the wide-spread, debris-free, steep slope sections, the debris slopes count amongst one of the most eye-catching forms of landscape in the NW Karakorum (Photo 1∇, 2 ∆ , 6 ∇ and 10 ●). Remarkable is their vertical extent of spread. In this way, they are not only represented in the snowline border as a zone of maximal debris production, but occur also in the altitude between 1000 and 5000 m (Iturrizaga 1996b: 2). A large part of the valley flanks is occupied by this loose material zone, which thus represents the slope-ward boundary of many settlement areas (Photo 10 ●). The wide spreading of the debris cones can be explained amongst others by the glacial preshaping of the relief. The glacial over- steepening of the valley flanks (Photos 1 and 2) delivered the abrupt, concave-formed angle of slope between the valley bottom and valley slope, which is necessary for the accumulation of debris in a conical form. At a moderately stretched slope, prepared from fluviatile processes, debris covers would arise. The supply of debris is favoured in the mountains, rich in incoming radiation, by the high frequency of freeze-thaw cycles which occur in every season in alternating altitude regions (Hewitt 1989: 14), as well as by the only sporadically occurring vegetation cover. Up to 1200 m high screes, extending over several kilometres, with an angle of slope of up to 40° border the valley flanks (Kalvoda 1992: 206).

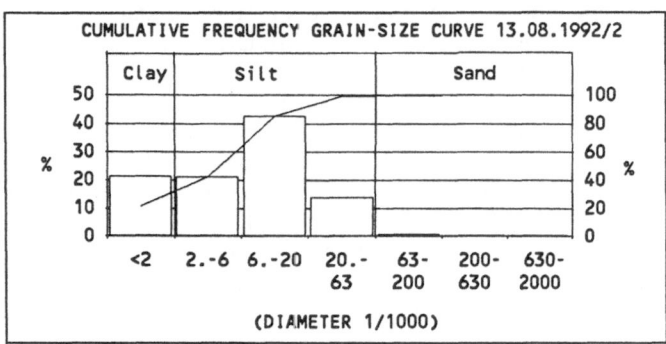

Fig. 4. Sample 13.08.1992/2. Orographic right bank formation of the Late Glacial to Neoglacial Momhil glacier (catchment area of the Shimshal valley) at 3520 m a.s.l; far above the surface of the present Momhil glacier tongue, approximately 3 km up-valley of its end (NW Karakorum, 36°26'N/75°04'E). The, on wetting (by snow or rain), sliding and mudflow susceptible moraine character of the many times redeposited and finally in the form of a debris cone as a kames deposited substrate, can be immediately identified by the preponderance of clay and silt in its fine-grained matrix or: the coarser sand fraction in the moraines is conspicuously of secondary importance in comparison to the primary debris cones (cf., however, Fig. 5).

Rockfalls represent the most frequent source of danger in the Karakorum for the settlement areas adjacent to the slope (Photo 6 ▽). The multitude of screes is in a saturated, critical state, i.e. they have reached their maximum angle of slope, which lies between 35 and 40°. That means, that each of the rocks falling down the slope has a high probability of reaching the distal areas of the talus cones and to penetrate into the settlement zone. In Shimshal, rockfalls reach up to approximately 300 m into the village land. The sallow thorn plantations, adjacent to the slope, are littered with fresh pieces of broken rock. Here, in contrast to the village land areas, they are not immediately removed from the Shimshalis. In the case of small fields, boulders that have fallen, represent already distinct harvest loss. Large boulders, which already on the foundation of the village lay on the settlement land, are frequently integrated in the building of houses. With its air-tight surface, the boulders provide protection against the attacks of wind and rockfall. By means of the planting of sallow thorn, the debris cones can be partly consolidated. With the construction of this natural buffer zone, the initial phase of this protection measure appeared the most problematical, i.e. the laying of the canals in the screes (Photo 10 ↓). In Shimshal it has been attempted for a long time, in vain, to lay several irrigation canals in loose material. Mass movements on the talus cones, such as mudflow and the sliding of debris prevent the plan. Not least, the canal itself undercuts the slope and initiates movements on the debris cone surface. Rockfalls are recorded throughout the whole year while, in spring, at the time when the snow melts, they increasingly appear when precipitation occurs, as well as storms. On this occasion, the predictability of the process already ends. Neither the direction of travel and the radius of action, nor the point in time and the size of the falling rock pieces can be predicted.

The debris cones in Shimshal are characterized through a high overprinting by mudflows, which run like an arrow and have dam-like borders, with a basal debris lobus. The predisposition of the debris cones in respect to mud flows is explained by their high content of fine material from clay and silt (Fig. 4). The fine material of high-lying moraine deposits is transported little by little down the slope. In particular at the time of the melting of snow, the high-lying moraine deposits on the valley flanks are mobilised and send mudflow movements, which run out over the debris cones. Most favourable for the mudflow movements are, in the extreme case, the moraine debris cones. These are debris cones, which derive directly from the redeposition of moraine material, such as those which can be observed e.g. on the orographic right-hand side of the Momhil valley along of a Late Glacial lateral moraine (Fig. 4). The high capacity to swell of the clay components transforms the moraine substrate on soaking into having a plastic flow behaviour. Pure debris cones (Fig. 5) are interlocked by means of their isolated components. This cone construction, following the "modular concept", makes them more stable, so that they react, as a tendency, on precipitation rather with rockfall. The two diagrams clearly show the differences in fine material content of both types of debris cones. The debris cones, characterized by moraine material, often show on their surfaces flat areas of debris slides and incipient cracks. Also, pieces of rock rounded at the edges, deriving from the moraine material are to some extent to be found on the debris cones. The mudflow lobes in Shimshal strike directly into the settlement areas. A several years old mudflow travels close to the settlement unit Barkut (Shimshal, Photo 10). The mudflow track created by the running off water masses has a diameter of 60-100 cm and is surrounded on the sides by mudflow walls. The mudflow track ends in a 6 m wide, hardly thick debris lobus. Today there are already fresh rock pieces on the mudflow track, which come from the slope. This mudflow movement, covering a vertical distance of more than 100 m, destroyed all six slope-parallel running canal branches of the main canal of the Shimshal settlement, the Hodber canal, on this debris cone. After a little meandering mudflow movement on the debris cone root, this broke through the canal branches mentioned, passed through the sallow thorn plantation, situated below, traversed a further moraine sequence, then several still cultivated terrace fields and ended a few metres in front of a Shimshal dwelling house in a lobular form. According to accounts of inhabitants of Shimshal, the mudflow took place at the end of April at the time of the spring melting of snow. A fracture of a canal at this point in time proved particularly precarious for the village inhabitants, as the start of the vegetation period is at this time. The likelihood of a repetition after a mudflow movement is, for the coming years low, as the delivery channel is cleared of loose material. Allowance is made - if possible - for the high activity of the debris bodies in an adapted housing distribution plan. One finds in the zone at the foot of the slope, in addition to green areas for young animals, chiefly threshing places, storage houses and cattle-sheds, arranged closely one behind another in the line of the debris cone (Fig. 6). Terraces which have been given up and destroyed by slope processes testify for an unsuccesful attempt of cultivation in the slope region.

4.2.4 The occurence of heavy precipitation between the 7th and 9th September 1992

The extraordinary precipitation events in summer contribute today - as is typical for arid regions - through the shifting of the loose material deposits, in combination with high relief energy, most markedly to the change in landscape in the NW Karakorum. An invasion of bad weather from the 7th to the 9th September represented, according to statements of the Hunza valley inhabitants, the most devastating occurence of precipitation since the year 1929. The amount of precipitation, which continually fell distributed over 56 hours, is nearly comparable with the annual amount of precipitation of 100 mm. From an altitude of 2700 m, the precipitation already fell as snow. The damaging effect of this precipitation affected all infrastructure facilities, such as dwelling houses and buildings with an economic use, route lines and canal installations, as well as field areas.

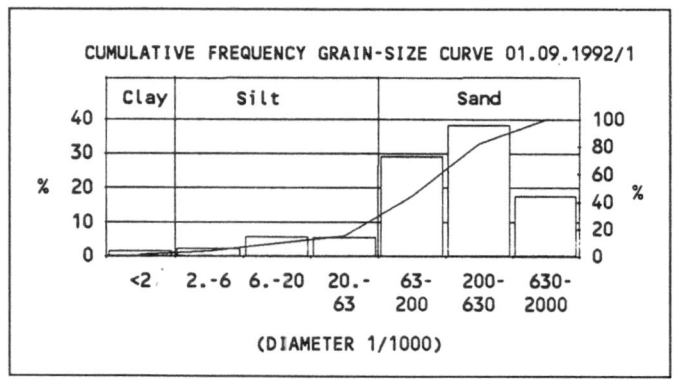

Fig. 5. Sample 01.09.1992/1. Slope debris, covering the bedrock at 3500 a.s.l., orographic right of the Shimshal Pamir valley, about 3 km to the north of the Chatmerk pass (NW Karakorum, 36°27'N/75°30'E). The matrix of this weathering material, producing primary debris cones which developed in situ, contrasts to the more slide- and mudflow vulnerable matrix of the moraine material (cf. Fig. 4) through a predominance in the coarser and, with that, more stable sand fraction.

4.3 THE ACTIVITY OF DEBRIS BODIES IN THE REGION OF THE ALPINE PASTURES IN THE GLACIAL LATERAL MORAINE VALLEYS AT SELECTED LOCATIONS: RAKHIOT VALLEY (NANGA PARBAT), BATURA VALLEY (BATURA I) AND HASSANABAD VALLEY (SHISPAR)

Sought-after locations for the temporary settlements are the lateral moraine valleys. They possess, by virtue of their own humid climate, a higher stock of vegetation than the non-glaciated valleys at the same altitude and offer access to extensive pasture areas. The climatic relations consequently determine the change from the permanent to the temporary type of settlement. This temporary settlement zone extends in an altitude between 2800 m and 3900 m. The lateral moraine valleys, i.e. the space between the glacier or its lateral moraine and the adjacent valley flanks, vary in their width from several metres up to hundreds of metres. Lateral moraine valley lakes, filled up by sedimentation, side valley alluvial fans and moraine deposits offer here a favourable location for settlements. Often, however, the lateral moraine valleys are so narrow that settlement must take place on the adjacent slopes (e.g. Yazghil pasture 3900 m/Yazghil glacier). But there also exist situations, where the lateral moraine valleys abruptly come to an end and the glacier directly undercuts the bedrock. An example for this is given on the orographic left side of the Batura glacier at approximately 2700 m.The glacially undercut, perpendicular-standing limestone flanks with partly overlying moraine material are highly dangerous, in terms of the danger of rockfall and rockslide, for the footpath leading directly under it. A blocking of the lateral moraine valley paths would make an improvised crossing of the glacier necessary. If the lateral moraine valleys are wide enough, alluvial and mudflow cones can develop. Two localities are chosen here as examples: the alpine settlement Bechal (3500 m)/Rakhiot valley, as well as Fatmahil (3330 m)/ Batura valley.

4.3.1 The lateral moraine valley on the Nanga Parbat-N-side

In the lateral moraine valley on both sides of the Rakhiot glacier on the Nanga Parbat N side (Nanga Parbat 8125 m, Fig. 2), the sediment fans of the tributary valley are rather active. Their catchment area lies in the nival and glacial zone. Mudflow as well as snow and ice avalanches run over the fan surfaces. On the orographic left side of the Rakhiot valley, the supply region of a selected mudflow cone ends in the Jiliper pass, located at 4837 m, which is framed by mountain crests of about 5000 m. The valley flanks are occupied by numerous debris cones, which in winter are covered by thick snow deposits. In 1994, according to the statement of a local resident, Mr. Rehmet Nabi Raes, snow avalanches discharged, which even the thick birch and pine forests on the foot of the fan could not withstand and were completely destroyed. Consequently, the alpine pasture Bechal (3500 m, 35°21'N/74°35'E) is not positioned immediately on the mudflow cone, but lies on the opposite-lying lateral moraine outer slope of the Rakhiot glacier. Further down the valley, numerous mudflows stretch along under the forest, which were sent from a nival catchment area. Similar mudflow processes were also observed in the Jaglot valley (Rakaposhi-W-side). On the orographic right side of the Rakhiot valley, steep glacier tongues flow down up. to a height of approximately 4800-5000 m. High rates of meltwater clear the steep side valleys of debris and provide mudflows. It is demonstrated in the examples mentioned very clearly that the side valleys, which join the lateral moraine valleys, because of their catchment areas in the range of fluctuations of the snow line and the related high snow depositing and melting processes, show a very high activity.

4.3.2 The lateral moraine valley at the Batura glacier (Karakorum N slope)

The position of the alpine pasture Fatmahil (3330 m, 36°34'N/ 74°40'E) is on an alluvial fan in the orographic left, extended Batura lateral moraine valley. The catchment area of the 57 m long Batura glacier reaches with the Batura I up to 7785 m; its glacier tongue comes to an end at a height of 2540 m. The arrangement of the buildings of the alpine settlement Fatmahil is line-like, so that eight housing units are put together in a row, to which a cattle kraal is connected. This type of building saves, on the one hand, building material and, on the other, it offers a more favourable heat budget than is the case with houses, standing isolated. The alpine pasture Fatmahil is located at the foot of a moderately inclined, c. 20-25° steep slope, which is supplied with closely-standing earth pyramids of up to 15 m in height. These earth pyramids are crowned by up to room-sized blocks. House-sized blocks lie directly on the slope. The earth pyramids represent the remainings of Late Glacial lateral moraine generations and have been observed, in a similar form, on the Yazghil glacier as well as on the Hassanabad glacier. Earth pyramids are to be regarded as potential sources of danger for rockfall during soaking processes and strong winds. However, the rockfall, which is caused daily by the driving down or driving up of the goat herds, is to be evaluated as

being disproportionately higher in comparison with "natural rockfall". But also unconsolidated debris slopes endanger the settlement area, situated further down-vally,
through the occurence of rockfalls.

*4.3.3 Late Glacial lateral moraines and narrow neoglacial to historical lateral mo-
raine valleys at the Shispar glacier*

The glaciers are not always accompanied by lateral moraine valleys. In the middle
of the Shispar valley (36°25'N/74°38'E) - the upper valley part of the Hassanabad valley (see section 2) - the Shispar glacier borders on the orographic right side of the
valley directly on the valley flanks, so that debris bodies are not formed. On the orographic left valley side a 200 m high Late Glacial moraine terrace serves as alpine
settlement location. The settlements are located on the top of the terrace at a height of
3500-3600 m and are called Daltar and Kaltar. The up to over 7000 m steeply proje-
cting gneiss-pillars - towering above the moraine terrace - already exceed the
maximum angle of slope for the debris cone formation (larger than 40°), so that only at
the foot of the wall rock fall cones can be observed with coniferous trees. In addition to
the vegetation stock, the absence of freshly broken rock- pieces points to the inactivity
of the wall sections. A gorge-like valley divides the moraine terrace into two parts.
This steep valley is cleaned throughout the year by mudflows and avalanches which
undercut the moraine terrace. In this case, the alpine settlements of Daltar and Kaltar
represent relatively safe settlement locations. However, sporadic rock slides from the
rock face, made unstable by glacial polishing and abrasion, must be expected. The
neoglacial to historical lateral moraine ramparts form, at the foot of the Late Glacial
moraine deposits, a narrow lateral moraine valley-like channel, which offers space for
alpine paths and settlement sites. However, with respect to the glacier oscillations in
this marginal zone, these are rather short-termed settlement locations, i.e. for building
rest houses. The backward erosion of the moraine walls endangers the locations at its
base due to rockfall and mud flows much more than the above-mentioned, high lying
alpine settlements Daltar and Kaltar.

4.4 CONCLUDING CONSIDERATIONS OF THE RISK POTENTIAL FROM DRY AND WET MASS MOVEMENTS

Starting from the perspective of landscape history of the NW Karakorum, i.e. an
almost relief-filling glaciation, a correspondingly large-scale diversity of sedimentation
of morainic deposits results. Consequently, the currently existing bedrock is no longer
decisive for the assessment of danger, but the type and location of the morainic deposits. In the dry state, the moraine material appears "baked" as firm as cement, but
already with a low amount of soaking it is extremely vulnerable to sliding. The glacial
as well as pure slope debris bodies occur in close spacial proximity. Their resedimentation leads to a high portion of polygenetic debris body forms. The shifting of the
moraine fine material into the debris cone zone additionally overprints the debris cones
- which really are rather ready to rock slide - with mudflows. Figure 6 graphically
summarizes the dangers associated with slopes appearing in the NW Karakorum.

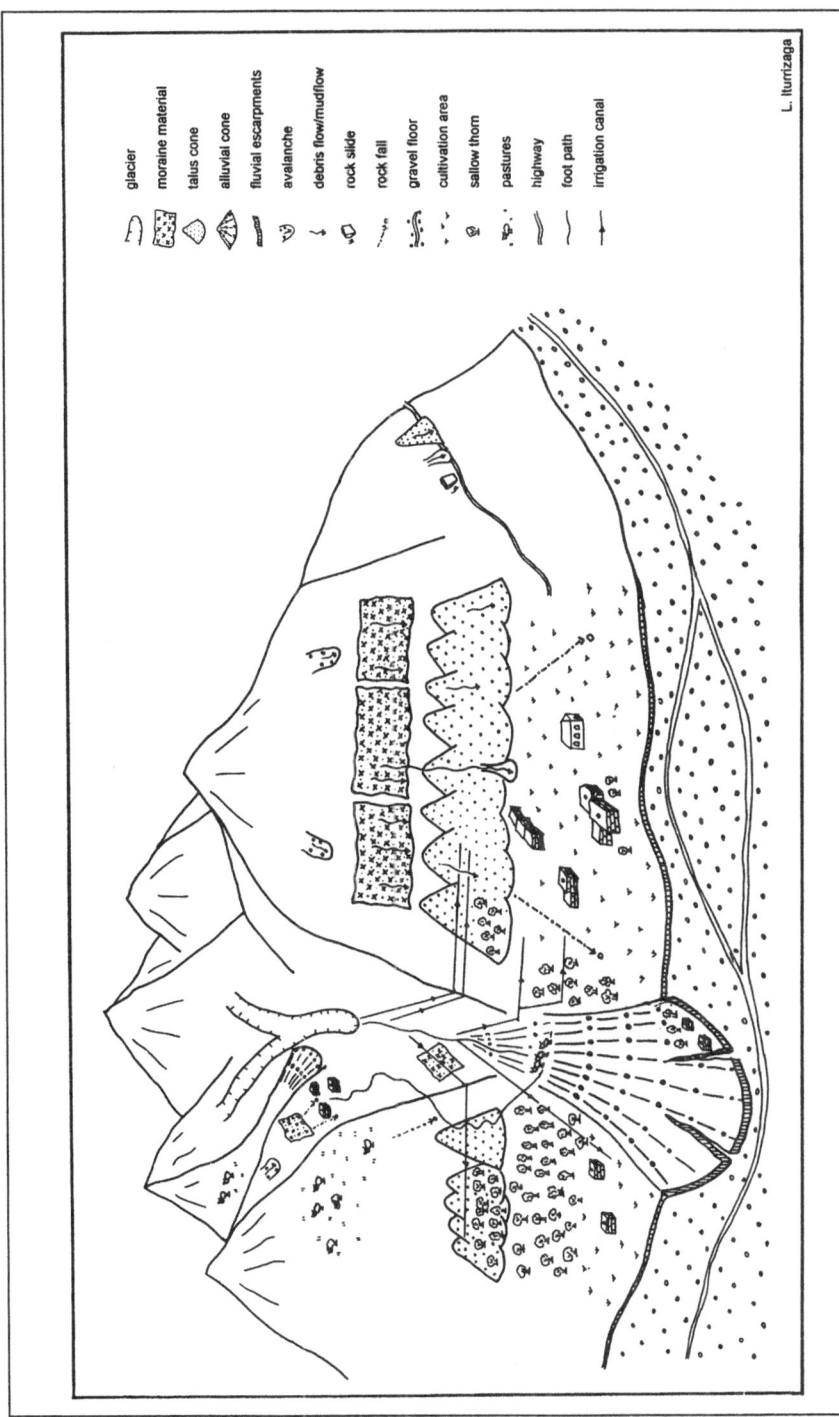

Fig. 6 Natural hazards caused by slope processes in the NW Karakorum

5. Summary

On the basis of a total of thirteen case examples from the Tien Shan, Karakorum, Himalaya and Tangula Shan (central Tibet), the risk potential and hazards are inferred from the development of landscape during the Quaternary. The history of glaciers can be seen as of central importance for this. The Ice Age glacial erosion created U-shaped valleys, which with their steep flanks - as a consequence of the interglacial formation of V-valleys - have prepared and brought about landslides as well as rockslides and the hazards, combined with them. The same is true for the moraines, which the glaciers have deposited high-up in the valley flanks and related loose stone deposits. Dry and wet mass movements follow after heavy precipitation, especially in the semi-arid investigation areas, and are catastrophes for the settlements and the communication routes in the valley floors. Their key-forms are debris cones and debris slopes, as well as mudflows and alluvial fans.

In addition to the Ice Age glaciation history, as a preparatory, indirect factor, the Holocene to present glaciation history is, as a result of the damming-up of glacier- and moraine lakes and their outbursts, a direct risk factor.

The examples presented of acute and already occurred cases of damage were investigated in the years 1989-1994.

Acknowledgements

The authors wish to thank the Deutsche Forschungsgemeinschaft (DFG), the Max Planck-Gesellschaft (MPG), the Volkswagen-Stiftung (VW) and the Deutscher Akademischer Austauschdienst (DAAD) for the financial support for the field-work.

References

1. Derbyshire, E.; Li Jijun; Perrot, F.A.; Xu Shuying; Waters, R.S. (1984): Quaternary glacial history of the Hunza Valley Karakoram, Pakistan.-In: Internat. Karakcram Project, ed. Miller, K.J., Vol. 1, p. 457-495.

2. Dyurgerov, M., Usnurtsev, S.N., Chicagov, A.V.,Mikhalenko, V.N., Kunakovich, M.G., Larin, A.D. (1991): Sary-Tor Glacier, Ak-Schirak Range, Internal Tian Shan: Mass Balance, Run off and Meterological Conditions 1985-1989. Academy of Science of the UDSSR Soviet Geophysical Commitee 71, Moskau.

3. Goudie, A. S. et. al.(1984): The geomorphology of the Hunza Valley, Karakoram Mountains, Pakistan. In: Miller, K.J. (Ed.), The International Karakoram Project, Vol. 2. Cambridge: Cambridge University Press, 359-410.

4. Haserodt, K. (1982): Die quartäre Vergletscherung am pakistanischen Hindukusch (Chitral). Sitz. Ber. Braunschweig. Wiss.Ges., Sonderheft 6, Göttingen, 25-27.

5. Haserodt, K. (1989): Zur pleistozänen und postglazialen Vergletscherung zwischen Hindukusch, Karakorum und Westhimalaya. In: Beitr. u. Mat. z. Reg. Geogr. H.2, Berlin, p. 181-233.

6. Hewitt, K. (1982): Natural dams and Outburst Floods of the Karakoram Himalaya. In: Glen, J.; Hydrological Aspects of High Mountain Areas. International Association of Scientific Hydrology. IAHS publ. 138, 259-269.

7. Hewitt, K. (1989): The Altitudinal Organisation of Karakoram Geomorphic Processes and Depositional Environments. In: Derbyshire, E. & Owen, L. A. (Ed.): Quaternary of the Karakoram and Himalaya, Zeitschrift für Geomorphologie, N.F., Suppl.Bd 76, 9-32.

8. Iturrizaga, L. (1994): Das Naturgefahrenpotential der Talschaft Shimshal, NW-Karakorum. Unpublished Thesis. University of Göttingen, Germany, 2 vol., 210 p.

9. Iturrizaga, L. (1996a): Über das Naturgefahrenpotential der Hochgebirgssiedlung Shimshal (3080 m), NW-Karakorum. In: Die Erde (in press).

10. Iturrizaga, L. (1996b): Preliminary Results of Field Observations on a Typology of Postglacial Debris Accumulations in the Karakorum and Himalaya Mountains In: Culture Area Karakorum Studies (in press).

11. Kalvoda, J. (1979): The Quaternary history of the Barun Glacier, Nepal Himalayas. In: Vestnik Ustredniho ustavu geologickeho, 54, 1, 11-23, Prague.

12. Kalvoda, J. (1992): Geomorphological Record of the Quatenary Orogeny in the Himalaya and the Karakoram. Developments in Earth Surface Processes 3.,Elsevier, Amsterdam, London, New York, Tokyo. 317 p.

13. Kreutzmann, H. (1989): Hunza - Ländliche Entwicklung im Karakorum. Berliner Abhandlungen Bd. 44. Institut für Geographische Wissenschaften. Freie Universität Berlin. 272 p.

14. Kuhle, M. (1984): Hanglabilität durch Rutschungen und Solifluktion im Verhältnis zum Pflanzenkleid in den Alpen, den Abruzzen und im Himalaya. Entwicklung und ländlicher Raum, 18. Jg., H. 3, Frankfurt a. M., 3-7.

15. Kuhle, M. (1988a): Letzteiszeitliche Gletscherausdehnung vom NW-Karakorum bis zum Nanga Parbat (Hunza-Gilgit- und Indusgletschersystem). Tagungsber. Wiss.Abh., 46. Dt. Geographentag 1987, München, Bd. 46, Stuttgart, 606-607.

16. Kuhle, M. (1988b): Zur Geomorphologie der nivalen und subnivalen Höhenstufe in der Karakorum-N-Abdachung zwischen Shaksgam-Tal und K2-N-Sporn: Die quartäre Vergletscherung und ihre geoökologische Konsequenz. Tagungsber.Wiss.Abh., 46. Dt. Geographentag München 1987, Stuttgart, 413-419.

17. Kuhle, M. (1988c): Die eiszeitliche Vergletscherung W-Tibets zwischen Karakorum und Tarim-Becken und ihr Einfluß auf die globale Energiebilanz. Geogr. Z., Jg. 76, H. 3, 135-148.

18. Kuhle, M. (1988d): Die eiszeitliche Vergletscherung W-Tibets zwischen Karakorum und Tarim-Becken und ihr Einfluß auf die globale Energiebilanz. Geogr Zeitschrift., Jg. 76, H. 3, 135-148. Franz Steiner Verlag Wiesbaden.

19. Kuhle, M. (1988e): Die Inlandvereisung Tibets als Basis einer in der Globalstrahlungsgeometrie fußenden, reliefspezifischen Eiszeittheorie, In: Petermanns Geographische Mitteilungen, 133, 4, 265-285.

20. Kuhle, M. (1991): Die Vergletscherung Tibets und ihre Bedeutung für die Geschichte des nordhemisphärischen Inlandeises. In: Frenzel, B. (Ed.), Klimageschichtliche Probleme der letzten 130 000 Jahre, Akademie der Wissenschaften u. d. Literatur, Mainz, 293-306.

21. Kuhle, M. (1993): The Pleistocene Glaciation of the Himalaya and Tibet and its Impact on the Global Climate. A Relief-specific Ice Age Theory. Journal of the Nepal Research Centre (JNRC), Vol. IX, pp 101-140, Franz Steiner Verlag, Wiesbaden.

22. Kuhle, M. (1994a): New Findings on the Ice-Cover between Issyk-Kul and K2 (Tian Shan, Karakorum) during the Last Glaciation). In: Zheng Du et. al. (Ed.): Proceedings of International Symposium on the Karakorum and Kunlin Mountains (ISKKM). Kashi 1992. 185-197.

23. Kuhle, M. (1994b): Present and Pleistocene Glaciation on the north western margin of Tibet between Karakorum main ridge and the Tarim Basin supporting the Pleistocene Inland Glaciation. In: GeoJournal vol. 33 no. 2/3, 133-272.

24. Mason, K. (1935): The Study of Threatening Glaciers. In: The Geographical Journal 85, 24-41. London.

25. Meiners, S. (1991): The Upper Limit of Land Use in Central, South- and Southeast Tibet. In: Kuhle. M. & Xu Daoming (Ed.), Tibet and High-Asia, Results of the Sino-German Joint Expeditions (II), GeoJournal, Vol. 25, No. 2/3,285-295.

26. Meiners, S. (1996): Zur rezenten, historischen und postglazialen Vergletscherung an ausgewählten Beispielen des Tien Shan und des Nord-West-Karakorum. Dissertation, Göttingen 1995 (in press).

27. Schneider, H.-J. (1956): Geologische und erdmagnetische Arbeiten im NW-Karakorum. Erdkunde 10, 1-93.

28. Schneider, H. J. (1957): Tektonik und Magmatismus im NW-Karakorum. Geol.Rdsch. 46, 426-476.

29. Schroder, J.F., Saqid Khan, M. (1988): High magnitude geomorphic processes and Quaternary chronology, Indus Valley and Nanga Parbat, Pakistan. Z.f.G., Suppl.Bd."The Neogene of the Karakoram and Himalayas" Leicester, March 1988.

30. Shams,F.A. & Khan, K. (1994): Landslide Hazards in the Murree Region. In: Centre for Integrated Mountain Research. Punjab University, Lahore 20, Pakistan.

31. Trinkler, E. (1930): The ice-age on the Tibetan Plateau and in the adjacent regions. In: Geogr. Jour., 75, 225-232.

32. Visser, Ph.C. (1938): Wiss. Ergebnisse d. niederl. Expeditionen in den Jahren 1922, 1925, 1929/30 u. 1935. In: Visser, Ph. C. & Visser-Hooff, J. (Ed.), Glaziologie, Vol. 2, IV, Leiden.

33. Wanatabe, T., Ives, J.D., Hammond, J.E. (1994): Rapid Growth of a glacial lake in Khumbu Himal, Nepal: Prospects for a Catastrophic Flood. In: Mountain Research and Development Vol. 14, No.4, 329-340.

34. Wiche, K. (1958): Die österreichische Karakorum-Expedition 1958. Mitt.Geogr.Ges. Wien 100, 280-294.

35. Xu Daoming (1991): Quaternary Glaciation of the North Slope of Karakorum Mountains. In:Kuhle, M. & Xu Daoming (Ed.), Tibet and High-Asia, Results of the Sino-German Joint Expeditions (II), GeoJournal, Vol., 25 No. 2/3, 233-242.

36. Zheng Du (1983): Untersuchungen zur floristisch-pflanzengeographischen Differenzierung des Xizang-Plateaus (Tibet), China. In: Erdkunde Bd. 37, 34-47.

37. Map (1: 50 000): "Changes of Glaciers of the Ak-Shyrak-Range from 1943-1977". Ed. by: Geographical Institute, Russian Academy of Science, Moscow, 1990.

8. Authors

Matthias Kuhle, Sigrid Meiners, Lasafam Iturrizaga
University of Goettingen, Department of Geography
Goldschmidtstr. 5, 37077 Goettingen,
Germany

CHAOS THEORY OF SLIDES/MUD FLOWS IN MOUNTAIN AREAS

Example: Xiaojiang Basin, NE Yunnan, China

A. E. SCHEIDEGGER

1. Introduction

The development of landscapes can be described phenomenologically as the outcome of the operation of a few simple fundamental principles amongst which the Antagonism Principle is the most important one (Scheidegger, 1979). It states that there are *two* types of processes active in the formation of a landscape at the same time: the endogenic (tectonic) and the exogenic (meteoric) processes. Generally, these two processes more or less balance each other so that a landscape represents the temporary dynamic steady state in a complex open, nonlinear system. If one of the external parameters is changed, the steady state is usually re-established by a corresponding continuous change of the other parameters (process-response theory). However, it is well known that the dependences between various landscape parameters may become multivalued at singularities, such as junctions, cusps etc. (Thom, 1972); in that case, small random perturbations can send the system from one branch of the process-response curve to another: a natural disaster occurs. Thus, anytime the "equilibrium" in a landscape is disturbed, humans experience this as a disaster.

2. The concept of equilibrium in landscape evolution

Inasmuch as hazards embody a (potential) deviation from "equilibrium", it is necessary to say a few words regarding the latter term. Indeed, this term has been applied in different ways in geomorphology. In this context, Thorn and Welford (1994) have reviewed the various uses of the term "equilibrium", such as *the dynamic equilibrium* originally formulated by Gilbert (1877), implying a "balance of forces", *the steady state equilibrium* introduced by Hack (1960), implying a situation in which landforms do not change with time, and *the attractor state*, towards which a landscape system ultimately tends to evolve. Ahnert (1994) has discussed the equilibrium problem further and has focussed his attention on the *scale-factor*, which is surely of paramount importance. In effect, this author believes that any so-called "equilibrium" can only be temporary, a so-called stationary (or steady) state can only be *quasi*-stationary. Particularly in the context of hazards, we perceive as stationary, or steady a state which does not change much within a generation or so. "True equilibrium" can only be presented by a completely "dead" world: landscapes, however, are (owing to the antagonism principle) in *constant* flux; they are nonlinear open systems with an endogenic input of mass (and energy) and an exogenic output owing to mass denudation.

97

J. Kalvoda and C.L. Rosenfeld (eds.), Geomorphological Hazards in High Mountain Areas, 97-105.
© *1998 Kluwer Academic Publishers.*

Thus, we start here with the proposition that a (quasi-) stationary landscape-state does not correspond to any *equilibrium* at all, but rather to a (temporary) *self-organized order at the edge of chaos* in a nonlinear complex system.

3. Complex systems

Background to the above is the description of natural systems in terms of complexity theory (cf. e.g. Çambel, 1993): A system (Bertalanffy, 1932) consists of a set of elements identified with some variable attributes whose values characterize the state of the system. The evolution of the system is then the consequence of the existence of a set of relationships between attributes and between attributes and the environment; it is described by a trajectory in *phase space,* i.e. in the space containing one axis (dimension) for each attribute. The interaction of an *open* system with the "environment" is specified by a set of parameters (stating the boundary and initial conditions, such as the temperature or the energy flux at the boundary etc.). The parameter values (varying through a multi-dimensional *parameter space*) exercise an influence regarding which attractors in the phase space of the system are in control; various regions in parameter space may cause the system to evolve in completely separated regions in phase space.

The ultimate evolution of a system is described by the attractors to which the trajectories converge. Such attractors can be regular closed curves (circles, loops) in phase space;- then the system tends to absolute stability; or "strange" attractors: curves (sets of points) which are nowhere differentiable and which have a fractal dimension.

4. General features of self-ordered criticality

Innumerable observations have led to the recognition of the frequent presence of self-organized critical states in many complex systems at the edge of chaos (examples in astrophysics: galaxies and solar flares (Bak, 1993); in geophysics: fractal dynamics of earthquakes (Bak and Tang, 1989); in geomorphology: minislips in sand-piles (Bak et al. 1988); in the life sciences: "Darwinian" evolution (Kauffman, 1993); in geomorphology: the distributions of heights in a landscape (Turcotte, 1992). An ordered state at the edge of criticality establishes itself in a complex system solely on account of the (normally highly non-linear) interactions between the individual elements of the system and not because of the presence of an external ordering principle.

From such observations it was found that the characteristic observables in ordered quasi-stationary states generally are spatially and temporally scale-invariant: they have been found to be fractal. In a fractal set of dimension D, there exists a power law for subsets: The number N of subsets of (linear) "size" L is proportional to L exp(-D). Such power laws have indeed also been found in complex natural systems. Thus, the number N of earthquakes of magnitude $m \geq M$ follows the famous Gutenberg-Richter (1949) law (a and b are constants):

$$\log N(M) = a - bM \tag{1}$$

Written in terms of the energy E (M = c log E) this is for N(E) a "power-law". "Order" seems to occur only at the edge of chaos, and one conjectures that the establishment of ordered conditions at the edge of chaos might be a general, self-generating feature of Nature (Haken and Wunderlin, 1991; Nicolis and Prigogine, 1977). Thus Bak et al. (1988) noted that *the temporal fingerprint of such states is the presence of flicker (=1/f) noise* (Dutta and Horn, 1981), *its spacial signature is the emergence of a scale-invariant (fractal) structure*; with regard to life-systems, Kauffman (1993) has even gone so far as to state this conjecture as a hypothesis (l.c., p. 232): *"Living systems exist in the solid regime near the edge of chaos, and natural selection achieves and sustains such a poised state"*. Kauffman (1993) based his hypothesis on the empirical behavior of systems with Boolean attributes: he found, by extensive computer simulations, that systems of N elements with Boolean attributes, where each attribute is influenced (in a set random geometry) by exactly K=2 elements through randomly set Boolean functions (so that the system is represented by a random Boolean network), develop into highly ordered systems with few attractors that are fairly stable in the sense that deleting any single element or altering its Boolean function typically causes only modest changes in the attractors and phase trajectories.

5. A New Law of Nature

All of the above was originally an "inductive" inference from observations and computer simulations. No actual theory existed as to why self-ordered complex systems should be fractal and why they should occur only at the edge of chaos. However, the frequent establishment of self-organized critical states in Nature (and, indeed, also in anthropogenic [such as social or economic] systems strongly suggests the existence of a corresponding *Universal Law of Nature*:
"Nonlinear open complex systems develop quasi-stationary states at the edge of chaos whose temporal signature is flicker noise and whose spacial signature is fractality with a dimension vastly lower than that of the phase space".

Therefore, one would like to have more *fundamental* insights into the circumstances that govern the establishment of self-organized order at the edge of criticality, than computer simulations. For something to be recognized and accepted as a "Law of Nature", it is, according to Cohen and Stewart (1994), not enough that it is true; one has to *know* that it is true and, moreover, one has to be able to explain *why* it is true. The present author believes that this has now been achieved (Scheidegger, 1995, 1996).

Thus, there are first of all a few quite obvious conditions that are *necessary* for the spontaneous establishment of a quasi-stationary, ordered state: Most systems under consideration (landscapes, galaxies, life) are in some way evolutionary: Elements are "born" and "die". Evidently, for a stationary state to develop, the death rate must be equal to the growth rate in the system, otherwise one has complete obliteration of the system or an explosion. Furthermore, all "evolutionary" systems must be open and dissipative. Therefore, the laws of equlibrium thermodynamics do not hold: the processes that occur in such systems are fundamentally irreversible and the usual form of the second law of thermodynamics does not apply: Prigogine (1947) noted that in open systems, the entropy may well decrease during the approach to a steady state, i.e. its value at the nonequilibrium steady-state may well be smaller than at equilibrium (death). *In a system obeying linear laws, the spontaneous emergence of order in the form of spatial or temporal patterns differing qualitatively from equilibrium-like behavior (death), is ruled out.* Thus, a necessary condition for the spontaneous development of order is that the relations between the elements of the system under consideration *be nonlinear.*

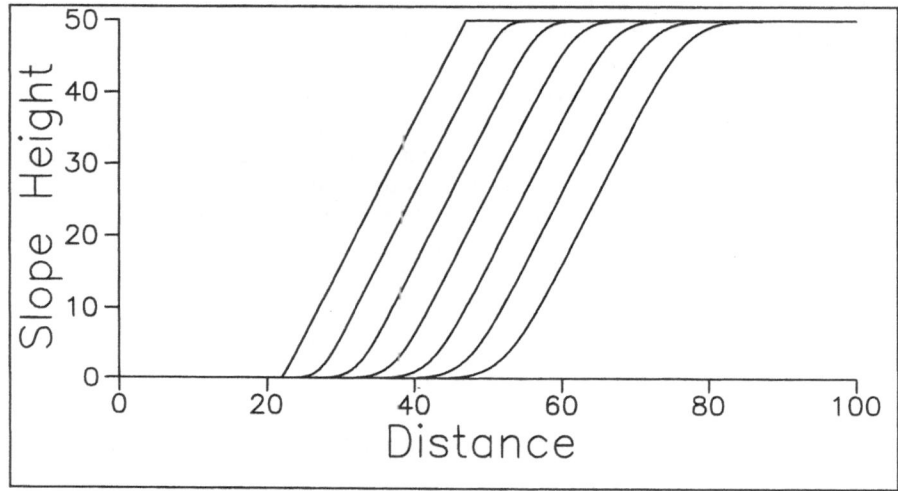

Figure 1. Regular uniform recession of a smooth slope bank (Scheidegger, 1961).

Next, one notices that the very concept of "order" requires that of all the possible states of a system, only a few are permitted: it requires the corresponding attractor to have a dimension which is much smaller than the dimension of the entire phase space; however, there is no condition that this dimension be an integer: if it is not an integer, it is fractal;- there are no other possibilities. This explaines the fractal structure of self-ordered states.

The order at the edge of chaos has to be (quasi-)*stationary* (at least for a while;-otherwise the ordered state cannot be "seen") with regard to small changes in the initial conditions. Thus, it is seen that ordered states are the *only* ones that have duration: there is a "Darwinian selection" of such states: Quite generally, in a system only those states can be "seen" which have some duration. Thus the problem of explaining *why* order develops out of chaos is solved; the new natural law formulated above can be considered as *established* according to the criteria of Cohen and Stewart (1994).

6. Example: Land- and Mudslides in Mountain Regions

System theory applied to slopes implies that the latter are self-organized ordered components of the system. A suddenly occurring instability expresses itself in a landslide. Such landslides are part and parcel of the geomorphic cycle, inasmuch as the continuously occurring uplift has to be balanced by a corresponding mass removal.

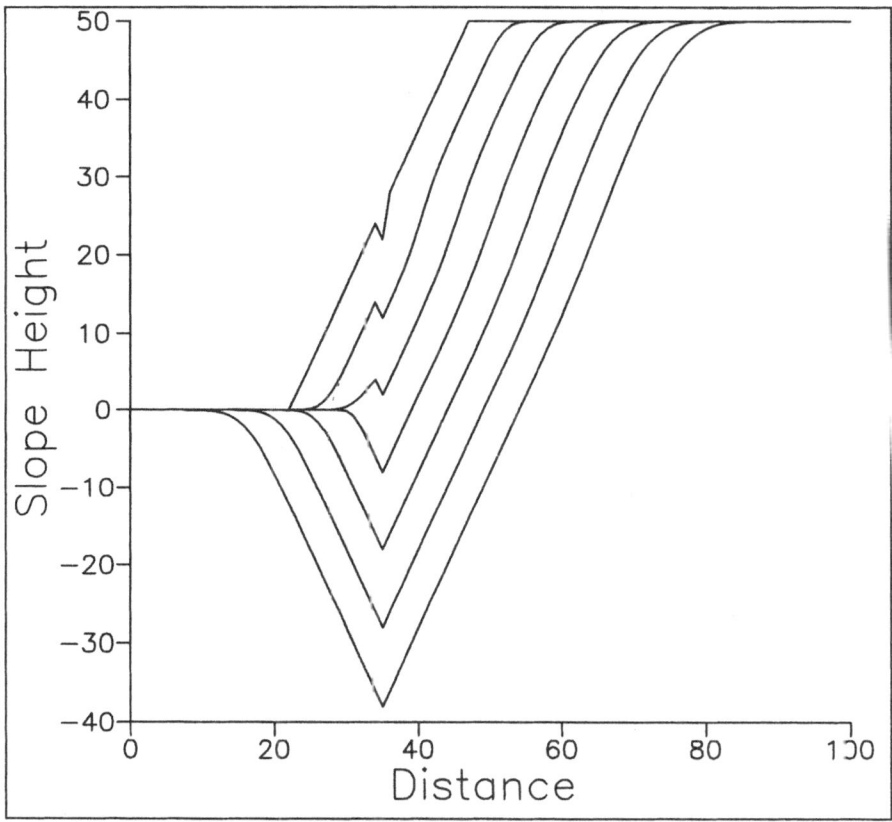

Figure 2. Unstable recession of a slope bank with a ditch in the middle;the ditch becomes deeper until instability is reached

The observation of the existence of a critical angle of repose on granular slopes shows that the latter are in a critical state at the edge of chaos. Whilst an ordered slope-development occurs on smooth slopes (Scheidegger, 1961), small "dents" in such a slope cause positive feed-back in the latter and therewith an instability in the evolution. This can be shown by integrating the fundamental non-linear slope-development equation (Scheidegger, 1961)

$$\partial y/\partial t = -|\partial y/\partial x|\ \sqrt{[1+(\partial y/\partial x)^2]} \tag{2}$$

where x is the abscissa (distance) and y the height of the slope on a computer by a finite-difference method for various initial conditions. Starting with a smooth slope bank leads to the well-known solution of Scheidegger (1961) shown in Fig.1, but any irregularity such as a knick or hole in a smooth slope leads to a positive-feedback mechanism. As an example, we show here the development of the profile of a regular smooth slope bank with a ditch (through which all eroded material is thought to be removed) half-way up (Fig. 2); in both figures the units are relative, i.e. $0 \leq x \leq 100$; $\Delta x = \Delta y$; $\Delta t = \Delta x$. It is clear that the development of a slope without a ditch is regular (or stationary); if there is a ditch, the latter becomes rapidly deeper, until the critical height (Fellenius, 1940) is inexorably (since the slope *angle* stays more or less the same) reached and a slide occurs.

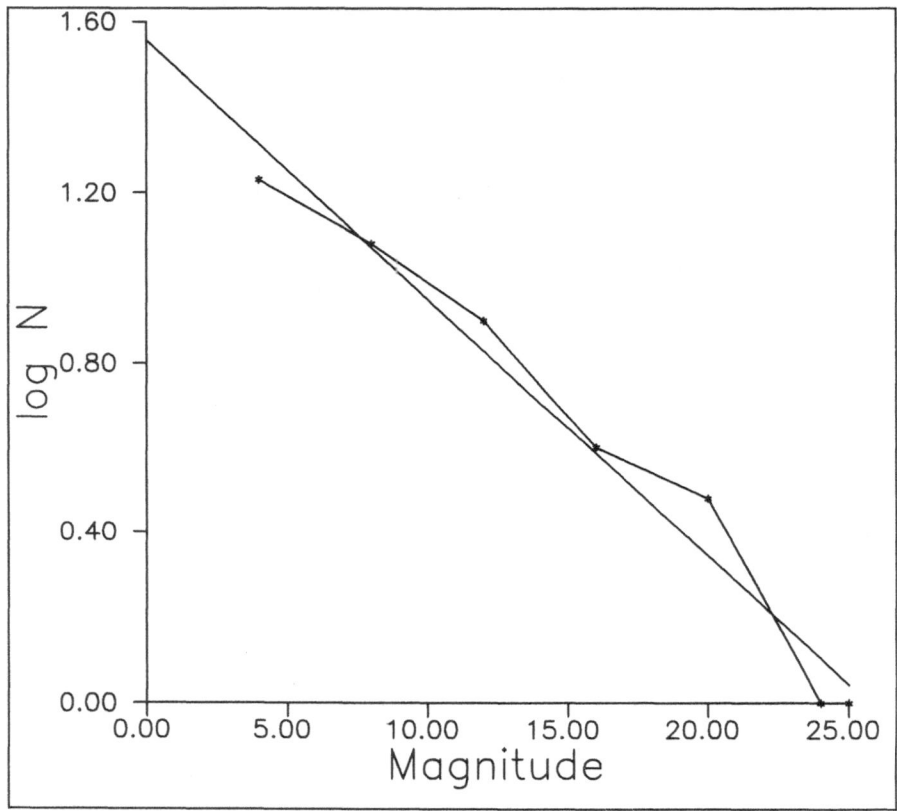

Figure 3. Magnitude-frequency relationship for mudslides in the Jiangjia Ravine of the Xiaojiang Basin in Yunnan, China.

Thus, a minute change in the initial conditions (i.e. introducing a small ditch) causes a fundamentally different long-term behavior. This is characteristic for complex systems at the edge of chaos.

The slope bank considered above has been assumed as somehow *a priori* given. Evidently, it exists at the edge of criticality; thus it is presumably the result of *self-organized criticality* as indicated earlier in this paper; according to the "sliding sand-pile" model of Bak et al. (1988) the number of slides in any given time interval (assuming the endogenic mass-addition rate as constant) as a function of size should follow a power-law distribution: the average frequency of a slide of a given size is inversely proportional to some power of its size. This prediction can be tested with data from nature. In a study of the Xiaojiang Basin, which lies between the ca. 3000m high Gangwang and Lingwang Ranges with a valley-floor-altitude of about 1500m, straddling the 103°E-meridian between 24°40' and 25°35' N latitude near Kunming (NE Yunnan, China), Li and Wang (1984) have published a summary of the mud-bursts per year for 17 years in a particular ravine of that basin (the Jiangjia-ravine, forming a deep gorge of 8km length on the E-side of the Xiaojiang at latitude 26½°N).

These mudflows are engendered by slides along the steep slopes of the ravine; one can take the number of outbursts as an integrated magnitude m of sliding in the ravine for each year. Then, it is possible to count the number N of years in which m≥M; the resulting numbers are shown in Table 1:

Table 1. Slide "magnitude" M against frequency N in the Jiangjia Ravine (103°E,26½°N) (Xiaojiang Basin, NE Yunnan) Original data after Li and Wang, 1984

M≥	4	8	12	16	20	24
N	17	12	8	4	3	1
log N	1.23	1.08	0.90	0.60	0.48	0.00

Furthermore, Fig.3 shows the plot of log N against M for the Jiangjia Ravine: one obtains a curve that is represented by a Gutenberg-Richter (1949) type of equation [Fig.3 shows the best-fitting straight line corresponding to Eq.(1)]. This shows that the land/mud slides in the mountain region investigated do indeed follow a power-law distribution.

The constants a and b in Eq.(1), as is the case in seismology, can be used to specify the risk present in any one region: from the linear best fit in Fig.3 one obtains a=1.558 and b=0.0607. The basis of the data of Li and Wang (1984) are observations for 17 years; thus the constant a given above refers to these 17 years. For a more standard period of 100 years, it has to be augmented by log(100/17)=0.76969; thus Eq.(1) becomes for this period

$$\log N_{100} = 2.328 - 0.0607 \, M \qquad (3)$$

This equation predicts for a period of 100 years the number N of years in which the "integrated slide-magnitude" m, i.e. the number of slides in a year in the Jiangjia Ravine, is m≥M. Correspondingly, the value of $100/N_{100}(M)$ would be the average recurrence time (in years) of sliding with m≥M in that ravine. Thus, a complete risk-characterization has been obtained for the area in question.

7. Conclusions

The above analysis shows that geomorphic systems are principally open and highly nonlinear with the possibility of positive feedback. Landslides are the result of sudden changes in long-term behavior caused by minute changes in the initial conditions. These can be the result of input from tectonic processes into the system and the resulting establishment of self-organized criticality with a fractal geometry. Landslides are therefore simply part and parcel of the normal geomorphic cycle: the steady-rate build-up of mass (input into the geomorphic system) by the plate-tectonic motions causes intermittent catastrophes with a power law characterizing the size-frequency distribution.

An example from the high mountain areas of South China (NE Yunnan) shows that this interpretation is indeed feasible and can be used for practical hazard-risk assessment.

References

1. Ahnert, F. (1994) Equilibrium, scale and inheritance in geomorphology, *Geomorphology* 11, 125-140.

2. Bak, P. (1993) Self-organized criticality in astrophysics, in A.Lejeune and J. Perdang (eds.) *Cellular Automata Models for Astrophysical Phenomena,* World Scientific Press, Singapore.

3. Bak, P., Tang, C., Wiesenfeld, K. (1988) Self-organized criticality, *Phys. Rev. A38, 364-374.*

4. Bak, P., Tang, C. (1989) Earthquakes as self-organized critical phenomenon, *J. Geophys. Res.* 94, 15635-15637.

5. Bertalanffy, L.v. (1932) *Theoretische Biologie,* Springer, Berlin, 170p.

6. Çambel, A.B. (1993) *Applied Chaos Theory, a Pardigm for Complexity,* Academic Press, Boston, 246p.

7. Cohen, J., Stewart, I. (1994) *The Collapse of Chaos,* Viking Press, New York, 495p.

8. Dutta, P., Horn, P.M. (1981) Low-frequency fluctuations in solids: 1/f noise, *Revs. Mod. Phys.* 53(3), 497-516.

9. Fellenius, W. (1940) *Erdstatische Berechnungen mit Reibung und Kohäsion (Adhäsion) und unter Annahme kreiszylindrischer Gleitflächen; 2. erg. Aufl.,* W. Ernst & Sohn, Berlin, 48pp.

10. Gilbert, G.K. (1877) *Report on the Geology of the Henry Mountains,* U.S.Geol. Survey; U.S.Govt.Printing Offc., Washington.

11. Gutenberg, B., Richter, C.F. (1949) *Seismicity of the Earth and Associated Phenomena,* Princeton University Press, Princeton, 211 p.

12. Hack, J.T. (1960) Interpretation of erosional topography in humid temperate regions. *Amer. J. Science* 258A, 80-97.

13. Haken, H., Wunderlin, A. (1991) *Die Selbstrukturierung der Materie,* Vieweg, Braunschweig, 466p.

14. Kauffman, S.A. (1993) *The Origins of Order Self-Organization and Selection in Evolution,* Oxford University Press, Oxford 709p.

15. Li, J., Wang, J.R. (1984) Mudflows in the Xiaojiang Basin, Yunnan Province, *Geooekodynamik* 5(3), 143-258.

16. Nicolis, G., Prigogine, I. (1977) *Self-Organization in Non-Equilibrium Systems. 8th printing,* Wiley, New York, 491p.

17. Prigogine, I. (1947) *Étude Thermodynamique des Processus Irreversibles,* Desoer, Liège, 143 p.

18. Scheidegger, A.E. (1961) Mathematical models of slope development. *Bull. Geol. Soc. Am.* 72, 37-50.

19. Scheidegger, A.E. (1979) The principle of antagonism in the Earth's evolution. *Tectonophysics* 55, T7-T10.

20. Scheidegger, A.E. (1995) Order at the edge of chaos in geophysics, *Abstracts IUGG XXI General Assembly* 1, A-11, UA51A-10.

21. Scheidegger, A.E. (1996) Ordnung am Rande des Chaos: Ein neues Naturgesetz. *Oesterr. Z. f. Vermessungskunde und Geoinformation* 84(1), 69-74.

22. Thom, R. (1972) *Stabilité Structurelle et Morphogénèse,* Benjamin, Reading,Pa., 362p.

23. Thom, C.E., Welford, M.R. (1994) The equilibrium concept in geomorphology, *Ann. Amer. Assoc. Geogr.* 84(4), 666-696.

24. Turcotte, D.L. (1992) *Fractals and Chaos in Geology and Geophysics.* University Press, Cambridge, 221pp.

Author

A. E. Scheidegger
Technical University
Section of Geophysics
A-1040 VIENNA,
Austria

THE SALT WEATHERING HAZARD IN DESERTS

ANDREW GOUDIE

1. Introduction: The hazard

The presence of salts in the environment, besides being a potent geomorphological process in coastal, urban and dry environments, is also a cause of severe deterioration in engineering structures. On the one hand, the fabric of some of the great new cities of the Middle East is being attacked, while on the other some of the world's great cultural treasures - the Pharonic temples and the Sphinx in Egypt, the Nabbatean site of Petra in Jordan, the Harappan city of Mohenjo-Daro in Pakistan, the Pueblo Buildings of the American South west, the Islamic treasures of Uzbekistan, and a range of great Christian cathedrals from the Mediterranean lands - are disintegrating.

Salinity, and associated salt weathering of natural rock outcrops and of engineering structures, is a characteristic of some high altitude mountainous desert environments. Examples include the Basin and Range Province of the western United States of America, the high altiplano of the Andes Cordillera in South America, the valleys of the Karakoram Mountains in Pakistan, and the many closed depressions of High Asia. Very rapid rates of salt weathering have been witnessed in the Hunza Valley in the Karakorams, where twentieth century moraines and rock fall debris have been attacked by a range of salts, of which magnesium sulphate is the most important. Likewise, there have been many studies of salt weathering in the dry valleys that dissect the mountainous terrain of Antarctica.

On a global basis the presence of salt in the environment is a widespread phenomenon, especially in some coastal and dryland situations. However, salts are also accumulating in urban environments because of air pollution. Indeed, the salt hazard may be increasing in extent and severity because of a whole range of human activities. Of these the most significant is probably the spread of irrigation (Table 1), which causes groundwater levels to rise and brings moisture to the surface where evaporation takes place leading to the precipitation of solutes. A good example of the affects of rising groundwater levels caused by irrigation is provided by the weathering of very important Islamic sites in Central Asia (Cooke, 1994) and in the Indus Valley (Goudie 1977).

A second prime cause of the accelerating spread of saline conditions (Table 2) is sea-water incursion brought about by over-pumping of groundwater in coastal marginal areas. Ocean water displaces less saline groundwater through a mechanism called the Ghyben-Herzberg principle. A third prime cause of soil salinity spread in drylands is vegetation clearance. The removal of native forest vegetation (e.g. Eucalyptus in the wheatbelt of western Australia) allows a greater penetration of rainfall into deeper soil layers which causes groundwater levels to rise, thereby creating conditions for the seepage of sometimes saline water into low-lying areas (e.g. Peck, 1983). A fourth

107

J. Kalvoda and C.L. Rosenfeld (eds.), Geomorphological Hazards in High Mountain Areas, 107-120.
© 1998 *Kluwer Academic Publishers.*

reason for increases in levels of salinity is the changing state of water bodies caused by inter-basin water transfers. A classic example of this is the shrinkage and subsequent increase in the mineralisation of the Aral sea, but the same has happened in the Owens valley area of California. Fifthly, urbanisation leads to changes in groundwater conditions that can aggravate salt attack. In some large dryland cities the importation of water and its subsequent usage, wastage and leakage can lead to deterioration of structures, as in various parts of Egypt (Hawass, 1993 ; Masuch - Oesterreich,1993; Smith, 1986). In addition, the urbanisation process itself can cause a rise in groundwater levels by affecting the amount of moisture lost by evapo-transpiration. The spread of impermeable surfaces (e.g. roads, buildings, car parks) in low-lying sabkha areas along the margins of the Arabian Gulf has caused groundwater levels to rise (Shehata and Lotfi, 1993).

Table 1. Global estimate of secondary salinisation in the world's irrigated lands.

Country	Cropped area (Mha)	Irrigated area (Mha)	Share of irrigated to cropped area (per cent)	Salt-affected land in irrigated (Mha)	Share of Salt-affected to irrigated land (per cent)
China	96.97	44.83	46.2	6.70	15.0
India	168.99	42.10	24.9	7.00	16.6
Commonwealth of Independent States	232.57	20.48	8.8	3.70	18.1
United States	189.91	18.10	9.5	4.16	23.0
Pakistan	20.76	16.08	77.5	4.22	26.2
Iran	14.83	5.74	38.7	1.72	30.0
Thailand	20.05	4.00	19.9	0.40	10.0
Egypt	2.69	2.69	100.0	0.88	33.0
Australia	47.11	1.83	3.9	0.16	8.7
Argentina	35.75	1.72	4.8	0.58	33.7
South Africa	13.17	1.13	8.6	0.10	8.9
Subtotal	842.80	158.70	18.8	29.62	20.0
World	1473.70	227.11	15.4	45.4	20.0

Source: from Ghassemi et al 1995, Table 18

In addition to the various hydrological changes outlined above, the salt hazard is being accelerated by air pollution, which brings sulphate and nitrate salts into the environment, and by the use of deicing salts on roads. The latter process leads to concrete decay, particularly on bridge decks (see Mallet,1994).

With regard to the future, global changes in climate and sea level could have implications for salinity conditions. For example, increasing drought risk in Mediterranean Europe could lead to an expansion of salt prone areas (Szabolcs,1994), while higher sea levels in susceptible geomorphological locations (e.g. coastal deltas, sabkhas, etc.) could change the position of the all important salt/freshwater interface and the height of the water table.

In recent years the significance of the salt hazard has become increasingly important and geomorphologists have undertaken a range of studies of the phenomenon in dryland environments. Table 3 lists a selection of the more important examples.

Table 2. Causes of aggravated salt attack

1.	Irrigation
	– rise in groundwater
	– evaporation of water from fields
	– evaporation of water from reservoirs
	– waterlogging produced by seepage losses
2.	Sea water incursion
	– caused by overpumping or by reduced freshwater
	– recharge from dammed rivers
3.	Vegetation clearance for farming and urbanisation
4.	Inter-basin water transfers
	– mineralisation of lake waters
	– deflation of salts from desiccated lakes and deposition downwind
5.	Urbanisation
	– water importation, faulty drains, production of impermeable surfaces, etc.
6.	Atmospheric pollution (especially sulphates and nitrates)
7.	Salting of roads etc. for deicing
8.	Changes in the internal microclimates of buildings

Table 3. Examples of work by geomorphologists and related scientists on the salt weathering hazard in drylands

Source	Date	Location
Goudie	1977	Móhenjo-Daro, Pakistan
Brown et al.	1979	Pueblo sites, Arizona
Doornkamp et al.	1980	Bahrain
Blackburn and Hutton	1980	Adelaide, Australia
Cooke et al.	1982	Suez City, Egypt
Horta	1985	Sahara (Algeria)
Ibrahim and Doornkamp	1991	Nile Delta
Akiner et al.	1992	Uzbekistan
Fitzner and Heinrichs	1994	Petra, Jordan
McNally	1995	Australia
Robinson	1995	Los Angeles, USA

2. Groundwater-related salt attack

As will be evident from the discussion so far, the presence of high groundwater levels, whether natural or caused by human activities, is one of the prime determinants of the efficacy of salt weathering of engineering structures in arid zones.

As Cooke et. al. (1982, p.168) point out:

"Aggressive soil conditions as an engineering hazard are produced when saline groundwaters are drawn upward through the soil to produce a capillary fringe that either reaches the ground surface or approaches sufficiently close to it to affect foundations. This upward movement of capillary water through surface and near-surface materials is sometimes known as "evaporative pumping" and is essentially the product of high surface temperatures. The height of capillary rise (i.e. the thickness of the capillary fringe), varies with the soil temperature gradient and the nature of the soil materials, normally being less than a metre in clean gravel but, according to laboratory studies... extends up to 10m in clay."

Terzaghi and Peck (1948) suggested that the height of capillary rise (hc) that might take place in different materials could be estimated by using the formula:

$$Hc = \frac{C}{e\,D10}$$

Where C = an empirical constant that depends on grain shape and surface impurities, and ranges between 0.1 and 0.5, e = void ratio, D10 = 'effective size' in which 10% of particles in the grain-size analysis are finer and 90% are coarser than this value.

Cooke et al's own experiences in Bahrain, Dubai and Egypt suggested to them that capillary rise under desert conditions is normally no more than 2-3m, and rarely if ever, exceeds 4m. Capillary rise may take place in buildings and the height of such rise, by what is sometimes termed "the wick effect", can be determined with a simple moisture meter, as was done in Ras Al Khaimah by Cooke and Goudie (1992, unpublished), where walls built as recently as the later 1980s show clear evidence of severe salt attack. Depending on the nature of the substratum, moisture rises up breeze block walls to as much as 1.8m (figure 1).

The position of the upper surface of the capillary fringe (known as "the limit of capillary fringe") with respect to the ground surface is of considerable importance in terms of building construction and land use planning and zoning. Cooke et al. (1982, p. 169- 170) believe that of particular importance is the line where the limit of capillary fringe intersects the ground surface, as this line marks the boundary between the zone where only foundations are at risk and that where both foundations and above-ground structures may be subjected to salt attack. They identified four zones (figure 2) with different orders of hazardousness:

Zone I - No hazard from groundwaters as the limit of capillary fringe is so deep within the ground as to be below the base of foundations.

Zone II - The limit of capillary fringe is below the ground surface but sufficiently close to it to affect foundations.

Zone III - The limit of capillary fringe is potentially above ground level so that both foundations and superstructures are at risk. Such areas usually show up as a dark tone on air photographs and have 'puffy' surfaces on which footprints and tyre-tracks are clearly visible.

Zone IV - Water-table within half a metre of the ground surface for most of the year so that foundations are emplaced in water and there is potential for capillary rise to well above ground level (sometimes in excess of 2m). Such areas are very low in elevation and may be periodically inundated as a consequence of rainfall or sea surges. These areas appear as a dark tone on air photographs, often with patches of standing water. On the ground, the surface is usually either `puffy' or crusted. Patches of salt efflorescences are often developed and, in certain instances, salt polygons may also be visible. Standing water areas are invariably rimmed with salt crystals.

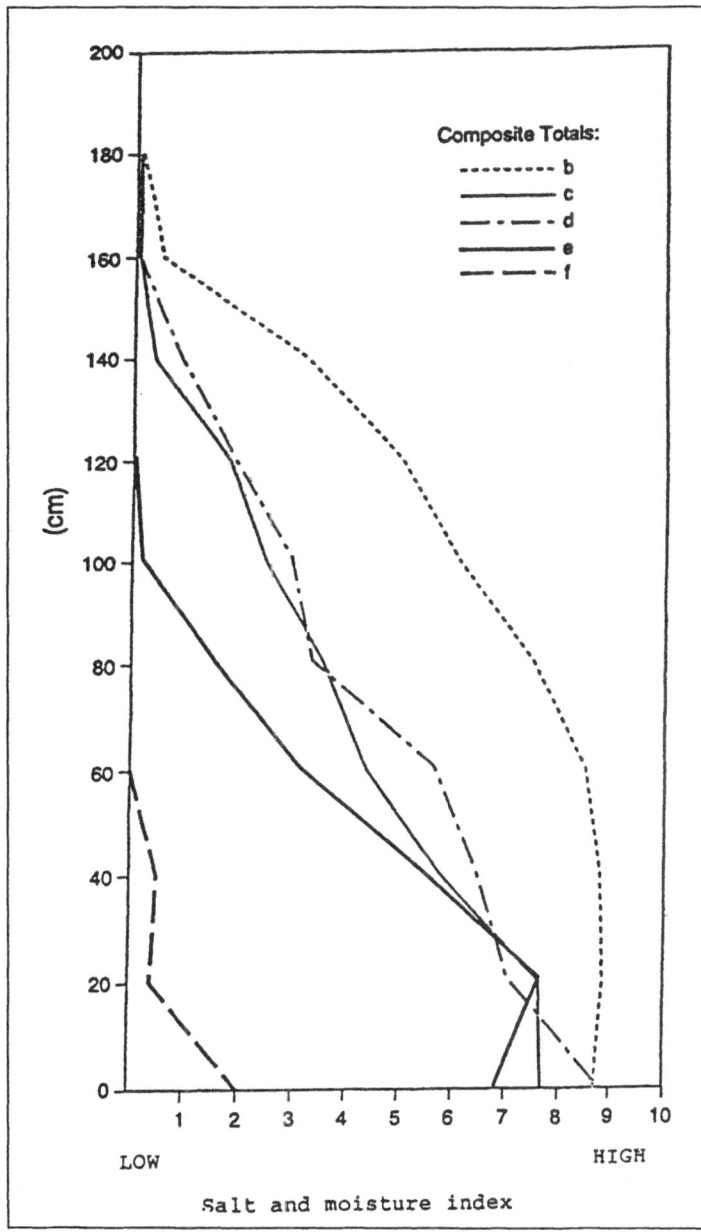

Figure 1. Salt and moisture profiles from walls from Ras-Al-Khaimah, United Arab Emirates. The scale on the X axis represents the salt and moisture index determined by the protimeter and ranges from 0 to 10. (source: Cooke and Goudie, 1992, unpublished).

Table 4. Mechanisms of salt attack

Physical Changes
        Crystallization
        Hydration
        Thermal expansion
        Electrical slaking double layer effects associated with hygroscopicity

Chemical Change
        Silica mobilization under alkaline conditions
        Changes to concrete mineralogy
        Corrosion of incorporated iron and steel
        Moisture related chemical weathering associated with hygroscopicity
        Gypsum/silicate replacement

In some buildings, however, the thickness of the capillary zone may be exceptionally and unexpectedly high, sometimes exceeding 6m (Cooke, 1994). Cooke reported a suggestion developed by R.A. Legg which might account for this phenomenon, which was observed in some of the decaying Islamic monuments of Uzbekistan. The idea is that (p.200-201) "There may be two zones of capillary rise one immediately above the water table (as normal), the other at or above ground level, the two separated by a relatively dry zone. The suggested mechanism for pumping water across the 'dry zone' is as follows: in winter, when the surface is relatively cool and there is a marked temperature gradient from the water table to the surface, water will evaporate from the groundwater at depth, and the warm vapour will pass upwards through the dry zone to condense in the relatively cool surface layers. Such condensed water could be supplemented by rainfall, snowmelt, local runoff, irrigation water and floodwater. In the condensation zone, capillary rise could proceed into finer-grained material of the buildings from the surrounding soil".

3. The mechanism of salt attack

The action of salt on rocks and building materials has conventionally been divided into those involving a predominantly physical change in the state of the material and those involving some change in its chemical state (Table 4).

Of the physical processes, the most cited cause is salt crystallisation - the process of salt crystal growth from solution. The growth of crystals can occur if solution temperatures fall (Kwaad, 1970) or as a result of evaporative concentration. Air humidity is a highly important control of the effectiveness of salt crystallisation, particularly for hygroscopic/deliquesecent salts like sodium chloride (Arnold 1981). Lewin (1981, 1990) provides a detailed consideration of the mechanism of crystallisation in pores and Winkler and Singer (1972) provide data on the crystallisation pressures that can occur. These pressures appear to exceed comfortably the tensile strength of most rocks and concrete (Table 5).

A second major mechanism is hydration, for certain common salts hydrate and dehydrate easily and quickly in response to changes in temperature and humidity. As a change of phase takes place to the hydrated form, water is absorbed and volume expansion occurs, thereby developing pressure against pore walls. The volumetric expansion for sodium carbonate and sodium sulphate can be in excess of 300% and the change of phase may occur at the sorts of temperatures encountered widely in nature (Mortensen, 1933). Winkler and Wilhelm (1970) have calculated the hydration pressures of some important common salts. Like crystallisation pressures, these also may exceed the tensile strengths of building materials (Table 5).

A third possible mechanism of rock disruption through salt action, proposed by Cooke and Smalley (1968), is that of differential thermal expansion, whereby disruption of rock may occur because of the fact that certain salts have higher coeffcients of linear expansion than do the minerals of the rocks in whose pores they occur. Supporters of the role of this mechanism include Johannessen et al. (1982).

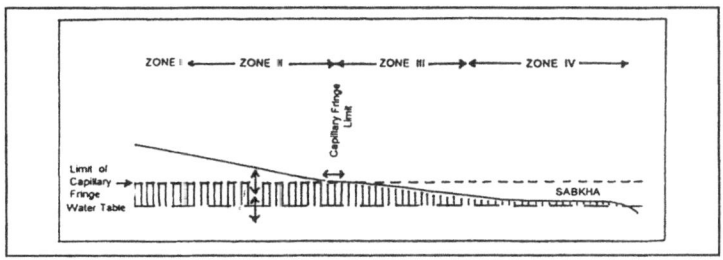

Figure 2. Hypothetical section to show zonal subdivision of aggressive ground in a sabkha environment (from Cooke et al.1982, fig. v.12).

Another possible mechanism of rock decay associated with the presence of salt is accelerated slaking of shales and mudstones. The presence of high amounts of exchangeable sodium has long been recognised as a cause of swelling and structure change in such materials (Seedsman, 1986; Taylor and Smith, 1986).

Chemical mechanisms are varied in type and include corrosion effects on iron reinforcements in concrete structures. The formation of the corrosion products of iron (i.e. rust) causes a volume expansion to occur. If one assumes that the prime composition of such corrosion products is $Fe(OH)_3$, then the volume increase over the uncorroded iron can be fourfold. Thus when rust is formed on the iron reinforcements, pressure is exerted on the surrounding concrete. This may cause the concrete cover over the reinforcements to crack, which in turn permits the ingress of oxygen and moisture which then aggravates the corrosion process. In due course, spalling of concrete takes place, the reinforcements become progressively less strong, and the whole structure may suffer severe deterioration.

Rates of corrosion are accelerated in the presence of chlorides. Chloride ions may occur in a concrete because of the use of contaminated aggregates or because of penetration from a saline environment (Soroka, 1993; section 10.5). However, the electrochemical corrosion of metals can also be produced by sulphates (Hong Naifeng, 1994, p. 33) for there are often sulphate-reducing bacteria in a saline soil containing sulphates, which can cause strong corrosion to metals.

Of no lesser importance in terms of the long-term durability of buildings is sulphate attack on concrete. Sulphates can cause severe damage to, and even complete deterioration of Portland cement concrete (Mehta, 1983; Bijen, 1989). Although there is still controversy as to the exact mechanism of sulphate attack (Cabrera and Plowman, 1988), it is widely appreciated that the sulphates react with the alumina-bearing phases of the hydrated cement to give a high sulphate form of calcium aluminate known as ettringite ($3CaO.Al_2O_3.3CaSO_4.32H_2O$).

Magnesium sulphate is particularly aggressive because in addition to reacting with the aluminate and calcium hydroxide as do the other sulphates, it decomposes the hydrated calcium silicates and, by continued action, also decomposes calcium sulphoaluminate (Addleson and Rice, 1991, p.407).

Table 5. Material strengths and salt pressures

A) Tensile strengths of materials in Mpa

Extremely high strength rocks[1]	> 10
Very high strength rocks[1]	3 - 10
High strength rocks[1]	1- 3
Medium strength rocks[1]	0.3 -1
Low strength rocks[1]	0.1- 0.3
Very low strength rocks[1]	0.03 - 0.1
Extra low stength rocks[1]	< 0.03
Concretes (typical values)[2]	2- 4

B) Pressures produced by salt processes in Mpa

Expansion of steel reinforcements on rusting[2]	up to 30
Crystallisation pressures of gypsum[3]	28.2-190
Crystallisation pressures of halite[3]	5.54- 373.7
Crystallisation pressures of thenardite[3]	29.2 -196.5
Hydration pressures of gypsum[4]	up to 254
Hydration pressures of $MgSO_4$[4]	up to 42
Hydration pressures of Na_2SO_4[4]	up to 48

Sources:
1 Bell (1992)
2 Murdock et al (1991)
3 Winkler and Singer (1972)
4 Kirchner (1995)

The formation of ettringite involves an increase in the volume of the reacting solids, a pressure build up, expansion and, in the most severe cases, cracking and deterioration (Soroka, 1993; section 9.3.1). The volume change on formation of ettingite is very large, and is even greater than that produced by the hydration of sodium sulphate.

Another mineral formed by sulphates coming into contact with cement is Thaumasite ($CaSiO_3.CaCO_3.CaSO_4.15H_2O$). This causes both expansion and softening of cement (Crammond, 1985) and has been seen as a cause of disintegration of rendered brick work and of concrete lining in tunnels (Lukas, 1975). A useful review of this mineral and its effects on Portland cement is provided by van Aardt and Visser (1975).

Some salts may cause decay of materials by producing alkaline conditions with elevated pH values. Saline lakes in desert areas may have pH values in excess of 9. This is significant because silica mobility tends to be greatly increased at pH levels above that figure. The presence of sodium chloride may also affect the degree and velocity of quartz solution (van Lier et al.,1960).

Hygroscopicity is another important contributor to chemical attack, and the phenomenon is reviewed by Piqué et al. (1992). Through it, the presence of salts like sodium

Table 6 Suggested limits of salt in aggregates (salts expressed as weight percentage anion of dry aggregate] by Fookes and French (1977)

	Bituminous wearing course/basecourse		Unbound base under		Unbound sub-base under	
	Thick/Dense (A)	Thin/Porous (B)	(A) or	(B)	(A) or	(B)
Soluable chloride anions in the aggregate, in possible presence of groundwater with very low dissolved salts.	0.5 (0.2)	0.4 (0.1)	1.0 (0.5)	0.4 (0.1)	1.0 (0.5	0.4 (0.1)
Soluble anions in the aggregate where more than 10% anion is sulphate, in possible presence of groundwater with very low dissolved salts.	0.3 (0.2)	0.2 (0.1)	0.3 (0.2)	0.3 (0.1)	0.5 (0.4)	0.2 (0.1
If local groundwater contains salts in solution	The design must also minimize moisture migration in road section.	The design must not allow capillary moisture to reach the road surface.				
Soluable chloride anions in the aggregate, in possible presence of transient moisture (rain, dew, etc.) only.	0.6 (0.3)	0.5 (0.2)	2.0 (0.8)	1.0 (0.4)	3.0 (2.0)	
Soluble anions in the aggregate where more than 10% anion is sulphate, in possible presence of transient water (rain, dew, etc.) only.	0.5 (0.2)	0.3 (0.1)	0.5 (0.2)	0.3 (0.1)	2.0 (1.0)	

(i) *Figures in parenthesis refer to soft or friable aggregates of high hazard potential, e. g. clayey or chalky limestones (say water absorbtion (BS812)>5%; soundness loss (ASTM C88) by MgSO4>30: ACV (BS812)>25.*
(ii) *Groundwater with very low dissolved salts is arbitrarily defined as containing less than 0,1 TDS.*
(iii) *The thick/dense surface is assumed to be a non-cement bound asphalt as least 40 mm thick.*
(iv) *Moisture movement in the road section from groundwater sources can be minimised by raising the road on embankment, use of impervious surface course, use of an impervious membrane (e. g. heavy duty polythene) or sand/bitumen layer or prime coat in the lower part of the road.*

chloride can attract moisture into the pores of rock or concrete, thereby increasing the potential for chemical attack (Hudec and Rigbey, 1976; MacInnis and Whiting, 1979). Campbell and Claridge (1987, p.140) identify this as being an important contributor to weathering in the otherwise very arid environments of the dry valleys of Antarctica, while Wallace (1915) suggested it contributed to the erosion of the Manitoba escarpment in Canada.

Gypsum/silicate replacement has been noted in granites in the presence of gypsum, one of the commonest of arid zone salts. According to Schiavon et al. (1995 p. 94):

"A likely mechanism for gypsum replacement of silicate minerals may involve the localisation of silicate dissolution and gypsum precipitation to thin solution films between the silicate and sulphate phases; according to this model, the dissolution of silicates and the precipitation of gypsum should occur simultaneously and original textural features in the primary mineralogy can be preserved after replacement."

4. Management solutions to the salt hazard

The methods that are available to attempt to manage the salt hazard in arid lands can be categorised into a series of groups.

1. Locate new structures away from zones with the environmental ingredients that can cause salt weathering (avoidance and zoning)

2. Reduce the severity of environmental factors that can cause salt weathering

e.g. by reducing groundwater levels

e.g. by reducing air pollution

e.g. by modifying micro-climate conditions

e.g. use of salt free materials in aggregates, etc.

3. Employ engineering solutions to the buildings themselves

e.g. use of salt and moisture retarding membranes

e.g. leaching out of salts

e.g. use of surfactants

e.g. impregnation of stone-work

e.g. use of salt resistant materials.

One of the most effective ways to cope with salt attack is to build structures away from aggressive terrain. This is particularly true of those situations where groundwater level and salinity are the crucial controls, as in the Middle East's low-lying coastal cities. Jones (1980) summarises some of the work undertaken by British geomorphologists to produce hazard-zone maps of value to both engineers and planners. In the Middle Eastern studies, the potential intensity of the salt weathering hazard is basically a function of the elevational relationship between the ground surface and the limit of capillary rise, and the salinity of the rising water. Areas with foundations above the capillary fringe are likely to be a relatively safe location for building.

Given that groundwater level is such an important control of salt attack, efforts need to be made to keep groundwater levels from rising (e.g. by control of irrigation developments), or, if they have already reached critical levels to make them fall. In dryland farming areas, such as Western and Southern Australia, where groundwater rise has followed deforestation, afforestation might be a reasonable ameliorative strategy, whereas in an area that has become salinised and waterlogged because of over-irrigation, the best strategy is probably to pump out groundwater by tube-wells and to evacuate it through drainage canals (disposal channels). This is the method proposed for dealing with the rapid decay at Mohenjo-Daro in Pakistan (van Lohuizen-de Loeuw, 1973).

Finally, there is a range of engineering solutions that may be applied to the salt problem. This can be illustrated by reference to means of preventing damage to bitumen pavements and to concrete bridges. Let us first consider the former, for salt can be a major cause of damage to bitumen roads and runways (see, for example, Januszke and Booth, 1984). The main means of attempting to reduce this problem have been summarised by Obika et al. (1989).

The first aim should be to keep the salt content of compacted basecourse and sub-grade to low limits, and table 6 shows some suggested limited developed by Fookes and French (1977) and reproduced by Okiba et al. (1989, Table 3).

However, it may not always be possible or economically feasible to avoid the use of saline materials and so other procedures may be necessary. One partial solution is to use a thick and dense asphalt, with the intention of greatly reducing evaporative loss of moisture and hence the migration and crystallisation of salt at the surface. Additionally, certain types of bituminous prime may help to reduce permeability, while salt migration by upward capillary rise of soil moisture can be prevented by placing an impermeable or semipermeable membrane in the base course. It is also important to use aggregates of high durability and soundness.

Finally, let us consider briefly how one might seek to protect bridges and other concrete structures against salt attack. Mallett (1994, Chapter 5) presents a useful review of some of the methods that are available to deal with salt attack on bridges while a more general review of means of dealing with the effects of chlorides and sulphates on concrete structures more generally is provided by Bijén (1989). Important measures include water-proofing and good drainage and care not to over-apply deicing salts on bridges. Various concrete sealing materials are available to control chloride intrusion, including epoxies, methacrylate, urethane and silane. It is also possible to remove salt from concrete by electrochemical means, while cathodic protection is a well established anti-corrosion method for protecting steel reinforcements that may be exposed to aggressively saline environments. Full discussion is, however, beyond the scope of this paper.

5. Conclusions

Salt weathering is a severe geomorphological hazard, particularly in dry environments, and has led to the decay of large numbers of engineering structures both ancient and modern. The process has become of increasing severity and importance in recent years, partly because of the growth of settlements in arid areas, but also because of human impacts on hydrological conditions through the extension of irrigation, dryland farming and urbanisation. Groundwater level and salinity is a particularly important control of salt attack, which may be physical or chemical in nature. There is a wide range of management strategies available to reduce or reverse salt attack, and these include procedures involving avoidance and zoning, procedures involving a reduction in the severity of environmental factors that can cause salt weathering, and procedures involving engineering solutions.

References

1. Addleston, L., and Rice, C. (1991) Performance of materials in buildings Oxford: Butterworth-Heinemann

2. Akiner, S., Cooke, R. U. and French, R. A. (1992) Salt damage to Islamic monuments in Uzbekistan. Geographical Journal, Vol.158, pp. 257-72

3. Arnold, A. (1981) Nature and reactions of saline minerals in walls. Conservation of Stone II, pp.13-23

4. Arnold, A. (1981) Nature and reactions of saline minerals in walls. Preprint of the Contributions to the International Symposium on the Conservation of Stone, II, Bologna, pp.13-23

5. Bell. F. G. (1992) Engineering in rock masses. Oxford: Butterworth Heinemann.

6. Bijen, J.M.J.M. 1989 Maintenance and repair of concrete structures. Heron (Delft), 34(2), pp. 1-82.

7. Blackburn, G., and Hutton, J. T. (1980) Soil conditions and the occurrence of salt damp in buildings of metropolitan Adelaide. Australian Geographer 14, pp. 360-365

8. Brown, P. W., Robbins, C. R., and Clifton, J. R., (1979) Adobe II: factors affecting the durability of adobe structures. Studies in Conservation 24, pp. 23-39

9. Cabrera, J. G., and Plowman, C. (1988) The mechanism and rate of attack of sodium sulphate on cement and cement/pfa pastes. *Advances in Cement Research* 1 (3), pp. 171-9

10. Cooke, R. U. (1994) Salt weathering and the urban water table in deserts in D. A. Robinson, and R. B. G. Williams (eds.) Rock Weathering and Landform Evolution, Chichester: Wiley, pp. 193-205

11. Cooke, R. U., and Smalley, I. J. (1968) Salt weathering in deserts. Nature 220, pp. 1226-1227

12. Crammond, N. J. (1985) Thausamite in failed cement mortars and renders from exposed brickwork. Cement and Concrete Research 15, pp. 1039-50

13. Doornkamp, J. C., Brunsden, D., and Jones, D. K. C. (1980) Geology, Geomorphology and Pedology of Bahrain. Norwich: Geobooks.

14. Fitzner, B., and Heinricks, K. (1991) Weathering forms and rock characteristics of historical monuments carved from bedrock in Petra/Jordan in N. S. Baer, C. Sabbioni and A. I. Sors (eds) Science, Technology and European Cultural Heritage. Oxford: Butterworth-Heinemann. pp. 908-911

15. Fookes, P.G. and French, W.J. 1977 Soluble salt damage to surfaced roads in the Middle East. The Highway Engineer 24 (December), 10-20

16. Ghassemi, F., Jakeman, A. J., and Nix, H. A. (1995) Salinisation of land and water resources. Wallingford: CAB International, pp. 536

17. Goudie, A. S. (1977) Sodium Sulphate Weathering and the Disintegration of Mohenjo-Daro, Pakistan. *Earth Surface Processes*, Vol. 2, pp. 75-86

18. Hawass, Z. (1993) The Egyptian monuments: problems and solutions in the M.J. Theil (ed.) Conservation of Stone and other materials. London: Spon. Vol. 1, pp.19-25

19. Hong Naifeng (1994) Corrosiveness of the arid saline soil in China. In Fookes P., and Parry, R.H.G. (Eds) Engineering characteristics of arid soils. Rotterdam: Balkema, pp. 29-34

20. Horta, J. C. (1985) Salt heaving in the Sahara. *Géotechnique* 35, No. 3, pp. 329-337

21. Hudec, P. P., and Rigbey, S. G. (1976) The Effect of Sodium Chloride on Water Sorption Characteristics or Rock Aggregate. *Bulletin of the Association of Engineering Geologists*, Vol. XIII, No. 3, pp. 199-211

22. Ibrahim, H. A. M., and Doornkamp, J. C. (1991) Towns of the Nile Delta and the potential for damage from aggressive saline groundwater. Third World Planning Review 13, pp. 83-90

23. Januszke, R.M. and Booth, E.H.S. 1984 Soluble salt damage to sprayed seals on the Stuart Highway. Australian Road Research Board Proceedings 12(3), 18-31

24. Johannessen, C. L., Feiereisen, J. J., and Wells, A. K. (1982) Weathering of ocean cliffs by salt expansion in mid-latitude coastal environment. Shore and Beach. pp. 26-34

25. Kirchner, G. (In print) Salt in weathering forms of the Basin and Range Province (USA and Mexico) and their relevance to the design of salt weathering experiments. Earth Surface Processes and Landforms

26. Kwaad, F. J. M. (1970) Experiments of the disintegration of granite by salt action. University Amsterdam Fys. Geogr. en Boden Kundiq. Lab. In.16, pp. 67-80

27. Lewin, S. (1990) The susceptibilty of calcareous stones to salt decay in F. Zezza (ed) The conservation of Monuments in the Mediterranean Basin Brescia: Grafo, pp. 59-63

28. Lewin, S. Z. (1981) The Mechanism of masony decay through crystallization. Conservation of Historic Buildings, pp. 120-144

29. Lukas, W. (1975) Betonzer störung durch SO₃ - Angriff unter bildung von Thaumaist und Woodfordit Cement and Concrete Research 5, pp. 503-18

30. Mallett, G.P.1975 Repair of concrete bridges in London: Thomas Telford

31. Masuch-Oesterreich, D. (1993) The groundwater rise in the east of Cairo and its impact on historic buildings in ?? Thorweihe and ?? Schandelmeier (eds.) Geoscientific Research in North east Africa. Rotterdam: Balkema

32. Mehta, P. K. (1983) Mechanism of sulphate attack on Portland cement concrete - another look. Cement and Concrete Research,13, pp 401-6

33. Mortensen, H. (1933) Die "Salzprengung" und ihre Bedeutung für die regional klimatische Gliederung der Wüsten. Petermann's Geographische Mitteilungen 79, pp. 130-135

34. Murdock, L.J., Brook, K.M. and Jgwar, J.D. (1991) Concrete Materials and practice. Arnold: London

35. McInnis, C., and Whiting, J.D. (1979) The frost resistance of concrete subjected to a deicing agent. Cement and Concrete Research 9, pp. 325-336

36. Obika, B., Freer-Hewish, R.J. and Fookes, P.G. (1989) Soluble salt damage to thin bituminuous road and runway surfaces. Quarterly Journal of Engineering Geology 22, 59-73

37. Peck, A. J. (1983) Response to groundwater to clearing in western Australia. Papers, International Conference on Groundwater and Man 2, pp. 327-35

38. Piqué, F, Dei, L., and Ferron, E. (1992) Physiochemical aspects of the deliquescence of calcium nitrate and its implications for wall painting conservation. Studies in conservation 37, pp. 217-227

39. Robinson, D.M. (1995) Concrete corrosion and slab heaving in a Sabkha environment: Long Beach - Newport Beach, California Environmental and Engineering Geoscience l, pp. 35-40

40. Schiavon, N. Chiavari, G., Schiavon, G. and Fabbri, D.(1995) Nature and decay effects of urban soiling on granitic building stones. The Science of the Total Environment 167, pp. 87-101

41. Seedsman, R. (1986) The behaviour of clay shales in water. Canadian Geotechnical Journal 23, pp. 18-21

42. Shehata, W., and Lotfi, H. (1993) Preconstruction solution for groundwater rise in Sabkha. Bulletin of the International Association of Engineering Geology 47, pp. 145-150

43. Smith, B. J., and McAlister, J. J. (1986) Observations on the occurrence and origins of salt weathering phenomena near Lake Magadi, southern Kenya. Z. Geomorph. N. F. 30, 4, pp. 445-460

44. Soroka, I. (1993) Concrete in hot environments. London: Spon

45. Taylor, R. K., and Smith, T. J. (1986) The Engineering Geology of Clay Minerals: Swelling, Shrinking and Mudrock Breakdown. Clay Minerals 21, pp. 235-260

46. Terzaghi, K., and Peck, R. B. (1948) Soil mechanics in engineering practice. New York: Wiley

47. Van Aardt, J. H. P, and Visser, S. (1975) Thausamite formation: a cause of deterioration of Portland cement and Related Substances in the presence of sulphates. Cement and concrete Research 5, pp. 225-37

48. Winkler, E. M., and Singer, P. C. (1972) Crystallization Pressure of Salts in Stone and Concrete. *Geological Society of America Bulletin*, V. 83, pp. 3509-3514

49. Winkler, E. M., and Wilhelm, E. J. (1970) Saltburst by hydration pressures in architectual stone in urban atmosphere. Bulletin Geological Society of America 81, pp. 567-572

Author

Andrew Goudie
University of Oxford
School of Geography
Mansfield Road
Oxford, OX1 3TB
United Kingdom

IMPACT OF CONVERSION OF UPLAND FOREST TO TOURISM AND AGRICULTURAL LAND USES IN THE GUNUNG KINABALU HIGHLANDS, SABAH, MALAYSIA

WAIDI SINUN, IAN DOUGLAS

1. Introduction

Accessible areas of equatorial tropical uplands close to cities and main roads are under increasing pressure of land use conversion from upland rain forests to temperate crops and recreational activities (Hamilton et al., 1995). In many developing countries of the humid tropics, rapid population growth, and of agriculture in tropical highlands and uplands. This pressure adds to the long developed exploitation of tropical uplands for tea, coffee and other commercial crops. Since 1980, expansion of temperate vegetable cultivation, both for the air freight markets in industrialised countries and for local supermarkets supplying the expanding middle class, has become a third agricultural pressure on the highlands of equatorial regions. Yet, although of such importance, farming in upland-steepland is said to be a neglected sector in Asian agriculture (Nangju, 1991).

The cooler climate of tropical uplands has long attracted people to the hills for leisure, from the hill stations of Simla and Ootacamund to the modern resorts of Genting Highlands and Bukit Tinggi. Golf courses and leisure complexes are adding to land cover and landform change in tropical uplands. Removal of the forest on steep slopes and the development of networks of paths and tracks lead to major modifications of rainfall: runoff relationships in areas where the lower montane and montane forests have a denser natural ground cover than their lowland counterparts. Litter decomposition is slower in the cooler upland temperatures and trees are less tall (Kitayama, 1995). Rain tends to be of longer duration and less intense than in the lowlands.

The vegetable cultivation of the Cameron Highlands of Peninsular Malaysia is long established and its environmental consequences are well known. The classic study by Shallow (1956) in the Cameron Highlands found annual sediment yields for catchments of between 25 and 35 % land use for tea or vegetable cultivation and the remainder under montane and lower montane forest to be 257 to 277 t km^2y^{-1}. A study 30 years later (Suki and Jaffar, 1990) based on a few suspended sediment determinations estimated a soil loss of approximately 200 t km^2 y^{-1} for the whole of the 712 km^2 Cameron Highlands. This study commented that due to the loss of top soil from farmland, farmers have cut into hitherto undisturbed hillsides to gain additional soil, thereby aggravating the erosion problem.

J. Kalvoda and C.L. Rosenfeld (eds.), Geomorphological Hazards in High Mountain Areas, 121-131
© 1998 *Kluwer Academic Publishers.*

2. Agricultural land development in the Kinabalu Park-Kundasang area of Sabah

In the Borneo state of Sabah, the greatest rate of steepland conversion to intensive agriculture has occurred since 1975 in the upland forest outside the Kinabalu National Park in the Kundasang area. Gunung Kinabalu (summit altitude 4,101 m) is the highest mountain in South-East Asia and "...because of its breath taking isolation, it has its own climate, a constant flux of cloud and wind, rain and cold, and the warmth of forest below... " (Harrison, 1978, p. 23). Although Gunung Kinabalu and surrounding areaswere designated a rainforest park in 1964, some land was degazetted into State land in 1980's to meet economic and social needs. Most of the degazetted area is still covered with forest, but is earmarked for development and agricultural purposes. In the 1980's, irrigation system improvements increased the area's productivity by making water available throughout the year (Matlan Gamalie, Kinabalu Farmers Association, Pers. Comm.). Probably the biggest and most important problem of Kundasang area is lack of 'proper' management during forest clearance, preparation of land for other uses, and under cultivation (Suhaimin, 1988). In late 1991, we observed land in hydrologically sensitive areas being cleared. Such clearance in fragile landscapes is known to accelerate mass movements and consequent land degradation.The massive expenditure to alleviate slope instability problems along the Kota Kinabalu - Ranau Highway, which passes the Kinabalu Park entrance, is but one indication of the economic costs of landuse change in this region.

Vegetable cultivation is continuous with several crops a year and now these highlands supply over 90 % of Sabah's vegetables, but at the cost of impacts on water supplies further down the Liwagu River and in the Labuk drainage system as a whole, in terms of flows, sediment loads and residues of agricultural chemicals. Tracks have to be built to provide access to the felds, but despite the terracing of the cropped areas, both the vehicular and walking access routes become severely eroded and in some cases develop into gullies up to 2 m deep. Within the Park itself there has to be vigilance to ensure that footpaths, roads and buildings do not alter the quality of water in streams.

In 1992, Sabah Parks and the University of Manchester in the UK., in collaboration with the Sabah Drainage and Irrigation Department, Yayasan Sabah, the Sabah Ministry of the Environment and Tourism and the then UKM (Sabah Campus), began a project to identify the benefits of maintaining a forest cover and assess the effects of loss of forest cover which set out to:

1) To evaluate the impacts of land cover changes on stream discharges; and

2) To compare sediment outputs from an undisturbed headwater stream, the Sungai Kalangaan, in Kinabalu Park with those from a stream, the Sungai Ayamut, draining the disturbed, vegetable-growing area just outside the Park and those from a stream, the Sungai Silau-Silau, partly affected by the tourist complex just inside the park gates (Figure 1).

3. Catchment characteristics

The three catchments (Table 1) had similar relief, slopes and lithology, although that affected by agriculture was at a slightly lower elevation than the other two. The annual rainfall of the Park Headquarters area at 1,680 m a.s.l. has fluctuated between 2,004 mm and 3,828 mm during the 30-year period records have been kept. Rainfall in the three catchments (Figure 2) is similar, but the Kemburongoh station further up the mountain records more in most months than those in the Silau-Silauor Ayamut catchments.

Table 1. Catchment characteristics and sediment yields

Catchment Name	Land Cover	Area km²	% of time discharge exceeds 0.125 m³ s⁻¹	sediment yield
Sungai Kalangaan	100 % lower and upper montane forest	2.45	1.9	15.1
Sungai Silau- Silau	95.6 % lower and upper montane forest,4.4% developed with tourist facilities (park buildings)	1.91	4.6	108.9
Sungai Ayamut	45 % lower montane forest, 55% cultivation, roads and buildings	0.91	4.3	1076.4

4. Methodology

Water level recorders were mounted on stilling wells a suitable sites with rock controls in the three catchments. A stone and cement weir was constructed in the Silau-Silau to improve the quality of the control in this steep boulder-strewn stream. Stream gauging by current meter and dilution methods produced rating curves with an accuracy of around ± 10%, but in the Silau-Silau boulder movement in major storms posed problems for stage: discharge relationships.

Float switch-activated automatic vacuum ALS water samplers collected 24 water samples during every storm event at 7.5 minute intervals for three hours. Suspended sediment concentrations were determined by filtration using 0.45 μm membrane filters in a vacuum pump assembly. Calibration of the samples and suspended sediment concentrations at low flows were assessed by samples collected with a USDH 48 depth-integrating hand suspended sediment sampler. Loads were calculated for every individual sample and then aggregated to give storm period sediment loads.

Nevertheless, the proportion of high discharge flows in the Kalangaan was far less than in the Silau-Silau and Ayamut. Discharge of the Kalangaan exceeded 0.125 m³ s⁻¹ km⁻² only 1.9% of the time but 4.6% and 4.3% of the time respectively in the Silau-Silau and Ayamut (Table 1). Discharges exceeding 0.225 m³ s⁻¹ km⁻² occurred 2.1% and 1.3% of the time in the Silau-Silau and Ayamut respectively but only 0.8% of the time in the Kalangaan.

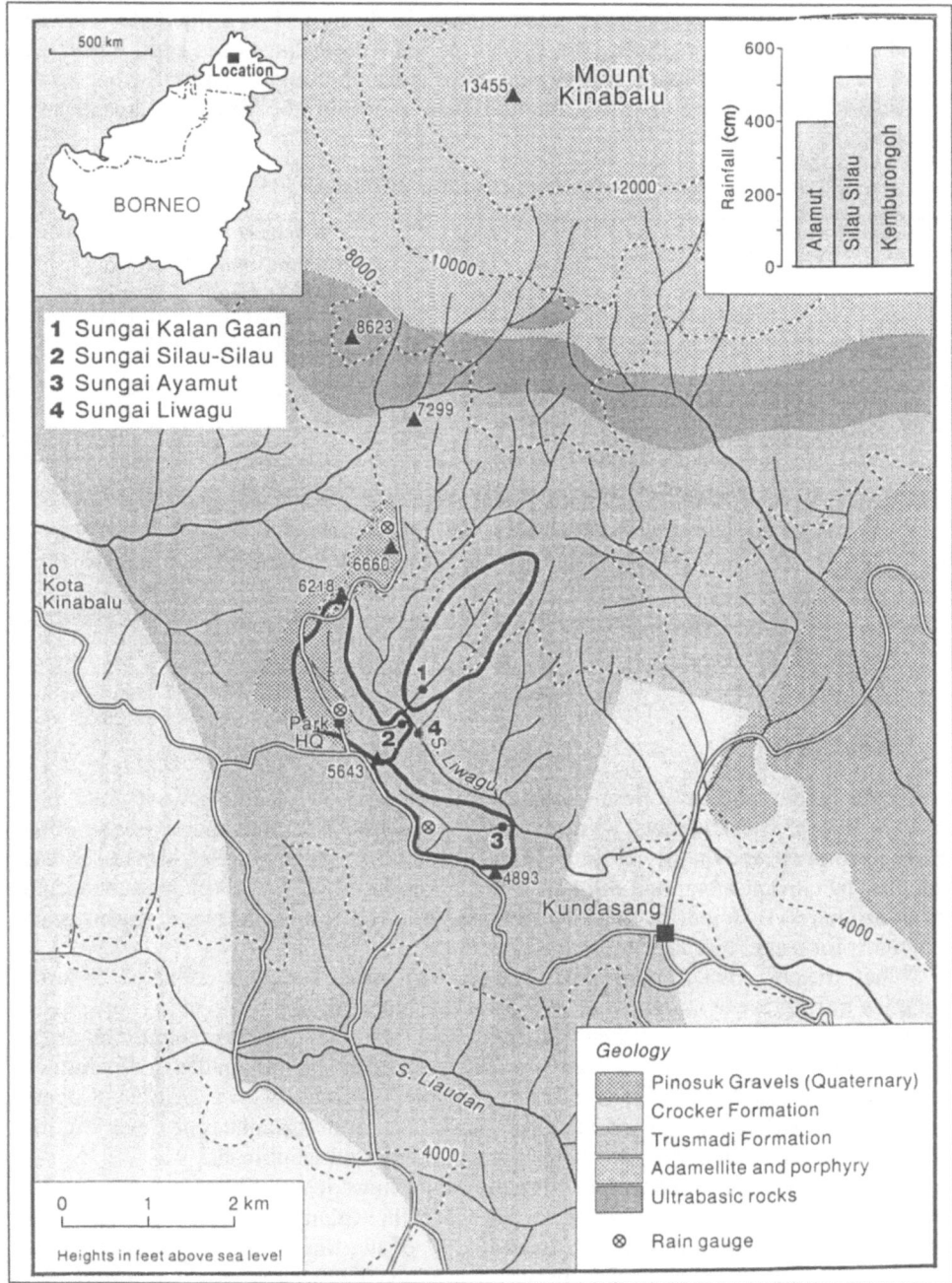

Figure 1. Location, geology and rainfall of the three study catchments at Mount Kinabalu

The regulatory influence of the montane steepland cloud-forest is shown by the way only 13.2% of the Kalangaan total runoff is derived from storm flow (quick return flow) compared with 38.0% and 29.7% for the Silau-Silau and Ayamut respectively. The contrast between the Silau-Silau and the Ayamut probably arises from the distribution of the land cover in the two catchments in relations to the location of the gauging station. Although the Silau-Silau is 94% forest-covered, the disturbed area, with roads and drains, is close to the gauging station in the lower fifth of the catchment. On the other hand, the cleared and developed areas of the Ayamut are on the southern and western divides of the catchment and the steep slopes immediately above the gauging station remain forest covered. Thus the Silau-Silau has direct pathways for rapid surface runoff to the channel, while the Ayamut, although having much concentrated runoff from tracks and fields, has opportunities for infiltration and subsurface flow in the forested areas downslope from the fields. Quick return flow in the Silau-Silau thus exceeds that in the Ayamut.

5. Sediment discharge

When the relationship between river flow (discharge on Figure 3) and suspended sediment concentration is examined, we see that in the undisturbed Sungai Kalangaan has no concentrations above 1,000 mg l^{-1}, while the Silau-Silau has just two and the Ayamut has many and several exceeding 10,000 mg l^{-1}. This indicates that the agricultural activity is supplying much more sediment to the Ayamut than enters the other rivers. In storms, the contrast between the disturbed agricultural Ayamut catchment and the others becomes even more apparent. Graphs for 17. 11. 93 for the Ayamut and Kalangaan show that during the rain that, the sediment concentration in the Ayamut rose to nearly 6000 mg l^{-1}, while that in the undisturbed natural forest Kalangaan only reached 500 mg l^{-1}. On 9.5. 93, the concentration in the Ayamut rose to almost 25,000 mg l^{-1}, a figure characteristic of rivers draining severely eroded soils. These careful observations show that the Ayamut exported about 9 times more sediment than the Silau-Silau and about 100 times more than the Kalangaan. Storms accounted for the bulk of the sediment output, especially in the Silau-Silau and Ayamut where they evacuated 90% and 89% of the annual sediment load respectively. The big difference between the Ayamut and the other two catchments is understandable because of the great degree of agricultural disturbance.

The sediment yield for the Kalangaan is low for a natural forest stream in Malaysia, even though it has steep slopes. There may be infrequent major mass movements or tree fall events that supply debris to the stream but none were sampled during the first 15 months of observations reported here. The sediment yield of the Kalangaan being less than in less steep lowland undisturbed catchments (Table 2). The annual suspended sediment yield at 15.5 t km^{-2} yr^{-1} is considerably less than those of lowland steepland areas. For example yield of W8S5 at Danum Valley was 312 t km^{-2} yr^{-1}, approximately 20 times higher (Douglas et al., 1992, Table 2). In an area well to the south of Kinabalu Highlands, Malmer (1990) obtained sediment yields of 58.4 and 20.1 t km^{-2} yr^{-1} for two of his forested catchments, which are also significantly greater than that obtained for Kalangaan.

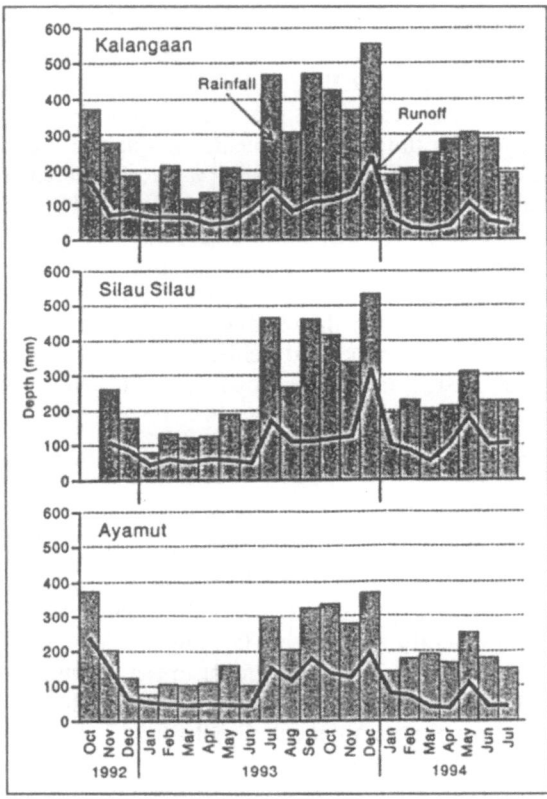

Figure 2. Rainfall and runoff for the three study catchments

Reasons for the large difference in suspended sediment yield between the undisturbed Kalangaan and the Silau-Silau, which despite the buildings and roads at the Park Headquarters, is 94 % covered with forest need discussion. The geological map (Figure I) provides one clue. Much of the Silau-Silau is cut into the Pinosuk Gravels, which erode easily when exposed on the river banks. However, the roads and roofs of buildings close to the gauging station in the Silau-Silau cause rainwater to flow rapidly to the river, giving it a rapid rise in flow and the energy to erode the unconsolidated banks and thus gain the extra sediment. Channel bank erosion and release of sediment from temporary debris dams associated with entry of material to the stream as a result of construction activity supply a large amount of material to the stream. In places new development leads to bulldozed earth close to the channel. Roadside runoff, even from sealed roads, adds to the detritus that contributes to the sediment load. Thus the Silau-Silau is affected by construction and earth movements related to the tourist activity, but the effect on erosion is much less than that caused by agriculture in the Ayamut.

In the Ayamut catchment, large quantities of clay are washed from the vegetable plots. Although vegetables are planted on ridges usually running approximately parallel to the contour, the lower ends of the areas between ridges usually allow water to flow on to the access path to the plot. The path then acts as an efficient conveyor of

Table 2. Suspended loads of selected Malaysian catchments

Catchments	Area (km²)	Vegetatiton/land use	Sediment Yield (t km⁻² yr⁻¹)	Methods	Comments
1. Bukit Berembun					
C1	0.13	Unsupervised logging 40% extraction	Before = 13.6 After = 27	Grab samples Weighted sediment	Baharuddin (1988) Zulkifli (1990)
C2	0.04	Forests	Before = 20 After = 19	concentrations calculated with daily discharge	
C3	0.30	Supervised logging 33% extraction	Before = 13.7 After = 12		
2. Sg. Tekam					
A	0.47	Forest to cocoa	Before = 22.5[a] After = 95[b]	USDH-48 and single stage rising	DID (1989);a-6 yr av.; b-3 yr av.; c-9
B[d]	0.96	Forest to oil palm	Before = 28[a] After = 225[b]	sampler. Sediment rating curve with mean daily	yr average; d-catchmentnested below A
C	0.56	Forest control	Before = 29[c]	discharge	
3. Sg. Gombak	140	Various	178.3	Sediment flow duration curves. USDH-48 sampling at various discharges.	Douglas (1968); sediment production calculated using sp Gravity of 2.65
4. Hulu Langat and Sg Lalang Forest Reserve					
Sg Batangsi	19.8	Logging on going	2,826.2	Selected base-flow samples and samples	Lai (1992)
Sg. Lawing (a) Aug 88 - Jul 89	4.7	100% upland forest	53.7	collected during storms using	Lai (1992)
Sg. Lawing (b) Jan 93 - Dec 93	4.7	Logging started Jan 93	1,129	an automatic sampler	Lai et al. (1995)
Sg.Chongkak(a) Mar 87 - Jun 88	12.7	Logging until Apr 87	2,476		Lai (1992)
Sg. Chongkak (b) Jul 88 - Oct 89	12.7	Post-logging recovery	1,335		Lai (1992)
Sg. Lui	68.1	80% forested 20% rural area	89.7		Lai (1992)
5. Danum Valley Sabah					
Sg. W8S5	1.1	100% Lowland forest	312	Daily samples using 500 ml bottles and	Douglas et al (1992)
Sg Steyshen Baru	0.56	Affected by logging	1,600	storm samples collected with Automatic liquid sampler (ALS)	

6. Lumaku Sepitang Sabah					
W1+W2	0.06	non-mechanised clearing, burning of slash and planting	71.4	Sampling at flume outlet. Automatic water samplers	SFI (1989) and Mal-mer (1990). 'Before' period
W3	0.19	Forest - control	58.4		averages of 10 months data; 'after'
W4	0.03	Manual felling, ma-nual extraction of logs - moving of slash into rows and planting. No burning	142.7		period average of 18.5 months data
W5	0.1	Manual felling, ex-traction by crawler tractors, burning of slash and planting	253		
W6	0.04	Forest - control	20.1		
7. Cameron Highlands					
Sg. Bertan	75.52	Forest - 64% Tea - 7% Vegetables - 7% Open areas - 8%	222	Winchester bottle lowered to 0.6 m stream depth. Sediment flow	Shallow (1956)
Sg. Kial	21.37	Forest - 70% Tea - 11% Vegetables - 19%	210	curves duration curves	
Sg. Telom	77.7	Forest - 94% Tea - 5% Vegetables - 1%	27.4		
8. Kinabalu Highlands					
Sg. Kalangan	2.45	Undisturbed Forest - 100%	15.5	Daily samples using 500 ml bottles and storm samples	Sinun (1995)
Sg. Silau-Silau	1.91	Undisturbed Forest - 95.60% Tourist development - 4.40%	108.9	collected with Automatic Liquid sampler (ALS) set at 15 and 7.5 minutes sampling interval	
Sg. Ayamut	0.91	Undisturbed Forest - 45% Vegetable Garden and Tourist Develop-ment - 45%	1076.4		

concentrated runoff and sediment downslope to the main track through the fields and then into a gully head which has been cutting back 2 to 3 m per year. Lack of any ero-sion control means that the only filter of water and sediment from agricultural land is the 100 m of forest on the steep convex slope immediately above the stream channel. Thus the channel be has huge accumulations of clay which even the largest storms have diffculty in removing. While sediment yields in the Kalangaan and Silau-Silau catchments are supply limited, that of the Ayamut is transport limited. The sediment delivery ratio for the Ayamut catchment is probably quite low. Small, bounded plot experiments in the agricultural fields revealed sediment losses equivalent to as much as 6,600 and 11,250 t km^{-2} yr^{-1} for two different slopes. While the soil loss from the plots depended on the type of ridge and furrow layout and the crop being grown, the sediment yield at the catchment outlet gauging station may be as little as one tenth of the soil loss on farmland. These losses from the fields themselves are of the same order as those reported for cultivated areas in various parts of China (Hill, 1994) and exceed those for some plots elsewhere in Malaysia (Table 3). As the gullying of the cultivated area intensifies, future sediment yields may be higher.

Table 3 Estimated* annual sediment yield of the Experimental Plots (t km^{-2} yr^{-1})

Stages	Period (days)	Plot A (t km^{-2} yr^{-1})	Plot B (t km^{-2} yr^{-1})
Phase I Cabbages: Beds downslope	91	634.7	1109.8
Phase 2 Chilli: Beds downslope	127	6601.94	11246.74
Phase 3 Cabbages: Beds across slope	98	469.830	555.07
Phase 4 Chilli Beds: across slope	77	773.76	1356.1

*Yield Calculation[(Total Yield (t km^{-2})/No. of days analysed) x 365.25 = Annual sediment Yield (t km^{-2} yr^{-1})

Sediment yield in the Ayamut, and in the vegetable growing areas of the Kinabalu Highlands as a whole, has a large social dimension through the way in which the production of crops is organised. The younger generations of the local Dusun peoples have little interest in the hard labour of vegetable growing and have leased their lands to middlemen who bring in migrant labour, often from eastern Indonesia, to do the actual work of cultivating the fields. Use of biocides and chicken-dung based fertilisers is high, but there is little long-term concern for the state of the land. The pressure on farmers is to respond to the growing external market demand, for the middle man to make a profit and the migrant labourer to earn an adequate wage. Here in the Kinabalu Highlands, the role of urbanisation, especially the consumption patterns of the urban middle class, as a driving force for land use/land cover change, and thus for global environmental change, is well demonstrated.

Table 4. Erosion levels and runoff under various vegetative covers on Regosol soil at Kemaman eastern Peninsular Malaysia (Ghulam et al. 1992)

Crop cover	Erosion t km^{-2} yr^{-1}
Kemaman (rainfall 4139 mm y^{-1})	
Cocoa + Gliricidia + Banana + no cover crop on 11° slope	9858.5
Cocoa + Gliricidia + Banana + cover crop on 8° slope	472
Cocoa + Gliricidia + no cover crop on 10° slope	1370.8
Cocoa + Gliricidia cover crop on 9° slope	130.2

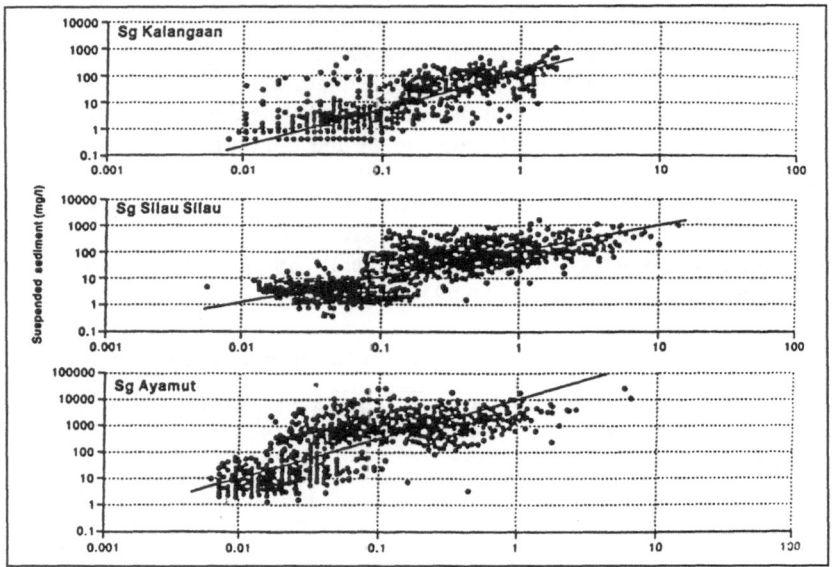

Figure 3. Suspended sediment: discharge relationships for the three catchments

Acknowledgements

We are grateful for the field assistance of Richard Siobi which has been essential to the project. The project was supported by NERC Grant GR and a Grant from the Conservation, Food and Health Foundation for which we are most grateful.

References

1. Baharuddin, K. 1988 Effect of logging on sediment yield in a hill dipterocarp forest in Peninsular; Malaysia. Journal of Tropical Forest Science, 1 (1), 56-66.

2. DID (1989): Sungai Tekam experimental Basin. Final Report July 1977 to June 1986. Water Resources; Publication No. 20. Drainage and Irrigation Department, Ministry of Agriculture, Kuala Lumpur, Malaysia, 107pp.

3. Douglas; I. 1968 Erosion in the Sungei Gombak catchment, Selangor, Malaysia. Journal of Tropical; Geography, 25, 1-16.

4. Douglas, I., Spencer, T., Greer, T., Kawi Bidin, Sinun, W. and Wong W.M.1992 The impact of selective commercial logging on stream hydrology, chemistry and sediment loads in the Ulu Segama Rain Forest, Sabah. Philosophical Transactions Royal Society, London B, 335, 397-406.

5. Hamilton, L.S., Juvik, J.O. and Scatena, F.N.1995 The Puerto Rico Tropical Cloud Forest Symposium: Introduction and Workshop Synthesis. In Hamilton L.S., Juvik, J.O. and Scatena, F.N. (eds); Tropical Montane Cloud Forests. (Ecological Studies 110) Springer, New York, 1-23.

6. Harrison, T. (1978): Kinbalu the wonderful Mountain of Change. In Kinabalu Summit of Borneo Sabah; Society Monograph. 24-44.

7. Hill, R.D. 1994 To the hills I shall lift my eyes...Inaugural Lecture November 17th 1994. The University of Hong Kong, Supplement to the Gazette Vol. XLI, (No. 2), 81-89.

8. Hudson, T.C. Sheng and San-Wei. Lee (eds) Development of Conservation Farming on Hillslopes. Soil and Water Conservation Society Ankeny, IOWA USA, 137-156.

9. Kitayama, K. I 995 Biophysical conditions of the montane cloud forests of Mount Kinabalu, Sabah Malaysia. In Hamilton, L.S., Juvik, J.O. and Scatena, F.N. (eds) Tropical Montane Cloud Forests. (Ecological Studies 110) Springer, New York, 183-197.

10. Lai, F.S., Lee, M.J. and Mohd. Rizal, S. 1995 Changes in sediment discharge resulting from commercial logging in the Sungai Lawing basin, Selangor, Malaysia. International Association of Hydrological Sciences Publication, 226, 55-62.

11. Lai, F.S.1993 Sediment yield from logged, steep upland catchments in Peninsular Malaysia.; International Association of Hydrological Sciences Publication, 216, 219-229.

12. Malmer, A. 1990 Stream suspended sediment load after clear felling and different forestry treatments in tropical rainforest, Sabah, Malaysia. International Association of Hydrological Sciences Publication, 192, 62-71.

13. Nangju, A. 1991 Management of Hillslope for Sustainable Agriculture. In W.C. Moldenhauer, N.W.

14. Shallow, P.G.D. (1956) River Flow in the Cameron Highlands. Hydro-electric Technical Memorandum No. 3, 27p

15. Sinun, W. (1995): Erosion in a tropical steepland environment: Mt. Kinabalu, Sabah. Unpublished Ph.D. Thesis. University of Manchester.

16. Suhaimin, J. (1988): Aspek Ekonomi dan Sosial Dalam Pembangunan Luar Bandar. Kajian Kes: Kundasang Ranau, Sabah. Unpublished BA Thesis. University Utara Malaysia.

17. Suki, A. and Jaffar, M.M.1990 A water qualitz survey of Cameron Highlands Watersheds. Paper presented at the Regional Seminar on Watershed Development and Management 19-23 February I 990, Kuala Lumpur.

18. Zulkifli Yusop (1991): Hydrologic nutrient osses following selective logging methods in a tropical rainforest. Paper presented at soil science conference of Malaysia, Genting Highland, Malaysia, 4-5 March 1991, 20 pp.

7. Authors

Waidi Sinun* and Ian Douglas
University of Manchester
United Kingdom
(*now at Forestry Division, Innoprise Corporation, Kota Kinabalu, Sabah)

LANDSLIDES IN THE ROCKY MOUNTAINS OF CANADA

D.M. CRUDEN, X.Q. HU

1. Introduction

Landslides from steep rock slopes can travel large distances and have catastrophic impacts on structures in their paths (Nicoletti and Sorriso-Valvo, 1991). Site reconnaissances in mountain regions seek to identify slopes which may move in the future and to locate facilities away from the consequences of these movements. So reconnaissance requires identifying potentially unstable slopes.

Terzaghi (1962) and Young (1972) described the stable slopes formed in rock masses with parallel, penetrative discontinuities and a joint set perpendicular to them. They assumed that the only movement possible was sliding along the discontinuities. Their diagrams of the stable slope forms are now reproduced in geomorphology texts (Gerrard, 1988, p174; Selby, 1993, p. 327; Trenhaile, 1987, p. 117). However, movements that occur in stratified rocks in the Canadian Rockies, in our experience, include toppling, buckling and sliding along combinations of discontinuities besides sliding along a single set of discontinuities. Stable slopes are also limited by these processes.

In this paper, we first provide ourselves with symbols from the terrain analysis literature to represent different types of steep sedimentary rock slopes (Figure 1). We then discuss the movements possible within each of the different types of slope, referring to examples from our field work in the Canadian Rockies (Figure 2). Young's (1972, p. 121) critical slope angles can be conveniently shown on a process diagram (Figure 3). We provide a new process diagram (Figure 4) which shows lower stable slopes than Young's. Cruden and Varnes' (1996) nomenclature describes the slope movements, symbols follow Barsvary et al. (1980) unless otherwise stated.

2. Terrain analysis

The Canadian Rockies, over 100 km wide, stretch 1400 km northwestwards from the U.S. border, 49°N, at 114°W. They are formed by Proterozoic clastics and Paleozoic carbonates which were moved northeastwards up listric thrusts and over Mesozoic clastics (Gabrielse and Yorath, 1992). The preferred orientation of slopes in the Rockies, perpendicular to the strike of bedding, is immediately obvious on LANDSAT photography (Taylor, 1981). The Canadian Rockies thus show "fabric-relief", in which elements of the topographic relief are controlled by the rock fabric (Sander, 1970, pp.210-212). This observation is neither new nor confined to the Canadian Rockies. Callaway (1879) commented "..., the strike of the mountain chain corresponds with the strike of the beds of which it is composed. Among examples of this law which will at once occur to geologists are the Pennine Chain, the Cotteswolds, the North and South Downs, the Swiss Alps and the Andes". Turner (1952) had previously

133

J. Kalvoda and C.L. Rosenfeld (eds.), Geomorphological Hazards in High Mountain Areas, 133-148.
© 1998 *Kluwer Academic Publishers.*

used Sander's German term, gefuge-relief, to describe schist-tor topography in New Zealand.

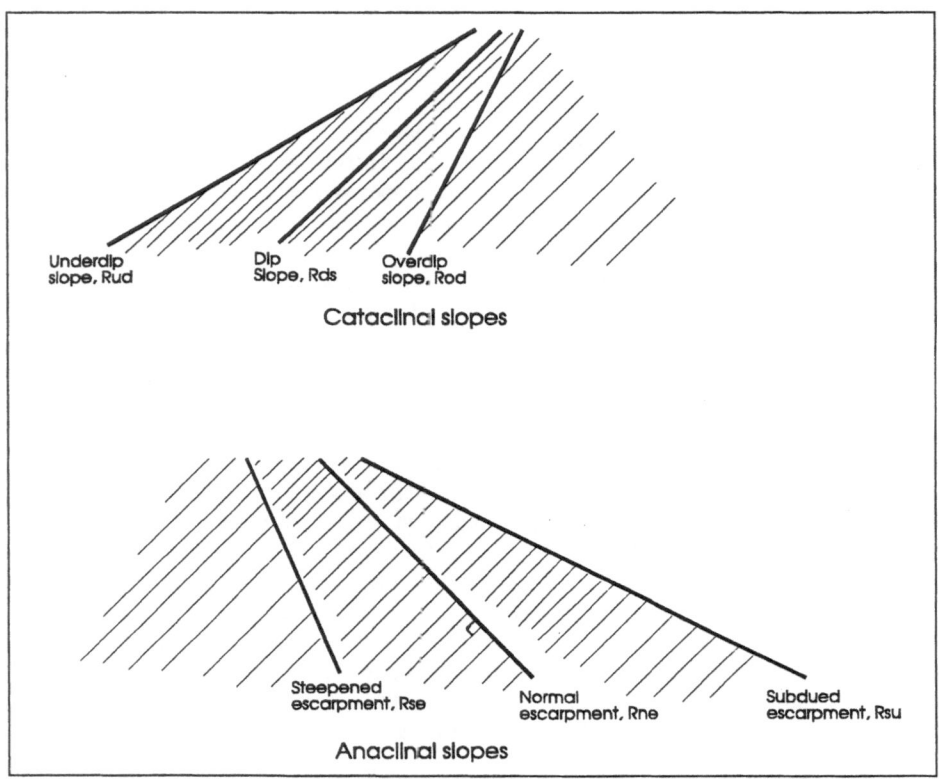

Figure 1.Sections through rock slopes classified by the relative orientation of the penetrative discontinuity (thin lines) and the slope (thick line).

When a rock mass contains penetrative discontinuities, such as bedding or schisto-sity, their orientation can be used to classify the fabric-controlled rock slopes formed on the rock mass (Figure 1). We use terms originally introduced by Powell (1875). In cataclinal slopes, the penetrative discontinuity dips in the same direction as the slope; in anaclinal slopes, the penetrative discontinuity dips in the direction opposite to the slope. A similar French terminology exists (Foucault, A., Raoult, J-F., 1988). Con-ventionally, these descriptive terms can be limited to slopes whose strikes are within 20 degrees of that of the penetrative discontinuity. Cataclinal slopes may be further divided into overdip slopes which are steeper than the dip of the discontinuity and underdip slopes which slope less than the dip of the discontinuity. Dip slopes follow the discontinuity. We propose to term anaclinal slopes which are perpendicular to the dip of the discontinuity "normal escarpments", slopes steeper than normal are "steepened escarpments" and slopes less steep than normal are " subdued escarp-ments".

In site reconnaissance studies, steep rock slopes, Rs, in the terminology of terrain analysis, (Canada Soil Survey Committee, 1978; Cruden and Thomson, 1987; Howes and Kenk, 1988) can be subdivided into the six types of slopes in Figure 1 with the assistance of airphotos and 1:50,000 topographic and geological maps readily available for the Canadian Rockies. The symbol, s, for steep slopes can be divided into subclasses, overdip slope, Rod, dip slope, Rds, underdip slope, Rud, steepened escarpment, Rse, normal escarpment, Rne, and subdued escarpment, Rsu (Table 1).

Table 1.

Type of steep rock slope	Modifying process predicted	Terrain symbol
overdip slope	rapid sliding	Rod-R^S
overdip slope	slow toppling	Rod-F^T
dip slope	rapid buckling	Rds-R^B
underdip slope	slow toppling	Rud-F^T
steepened escarpment	rapid sliding	Rse-R^S
steepened escarpment	rapid toppling	Rse-R^T
normal escarpment	slow toppling	Rne-F^T
subdued escarpment	slow toppling	Rsu-F^T

Howes and Kenk (1988) divided modifying processes on slopes into rapid, -R, and slow, -F, movements. Qualifying descriptors of these processes can be used to indicate buckling, B, sliding, S, and toppling, T. These can be placed as superscripts on the modifying processes. So the symbol, - F^S, represents a slope being modified by slow sliding. In Table 1, we collect the types of steep rock slopes we recognized together with the modifying processes we predict on such a slope from Figure 3 and the terrain symbol which would appear on a map or airphoto overlay to guide site reconnaissance.

We have overlaid the critical angles proposed by Young (1972, p. 121) on Figure 3. Toppling from kathetal joints and from bedding takes place on slopes less than Young's critical angles. Buckling may reduce dip slopes to underdip slopes, again reducing a critical angle. We have not pursued the distinctions Young made between various arrangements of cross (or kathetal) joints which resulted in 3 different critical angles. Joint spacings, typically less than a metre in the Rockies, are several orders of magnitude smaller than the displaced masses in large slope movements and contribute only to the roughness of surfaces of rupture (Hu and Cruden, 1992).

Cruden's (1988, Figure 8) process diagram (Figure 3) showed different slope movement modes determined by the relationship between the orientations of slopes and penetrative discontinuities. We update this process diagram as Figure 4 based on more recent studies. We define the boundary of each slope process in Figure 4 and for each slope process, using symbols from Table 1 to indicate both the slope classification and the slope process. Rapid movements here include both sliding from discontinuities

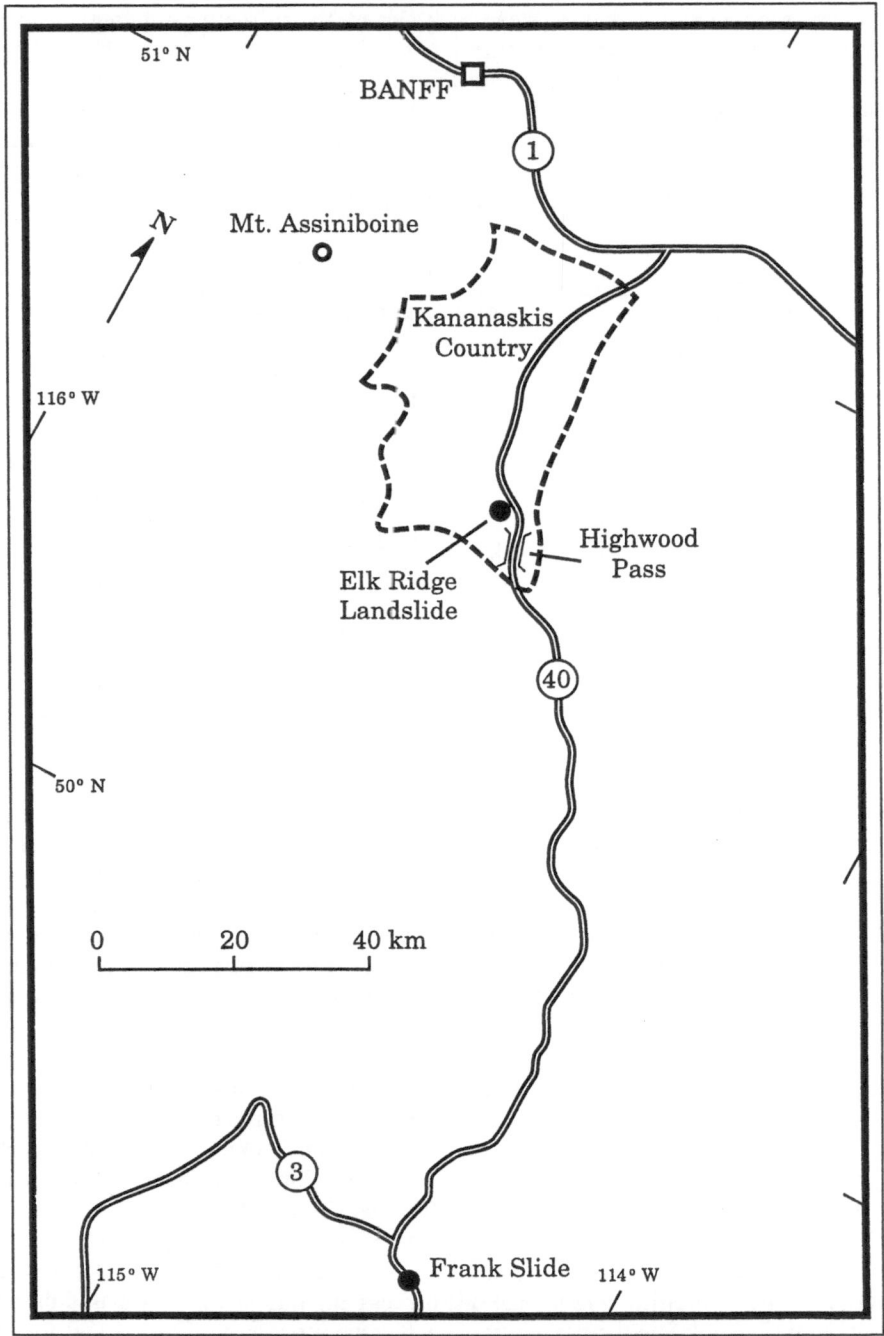

Figure 2. Location map

when the dips of the discontinuities exceed the friction angles of the discontinuities and rapid sliding following toppling. We assume that friction angles are $30°$ and cohesion zero to simplify discussion. If friction angles are different, slope process boundaries should be adjusted accordingly.

3. Cataclinal slopes

3.1. OVERDIP SLOPES, ROD

The behavior of overdip slopes can be modeled by the behavior of a sliding block of the rock mass. In the simplest model, the block slides when the dip, Ψ, of the discontinuity down the slope is greater than the friction angle, Φ, of the block on the slope.

Cruden (1985) demonstrated that the basic friction angle, Φ_b, of the rock is a useful estimate of the lower bound of rockslide avalanche activity in aseismic areas. When overdip slopes are less steep than this angle, the rupture surface remains covered with debris and rapid movements do not take place (Simmons and Cruden, 1980). It seems reasonable then to take Φ_b as an estimate of the friction angle on natural bedding surfaces except when the rock has softened and altered or the discontinuity has been polished by displacement. In these cases, shear tests should be carried out on samples of the actual discontinuities which may be at their ultimate friction angle, Φ_u, (Krahn and Morgenstern, 1979).

Cruden (1975) showed a slab of thickness, t, can be supported by a cohesion. c, where

$$t = c \cos \Phi/\gamma g \sin (\Psi-\Phi) \qquad (1)$$

Clearly major rock slides can be expected on cataclinal overdip slopes in cohesive rock masses along discontinuities dipping at relatively low angles which slightly exceed the basic angle of friction along the discontinuity. Bedding dipping at Φ_b thus forms a threshold for the occurrence of large slides in the Rockies. Steep overdip slopes (Rod- R^S, Fig. 4) give way, with the loss of cohesion over time, to dip slopes which, as characteristic landforms (Brunsden and Thornes, 1979), contribute to the fabric relief.

Rock slides can be expected on slopes plotting in Rod-R^S on Fig. 4. When bedding dips are over the basic friction angles on cataclinal slopes, the final characteristic landforms are dip slopes. They evolve from glacially steepened overdip slopes by sliding along bedding planes. The evolution may be sudden and movements can be rapid. Two historic rock slides along bedding have been documented (Cruden, 1982. Gadd, 1995, p. 156).

The lower bounds of apparent cohesion (Cruden and Hu, 1993, Fig. 5) were calculated using equation 1 for all 14 overdip slopes in Kananakis Country (Fig. 1) where the dip angles of the bedding surfaces exceed the basic friction angles. The basic friction angles were estimated by tilting table tests on sawn and lapped surfaces (Cruden and Hu, 1988). Roughness angles were taken as zero at the scales of blocks capable of sliding. The calculated lower bounds of cohesion include both the rock-bridge cohesive strength and the cohesion from surface roughness at scales smaller than the sliding

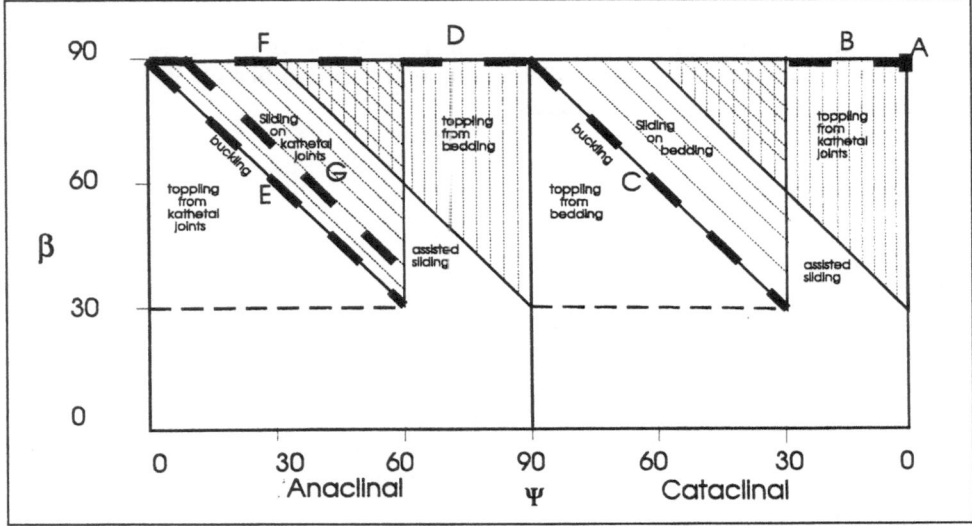

Figure 3.   Process diagram for anaclinal and cataclinal slopes in sedimentary rock masses with $\Phi = 30$, $c' = 0$. Heavy dashed lines indicate critical angles from Young (1972, p. 121). A: horizontal bedding; B: bedding dip less than friction angle; C: bedding dip greater than friction angle; D: dip of cross joints less than friction angle; E: cross joints not offset; F: cross joints offset h/b $\geq$ 1; G: cross joints offset, h/b < 1.

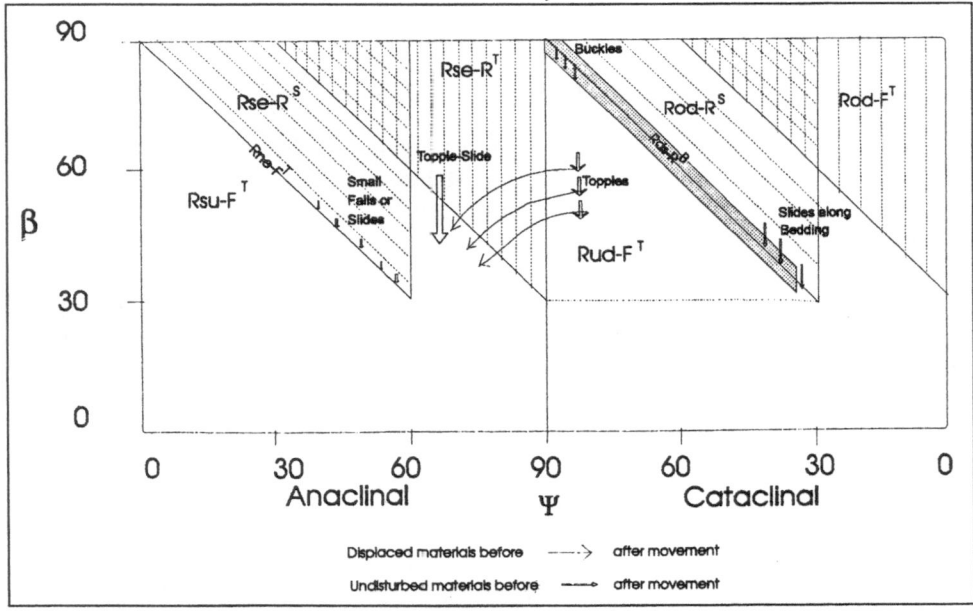

Figure 4.   Process diagram for steep rock slopes in the Canadian Rockies. All slopes are perpendicular to the strike of bedding.

blocks. For 12 of the overdip slopes, the lower bounds of cohesion are between 28 kPa and 47 kPa, equivalent to less than 0.1% of rock bridges along the bedding surfaces. One of the remaining slopes, with a back calculated cohesion of 100 kPa, also had a low basic friction angle which may underestimate friction over the whole potential rupture surface. The other high cohesion estimate for a slope, 280 kPa, may be due to a curved potential rupture surface. Generally then, the present topography of the Canadian Rockies requires little cohesion to support it. Cohesion may have been destroyed or reduced by flexure slipping along bedding surfaces during folding or by weathering.

After the last glaciation, many overdip slopes had been created by glacial erosion at their toes. On these overdip slopes, where bedding dips are greater than basic friction angles, shear resistance has been reduced gradually as cohesion has been destroyed over time. When the factor of safety on a slope is reduced to one, the mass on the slope slid. So the inventory of overdip slopes is being exhausted by one rock slide after another. The number of potential rock slides decreases with each event and the return period of rock sliding increases with time (Cruden and Hu, 1993).

3.2. DIP SLOPES, RDS

Recent work has contradicted McConnell and Brock's claim (1904, p.12) that "The Frank Slide was a Bergsturz", a breaking away of the mountain mass across the bedding planes. The Frank Slide demonstrated that large, extremely rapid movements were possible on dip slopes if other low angle discontinuities daylighted on the slope. McConnell and Brock's often reproduced section has been remapped (Cruden and Krahn, 1973). The Frank Slide now appears to have taken place on a rupture surface following bedding and a thrust fault at its toe, joints control the scarp of the slide.

When discontinuities suitably oriented for sliding do not exist, movements can occur in steeply dipping rock masses by buckling. Classical buckling theory suggested thinly-bedded, cohesionless, low-modulus rock masses would deform under low loads. Once rupture surfaces have been formed, the displaced masses behave in the same way as slides on overdip slopes, sliding on bedding and on the rupture surfaces of the buckles. Their continued movement creates underdip slopes rather than the dip slopes created by movements on overdip slopes.

Several buckles have been documented in the Highwood Pass (Figure 1), where the beds are almost vertical (Hu and Cruden, 1993). Two buckling models, the Euler model and the Timoshenko-Gere model, used to estimate the critical heights of the buckled beds, showed reasonable agreement between the calculated critical heights and the existing cliff heights.

One of the buckles (Hu and Cruden, 1993, Site 1) had deposited 2×10^6 m^3 debris at the toe of its subvertical surface of rupture probably as the result of the retrogressive buckling of layer after layer. So buckling of high rock walls can lead to large rock mass movements. One of the buckles was associated with toppling (Hu and Cruden, 1993, Site 3). The documented buckles are on slopes, Rds, at the boundary of Regions Rud-F^T and Rod-R^S (Fig. 4).

Euler's buckling model can estimate the limit on the relief of a dip slope imposed by buckling instability of a single rock slab, if the strength parameters and the bed thickness, t, can be estimated. Hu and Cruden(1993, Eq. 6) gave the critical length of a buckled bed, L, as

$$L = [(0.141\pi^2 Et^3)/(t\gamma\sin\Psi - t\gamma\cos\Psi \tan\Phi - c)]^{0.333} \qquad (2)$$

where E is the Young's modulus of the bed, Ψ is the bedding dip, γ is the unit weight of the rock, Φ is the friction angle and c is the cohesion. The critical height, H, is then L $\sin\Psi$.

Our analysis of overdip slopes suggested that cohesion can reasonably be ignored in these slopes. Based on the strength parameters that we used (Hu and Cruden, 1993), Φ is 30° and E is between 2 Gpa and 300 Gpa., the critical height of a planar dip slope calculated from Equation (2) increases from several metres to 800 metres as the effective bed thickness increases from 0.01 to 5 metres, the typical range of bedding thickness in the Canadian Rockies, and bedding inclination decreases from 90° to 40° (Figures 5, 6). Figures 5 and 6 also shows that the critical height of a dip slope is nearly independent of bedding inclination if other parameters remain unchanged and the bedding inclination is at least 5° greater than the friction angle. So, buckling stability of a rock slab is mainly controlled by rock stiffness and effective slab thickness. The range of the height is of the same magnitude as the existing relief on dip slopes in the Rockies, indicating that buckling may currently limit length and height of dip slopes. Equation (2) assumes that the top of the buckling column is free and the bottom is fixed, conditions which give the minimum critical length of the buckling slab. If the end conditions are different, the calculated critical length will be greater than that from Equation (2).

The buckling region is shown in Figure 4. Because critical buckling heights remain almost unchanged where bedding inclinations are at least 5° steeper than friction angles (Figures 5 & 6), a reasonable lower limit is at bedding inclinations 5° above friction angles (Figure 4). Based on our experience of bedding plane roughness on dip slopes in the Canadian Rockies, we propose slopes within 3 degrees of the bedding dip may buckle.

3.3. UNDERDIP SLOPES, RUD

Cruden (1989) pointed out that toppling may occur on underdip cataclinal slopes when

$$\beta + (90 - \Psi) > \Phi \qquad (3)$$

in the absence of cohesion on the bedding surfaces and assuming that the largest principal compressive stress is parallel to the slope. Equation 3 shows, that as $(90 - \Psi)$ is positive, toppling is possible on all underdip cataclinal slopes steeper than the angle of friction, Φ, on the bedding. Equation 3 also shows that toppling is possible on low angle cataclinal slopes when discontinuities dip at angles slightly greater than the slope, β. However, the analytical solution given by Savage et al. (1985) indicated that the maximum principal compressive stress may not parallel low angle slopes. Notice, too, that toppling on cataclinal slopes inclined at less than Φ_b leaves the slopes mantled with colluvial debris. So, a lower limit to toppling, $\beta = \Phi_b$ has been indicated on our process diagram (Figure 4). Cruden and Hu (1994) found two of the three types of toppling Goodman and Bray (1976) distinguished on anaclinal slopes occurred on underdip slopes. Block flexural toppling was typical of rock masses with closely - spaced discontinuities which move without severe cracking, and block toppling occurred where discontinuities were more widely - spaced.

Within the toppled rock mass in a block flexure topple, rock blocks separated by discontinuities maintain contact throughout the moving rock mass with transmission of the gravitational stress field. The bedding dips of different blocks change gradually. A block flexure topple is like a rounded fold.

In a block topple, cracking develops within the toppled mass and disrupts the gravitational stress field. Sliding surfaces may form as a result of cracking in the toppled mass and the rock masses overlying the sliding surfaces may be displaced. Development of sliding surfaces is associated with large rotational shear movements and bedding dips change abruptly across sliding surfaces. So, development of sliding surfaces in a topple is similar to the development of shear zones in the translational shear of clay-rich rocks (Skempton, 1985).

Topples on underdip slopes in the Highwood Pass (Cruden and Hu, 1994) clearly showed that toppling and crack development were controlled by the ratio of the spacing of the strike joints to that of the bedding thickness, the block ratio. In the sandstones and the shales of the Triassic Spray River formation, block flexure topples occurred where block ratios were less than 2. The toppled beds formed rounded folds (Cruden and Hu, 1994, Figure 8) without sliding surfaces. Block topples occurred where the block ratios were larger than 2, sliding surfaces formed oblique to both bedding surfaces and strike joints, extend less than 10 m and dip at around 35° downslope (Cruden and Hu, 1994, Figure 12). Bedding dips change abruptly across the sliding surfaces (Cruden and Hu, 1994, Figure 10).

As shown in the process diagram (Figure 4), the displaced rock masses in the topples in Highwood Pass rotated their bedding to change dip directions and reduce slope gradients (Rud-F^T in Fig. 4). When toppled beds rotate to dip into the slopes, the toppled beds behave as common topples on anaclinal slopes.

4. Anaclinal slopes

4.1. STEEPENED ESCARPMENTS, RSE

The development of toppling requires discontinuities perpendicular to the penetrative discontinues (Hu and Cruden, 1992, Fig. 3). Kathetal joints have been described from the Rockies by Muecke and Charlesworth (1964) among others. These joints have thresholds for sliding and toppling similar to those for bedding and they can be shown in a similar way on the process diagram (Figure 4).

Goodman and Bray (1976) have suggested that common toppling occurred on anaclinal slopes when

$$\beta > (90 - \Psi) + \Phi \qquad (4)$$

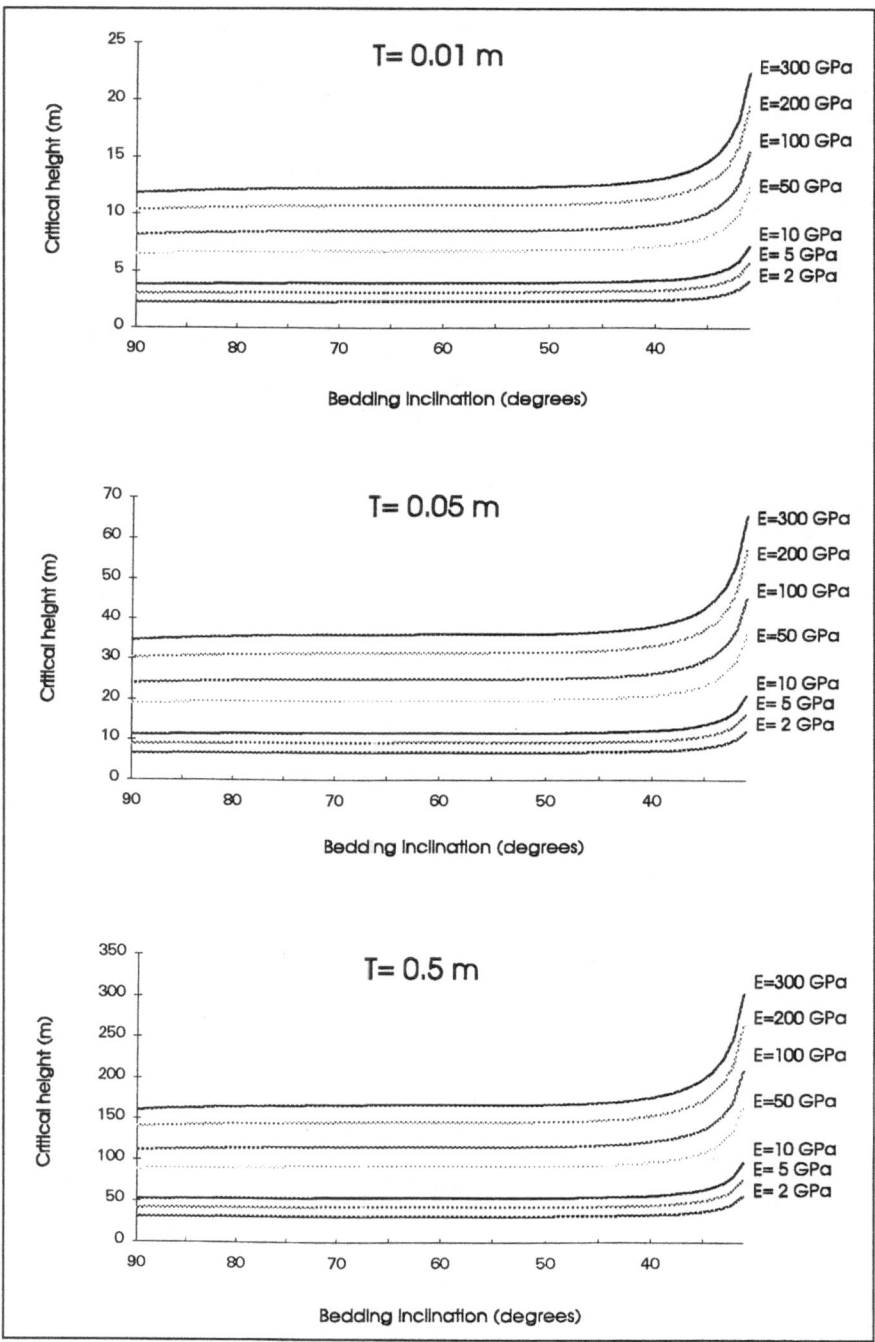

Figure 5. The relationship between the bedding inclination and the critical height when the buckled beds
are 0.01 m, 0.05 m and 0.5 m thick.

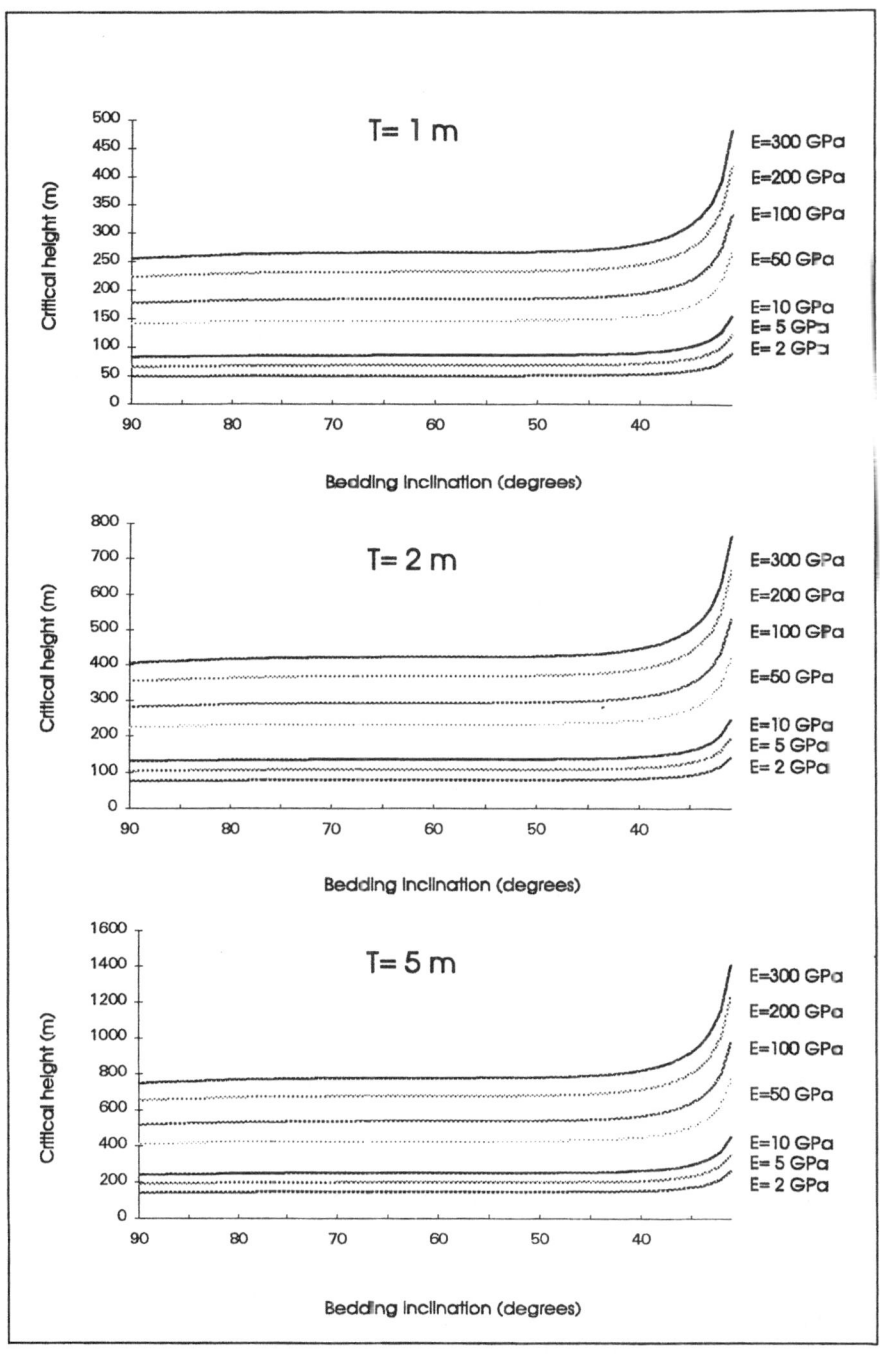

Figure 6.   The relationship between the bedding inclination and the critical height when the buckled beds
are 1 m, 2 m and 5 m thick.

This condition can be conveniently shown on the process diagram (Fig 4, Rse-R^T), given estimates of Φ and assuming zero cohesion. Because a normal escarpment has ß = (90- Ψ), Equation 4 shows that common toppling on anaclinal slopes is limited to substantially steepened escarpments. These might develop by glacial erosion of the toes of the rock walls of strike valleys. If the rock block topples, then Equation (4) shows the toppled block will slide. Wyllie (1980) has documented such a movement in a surface coal mine in the Rockies where bedding dipped at about 70 degrees into a 50 degree slope.

Both rock slides and rock falls are common on steepened escarpments. In an air-photo reconnaissance of the slope movements, Eaton (1986) identified 23 slope movements on anaclinal slopes. We examined 19 in Kananakis Country (Hu and Cruden, 1992). Sixteen of the displaced masses were smaller than 0.15×10^6 m^3 in volume. Fifteen of these smaller movements were slides along kathetal joints striking parallel to the strike of the bedding. These joints dipped at over 50°, exceeding the basic friction angle of both carbonates and sandstones in the succession. The one other small movement, on joints dipping at 25°, may have been a small topple from bedding. The displaced masses resemble those due to rock falls accumulating over long periods rather than in single events. The observed small slides or falls were in Rse-R^S (Fig. 4) and the small volumes were resulted from failure along non-penetrative discontinuities. The travel distance from these small events were also small, limited by the small volume of the displaced material (Nicoletti and Sorriso-Valvo, 1991).

On the other three slopes, the volumes of the displaced rock masses are considerably larger. The bedding surfaces dip into the slopes at 60°-70° while the strike joints dip out of the slope at 20°-30°. Sheeting joints develop parallel to slope surfaces in thickly bedded limestones. They are spaced from 1 to 2 metres apart but penetrate less than 5 metres into the slope. The rock masses slid along the strike joint surfaces in the thinly-bedded, intensely jointed units and along the sheeting joints in the thick bedded units.

The landslide on Elk Ridge, west of the Highwood Pass, provided a typical example of a composite topple slide (Hu and Cruden, 1992). The lower part of the surface of rupture is along sheeting joints; the upper part is oblique to the bedding surfaces and the strike joints (Hu and Cruden, 1992, figure 9). Identical rock sequences in the debris and in the rockwall (Hu and Cruden, 1992, figure 8) indicate that the movement was a single event. The process diagram shows (Figure 4, Rse-R^T) that, large topple-slides reduce slope angles as the slopes move to the stable region. The attitudes of bedding dipping at 60°-70° into the slopes and those of kathetal joints dipping at 20°-30° out of the slopes favour large topples from bedding and slides along combinations of bedding and kathetal joints following toppling .

4.2. NORMAL AND SUBDUED ESCARPMENTS, RSE, RSU

We have not observed any large movements on normal or subdued escarpments or slopes in Rsu-F^T and Rse (Figure 4). Sliding on kathetal joints and toppling from kathetal joints are probably on the scale of individual blocks, due to the non-penetrative nature of kathetal joints. So, displaced materials from these processes are not significant enough to be noticed individually.

Figure 7.   Mount Assiniboine (reproduced with kind permission of Whyte Museum).

High peaks in the Canadian Rockies develop in beds where bedding dips are gently inclined or close to horizontal, (Gadd, 1995, p. 157). The highest peaks of the Canadian Rockies, over 3500 m, are shrouded in ice and snow. The most southerly of these, Mount Assinboine, is however relatively well exposed and the flat-lying Palaeozoic sedimentary rocks can be clearly seen. The western face slopes at about 50 degrees, 750 metres from the peak to the northwest shoulder (Figure 7). Wilcox (1896, p. 178) commented "Mount Assiniboine, especially when seen from the north, resembles the Matterhorn in a striking manner... moreover, the horizontal strata have weathered away in such a manner as to form vertical ledges, which completely girdle the mountain ...". In rock masses in gently inclined beds, (Rsu, Rod, Figure 4) movements, such as slides and topples, need assistance. Slope gradients reflect the rock mass strength rating (Selby, 1993, p. 103) and in strong and thick sequences, slopes can be steep and high.

5. Discussion

There are still some uncertainties remaining in the process diagrams (Figures 3 and 4). Boundaries for assisted sliding and toppling will depend on the processes assisting slope movements. No falls are shown on the diagrams as, generally, rock masses move first as slides or topples before a local steepening of the slope projects the displaced mass into the air and it falls.

Processes have been analyzed assuming pore pressures are absent from the rock masses. Such assumptions may not be valid above the treeline in the presence of permafrost or if glaciers feed meltwater into the rock slopes. Two fabric elements have been considered, penetrative discontinuities and kathetal joints. Where other fabric elements such as faults or sheeting joints occur, kinematic possibilities become more complex and the simple generalizations proposed break down.

Simple rock mechanics then explains the broad outlines of the structurally controlled relief that characterizes the Rockies. Individual sites still require careful investigation and may offer challenging complexities.

The results of this paper have important implications for siting of structures. Because the hazards and magnitudes of slope movements differ on different types of slopes, attention paid to slope processes prevents loss and damage to structures and human lives. Structures should be safe distances away from slopes in Rod-R^S and Rse-R^T (Figure 4) where rapid slope movements may occur. Rock falls and rock slides should be expected below normal escarpments or toppled rock masses on underdip slopes. Steep high dip slopes may buckle, which may lead to damage to structures at the toes of these slopes.

6. Conclusions

Modes of hazardous rock slope movements in the Rockies are controlled by the relationship between attitudes of penetrative discontinuities and orientations of slopes. Major rapid rock slope movements include rock slides on overdip cataclinal slopes in cohesive rock mass and complex rock topple-slides on steepened escarpments on anaclinal slopes. Buckles are found on steep dip slopes, buckling of high walls can be rapid. Two types of topples, block flexure topples and block topples occur on underdip cataclinal slopes, the type of toppling is controlled by discontinuity spacings. While no discontinuity forms in block flexure topples, sliding surfaces develop in block topples and toppled materials can be displaced as rock slides along these sliding surfaces.

Landform development in the stratified sequences in the Rockies is controlled by the slope processes. When bedding dips on an overdip slope or dips of kathetal joints on a steepened escarpment exceed their basic friction angles, the slopes evolve to stable forms, dip slopes on cataclinal slopes or the normal escarpments on anaclinal slopes. Other processes, such as toppling on underdip slopes, buckling and toppling-sliding on steepened escarpments all reduce slope gradients and the rates of processes themselves.

Relief in steeply inclined beds is limited by sliding on overdip slopes, by toppling-sliding on steepened escarpments and by buckling on dip slopes. High mountain peaks develop in gently inclined beds or horizontal beds where sliding, toppling, and buckling are not active.

References

1. Barsvary, A.K., Klym, T.W., Franklin, J.A., 1980, List of terms, symbols and recognized SI units and multiples for geotechnical engineering, Canadian Geotechnical Journal. 17:89-96.

2. Brunsden, D., Thornes, J.B.,1979. Landscape sensitivity and change, Transactions Institute of British Geographers. 4: 463-484.

3. Callaway, C., 1879. On plagioclinal mountains, Geological Magazine. 15:216-221.

4. Canada Soil Survey Committee, 1978. The Canadian system of soil classification. Canada Department of Agriculture Publication 1646. Supply and Services Canada, Ottawa.

5. Cruden,D.M.,1975. The influence of discontinuities on the stability of rock slopes, Proceedings, 4th Guelph Symposium on Geomorphology, Geoabstracts, Norwich pp. 57-67.

6. Cruden, D.M., 1982. The Brazeau Lake slide, Jasper National Park, Alberta. Canadian Geotechnical Journal. 19:975-981.

7. Cruden, D.M., 1985. Rockslope movements in the Canadian Cordillera. Canadian Geotechnical Journal. 22: 528-540.

8. Cruden, D.M., 1988, Thresholds for catastrophic instabilities in sedimentary rock slopes, some examples from the Canadian Rockies, Zeitschrift fur Geomorphologie Suplementum 76, 67-76.

9. Cruden, D.M., 1989, The limits to common toppling. Canadian Geotechnical Journal. 26: 737-742.

10. Cruden, D.M., Hu X-Q, 1988, Basic friction angles of carbonate rocks from Kananaskis Country, International Association of Engineering Geology Bulletin, 38:55-59.

11. Cruden, D.M. and Hu, X.Q., 1993. Exhaustion and steady state models for predicting rockslide hazards in the Canadian Rocky Mountains. Geomorphology, 8:279-285.

12. Cruden D.M. and Hu, X.Q., 1994. Topples on underdip slopes in the Highwood Pass, Alberta, Canada, Quarterly Journal of Engineering Geology, 27:57-68.

13. Cruden, D.M., Krahn, J., 1973, A re-examination of the geology of the Frank Slide, Canadian Geotechnical Journal, 10:581-591.

14. Cruden, D.M. and Varnes, D.J. 1996. Landslide types and processes. In Landslides: Investigation and Mitigation. Transportation Research Board, National Academy of Sciences, Special Report, 247.

15. Cruden, D.M. and Thomson, S., 1987. Exercises in terrain analysis. University of Alberta Press, Edmonton..

16. Foucault, A., Raoult, J-F., 1988. Dictionnaire de Geologie. Masson, Paris, 352 p.

17. Gabrielse, H., Yorath, C.J., 1992, Geology of the Cordilleran Orogen in Canada, Geology of Canada, Volume 4, Geological Survey of Canada, Ottawa.

18. Gadd, B., 1995. Handbook of the Canadian Rockies, Corax Press, Jasper, Alberta.

19. Gerrard, A.J. 1988. Rocks and landforms. Unwin Hyman, London.

20. Goodman, R.E., Bray, J.W., 1976, Toppling of rock slopes, Proceedings of the Specialty Conference on Rock Engineering for Foundations and Slopes, American Society of Civil Engineers, Boulder, Colorado, pp. 201-234.

21. Howes, D.E. and Kenk, E., 1988. Terrain classification system for British Columbia. Ministry of Environment, Manual 10, Province of British Columbia, Victoria.

22. Hu, X.Q. and Cruden, D.M., 1992. Rock mass movements across bedding in Kananaskis Country, Alberta, Canadian Geotechnical Journal, 29: 675-685.

23. Hu, X.Q. and Cruden, D.M., 1993. Buckling deformation in the Highwood Pass, Alberta, Canada. Canadian Geotechnical Journal. 30: 276-286.

24. Krahn, J., Morgenstern, N.R., 1979. The ultimate frictional resistance of rock discontinuities, International Journal of Rock Mechanics and Mining Science, 16:127-133.

25. McConnell, R.G., Brock, R.W., 1904. Report on the great landslide at Frank, Alberta, Department of Interior, Annual Report for 1903, Ottawa, Part 8, 17 p.

26. Muecke, G.K., Charlesworth, H.A.K., 1966. Jointing in folded Cardium sandstones along the Bow River, Alberta, Canadian Journal of Earth Sciences, 3, 579-596.

27. Nicoletti, P.G. and Sorriso-Valvo, M. 1991. Geomorphic controls of the shape and mobility of rock avalanches, Geological Society of America Bulletin. 103:1365-1373.

28. Powell, J.W., 1875. Exploration of the Colorado River of the West and its tributaries, Government Printing Office, Washington.

29. Sander, B., 1970. An introduction to the study of the fabric of the geological bodies, Pergammon, Oxford.

30. Savage, W.Z., Swolfs, H.S., Powers, P.S., 1985. Gravitational stresses in long symmetric ridges and valleys. International Journal Rock Mechanics and Mineral Science, 22:291-302.

31. Selby, M.J. 1993. Hillslope materials and processes. Oxford University Press, Oxford.

32. Simmons, J.V., Cruden, D.M., 1980. A rock labyrinth in the Front Ranges of the Rockies, Alberta, Canadian Journal of Earth Sciences, 17:1300-1309.

33. Skempton, A.W., 1985. Residual strength of clays in landslides, folded strata and the laboratory. Geotechnique. 35:3-18.

34. Taylor, C.G., 1981. Fort St. John, in Slaney, V.R., LANDSAT images of Canada, a geological appraisal, Geological Survey of Canada, Paper 80-15, p. 88.

35. Terzaghi, K. 1962. Stability of steep slopes on hard unweathered rock. Geotechnique, 12:251-270.

36. Trenhaile, A.S., 1987. The geomorphology of rock coasts. Clarendon Press, Oxford.

37. Turner, F.J. 1952. "Gefugerelief" illustrated by "Schist tor" topography in central Otago, New Zealand. American Journal of Science. 250:802-807.

38. Wilcox, W.D., 1896. Camping in the Canadian Rockies. Putnam, New York.

39. Wyllie, D.C., 1980, Toppling rock slope failures, examples of analysis and stabilization, Rock Mechanics, 13:89-98.

40. Young, A., 1972. Slopes, Oliver and Boyd, Edinburgh.

Authors

D. M. Cruden
Department of Civil Engineering
University of Alberta
Edmonton, Alberta, Canada
T6G 2G7

X.Q. Hu
Department of Earth and Atmospheric Sciences
University of Alberta
Edmonton, Alberta, Canada
T6G 2E3

LATE HOLOCENE STURZSTROMS IN GLACIER NATIONAL PARK, MONTANA, U.S.A.

Hazardous Mass Movements Along the Eastern Front of the Lewis Overthrust Fault

DAVID R. BUTLER, GEORGE P. MALANSON,

FORREST D. WILKERSON, GINGER L. SCHMID

1. Introduction and Background

Recent decades have seen population expansion into, and pressure upon, marginal mountain lands. The inevitable disasters associated with such expansion have in many cases forced attention onto the distribution of, and processes associated with, rock avalanches (*sturzstroms*). Increased attention on rock avalanches has also resulted from the United Nations' declaration of the 1990s as the International Decade for Natural Disaster Reduction. In North America, particularly notable efforts directed at identifying and reducing rock-avalanche hazards have taken place in the Canadian Rockies of Alberta, British Columbia, and the Northwest Territories (Eisbacher, 1978, 1979; Cruden, 1982, 1985; Eisbacher and Clague, 1984; Cruden and Hungr, 1986; Evans, 1987, 1989a, 1989b; Evans et al., 1987, 1989; Clague and Evans, 1987; Van Gassen and Cruden, 1989; Jackson and Isobe, 1990; Ryder et al., 1990; Evans and Clague, 1990, 1994; Kaiser and Simmons, 1990; Hungr, 1995). In the U.S.A., the bulk of rock-avalanche-hazard research has been focused on the tectonically active states of Alaska and California (Griggs, 1920; Shreve, 1966, 1968a, 1968b; Eppler et al., 1987). Work in the Rocky Mountain states has also been undertaken (Voight, 1978), but few studies have focused on the distribution and age of rock-avalanche deposits in Montana, a state where the western third is seismically active and mountainous. Exceptions include works by Hadley (1964), who described the rockslide avalanche produced by the Madison Canyon earthquake of 1959, near Yellowstone National Park; Mudge (1965), who described rockfall-avalanche and rockslide-avalanche deposits of Pleistocene and Holocene age at Sawtooth Ridge, Montana, approximately 100 km southeast of Glacier National Park; and previous papers describing site characteristics and ages associated with two sturzstrom sites in Glacier National Park (Butler, 1983; Butler et al., 1986, 1991; Butler and Schipke, 1992; Butler and Malanson, 1993).

Our interest in the two aforementioned sturzstrom sites in Glacier National Park has led us to examine several more such deposits in the field. Here we describe and discuss these sturzstrom sites. We also present new data on the ages of several of these deposits, including information associated with new sturzstroms triggered by an ear-

149

J. Kalvoda and C.L. Rosenfeld (eds.), Geomorphological Hazards in High Mountain Areas, 149-156.
© 1998 *Kluwer Academic Publishers.*

thquake in July 1992, as well as some refinements on the ages of the previously described sturzstrom deposits.

2. The study area

Our study area is in Glacier National Park, Montana, U.S.A. Glacier National Park
(GNP) is a United Nations-designated International Biosphere Reserve comprising
about 0.4 million hectares astride the Continental Divide in northwest Montana
(Figure 1). The Continental Divide subdivides the Park into two climatic realms; the
western one-half illustrates characteristics typical of a modified Pacific maritime environment (Walsh *et al.*, 1992; Butler *et al.*, 1994). East of the Continental Divide, the
climate is characterized as continental in nature; it is drier, has stronger winds, and
lower winter temperatures (Walsh *et al.*, 1992). Terrain is steep and bears the imprint
of Pleistocene glaciation, and minor Little Ice Age glacial remnants still exist. This
steep, recently glaciated terrain is prone to a variety of mass movements, especially
debris flows and snow avalanches (Butler, 1979, 1989; Butler and Malanson, 1990,
1992, 1996; Butler and Walsh, 1990, 1994; Butler *et al.*, 1992; Walsh *et al.*, 1990,
1994); however, our focus here is on rock avalanches occurring in a geologically unstable portion of eastern GNP.

The principal study area for this research is a zone along the eastern edge of the
Lewis Overthrust Fault in eastern GNP. This fault extends the entire length of GNP
along the eastern front of the Lewis Range. The Lewis Overthrust Fault is an area
where older Precambrian sedimentary rocks (limestones and argillites) now overlie
younger Cretaceous rocks (Table 1). During the development of the Lewis Overthrust,
an extreme amount of folding in the Cretaceous rocks beneath the Lewis thrust occurred. Several subsidiary faults exist in conjunction with the Lewis thrust fault, adding
to instability of the area. The overarching importance of the Overthrust Fault for this
study is that highly jointed, fractured, and faulted Precambrian Altyn limestone and
Appekunny argillite overlie weak Cretaceous mudstones, creating an area prone to
landslide activity (Oelfke and Butler, 1985a). In effect, the entire Lewis thrust fault
zone in GNP represents a sturzstrom-prone area.

Oelfke and Butler (1985a) and Butler *et al.* (1986) published maps illustrating the
locations of landslides along the entire length of the Lewis Overthrust Fault in GNP.
Several areas of rockfall/rockslide deposits were mapped, but were not differentiated
into deposits produced by slow, isolated rockfall versus large rockfall/rockslide avalanche deposits. Two distinct sturzstrom deposits were identified, the Napi Point
deposit and an agglomeration of deposits at Slide Lake (Figure 1). Subsequent fieldwork and archival research has revealed additional information concerning the nature
and timing of deposition of the latter deposit. Additional detail concerning instability
at Napi Point was presented by Butler *et al.* (1991), who noted through dendrogeomorphological analyses of tilted trees that the sturzstrom there occurred in 1954, with
additional movement corroborated by eyewitness reports in 1959 and 1962.

Aerial photo analyses (U.S. Geological Survey panchromatic photographs from
1966 and 1991, and U.S. Geological Survey color infra-red photographs from 1974
and 1983); and field reconnaissance in 1972-1975, and 1981 (by DRB); in 1983, 1987-
1988, and 1990-1993 (by DRB and GPM); and in 1994-1995 (by DRB, GPM, FDW,

Figure 1.- Map of Glacier National Park. Montana, with the sturzstrom sites mentioned in the text. 1, Curly Bear older sturzstrom deposit; 2, Curly Bear younger deposit; 3, Napi Point sturzstrom; 4, Gable Mountain rockslide avalanche deposit; 5, Otatso Creek 1992 rock avalanche deposit; 6, Slide Pond adjacent sturzstrom deposit; 7, sturzstrom deposit that dams Slide Pond; 8, sturzstrom deposit that dams Slide Lake; C, Chief Mountain.

and GLS) have not identified similar widespread sturzstrom deposits in the southeastern quarter of GNP along the Lewis Overthrust Fault. This is not to suggest that such phenomena will not occur in the future in the southeastern quarter of GNP, but only that none have occurred thus far. Geologic and topographic conditions in that area are certainly amenable for the likelihood of sturzstrom in the future, and planners should be aware of such a distinct possibility.

3. Descriptions and dating of individual sturzstroms

Figure 1 illustrates the locations of eight distinct sturzstrom deposits in the north-eastern quarter of GNP along the Lewis Overthrust Fault, as well as the location of Chief Mountain, which is literally ringed with rockfall and sturzstrom deposits. We describe these deposits in a roughly south-to-north transect, beginning with the Curly Bear deposits along the southern flank of the St. Mary Valley.

3.1 THE CURLY BEAR MOUNTAIN STURZSTROM DEPOSITS

Along the northwest-facing slope of Curly Bear Mountain, two sturzstrom deposits of differing ages occur in close proximity to each other (Figure 2). We judge the deposit to the west to be older, based on its extensive tree cover (Figure 2a). Nevertheless, the freshness of the bedrock scar above the deposit (Figure 2a) suggests a time of deposition within the last 500 years, based on comparisons with bedrock scars above sturzstrom and glacial deposits of known age in eastern GNP. The forest cover on the older Curly Bear deposit may be up to 200 years old, but is more likely only 100-150 years old (an estimate derived by comparing the height of the trees to those of known age in the park). This deposit is *ca.* 700 m long and 300 m wide. We have no depth measurements from which to calculate volume.

The nearby younger Curly Bear deposit (Figure 2b) possesses some surface morphological characteristics similar to rock glaciers. However, Holocene-aged rock glaciers in GNP are substantially smaller than this deposit (Carrara, 1990), and because the deposit mantles late-Pleistocene glacial till, it must post-date the Pleistocene. In comparison to the older Curly Bear deposit, significantly fewer trees populate its surface, suggesting either a younger age, greater instability than the other deposit, or both. Given the similarity in slope angle between the two deposits, we deem a younger age the more likely reason for such a limited forest cover. The deposit pre-dates the 1966 aerial photographs of the study area, and contour lines on the early-1900s topographic map of the area also suggest that the deposit had been emplaced by 1900. The deposit is approximately 1.3 km in length, and varies in width from *ca.* 300-450 m. No depth data from which volume could be calculated are currently available.

3.2 NAPI POINT

The Napi Point sturzstrom deposit (Figure 1) flows northward from Napi Point toward Boulder Creek (Figure 3). The deposit illustrates excellent lobate flow morphology and distinct bounding lateral ridges (Butler, 1983). The surface of the deposit is comprised of highly angular clasts of Altyn limestone and Appekunny argillite of impressive sizes (Figure 4). Dendrogeomorphological dating of trees impacted by passage of the sturzstrom revealed that the rock avalanche occurred in 1954 (Butler *et al.*, 1986). Ground reconnaissance by J.G. and L.A. Oelfke provided data from which a volume of 2.6×10^6 m^3 was calculated for the Napi Point deposit. Other details of the site morphology are available in Butler *et al.* (1986). Here, we additionally note how close this sturzstrom came to blocking Boulder Creek at the base of Napi Point. The sturzstrom flowed a distance of approximately 1.3 km; a travel distance of only an additional 300 m would have impounded Boulder Creek and created a potenti-

Figure 2. - Curly Bear deposits. 2a (above), view of Curly Bear older rock-avalanche deposit, covered in trees (dashed line). Curly Bear younger deposit is at top of view. 2b (below), Curly Bear younger deposit. Note the distinct flow ridges and levees. Photos by DRB, 1995.

ally dangerous landslide-dammed lake. Given the extremely fractured nature of the bedrock along the crest of Napi Point, we deem additional failure and possible sturz-strom flowage to be very likely at this site. These conditions, in concert with a sparse vegetation cover produced by a recent forest fire, set the stage for a potentially disas-trous scenario of additional mass movement, possible remobilization, and blockage and impoundment of Boulder Creek. Such a scenario would produce a strong likelihood of a subsequent outburst flood along Boulder Creek that would destroy the highway ac-cess into the popular Many Glacier region of GNP.

3.3 STURZSTROM DEPOSITS IN THE OTATSO CREEK DRAINAGE

The mountains surrounding the Otatso Creek drainage in the northeastern corner of GNP provide the greatest number, representing the greatest hazard, of anywhere in the Park. Yellow Mountain on the south side of Otatso Creek is the source of sturzstrom deposits 5-8 on Figure 1. Nearby Gable Mountain to the north has been the source of a major rockslide avalanche, and numerous rockfall-avalanche deposits nearly encircle Chief Mountain.

Figure 5 illustrates the spatial relationships between the Gable Mountain and Yel-low Mountain (Slide Lake) sturzstrom deposits, as illustrated on a 1966 U.S. Geological Survey aerial photograph. The Gable Mountain rockslide, as well as three of the Yellow Mountain deposits, must therefore pre-date this photograph. An aerial photograph taken from a helicopter in 1995 illustrates the same Yellow Mountain deposits, looking southward from a point just east of the Gable Mountain deposit (Figure 6). In addition, the uppermost portion of a sturzstrom that occurred in 1992 is visible at the far left of the photograph.

3.3.1 Gable Mountain Rockslide-Avalanche

The Gable Mountain sturzstrom is comprised almost exclusively of Altyn limestone that slid down a southwest-facing dip slope (Figures 5, 7a). The resulting deposit came to rest at the base of Gable Mountain, with a portion of the deposit accelerating in a lobate fashion down the valley axis to the southeast (Figures 5, 7b).

The clasts of the deposit are very angular, and several of the boulders are the size of automobiles and even of small houses. We have no depth data from which to provide a volumetric estimate, but note that the deposit is about the same size as the younger Curly Bear deposit. The freshness of the clasts (Figure 7) suggests that this rockslide-avalanche is quite young, but no pre-1966 photographs are available for providing a bracketing date for its movement. The deposit is a long distance from the nearest hu-man trail, so it is not surprising that it has escaped notice until this first published description.

3.3.2 Sturzstrom Deposits at and Adjacent to Slide Lake and Slide Pond

The group of sturzstroms that drape the north-facing slope of Yellow Mountain (Figure 6) are collectively the most impressive, and the most potentially dangerous, of all the deposits described in this paper. The likelihood of dam failure and characteris-tics of the lakes produced by the Slide Lake and Slide Pond deposits have been described by Butler *et al.* (1991), Butler and Schipke (1992), and Butler and Malanson (1993). Photographs taken in 1901 by the U.S. Geological Survey during their topogra-

Figure 3 - Napi Point sturzstrom. View is from headscarp of Napi Point (left) to flow levees and sturzstrom deposit extending below. Deposit bifurcates at arrow. Photo composite by FDW and DRB, 1995.

Figure 4. - Surface view of typical clasts comprising the Napi Point deposit. Photo courtesy of J.G. Oelfke, 1983.

phic mapping of the area illustrate that no rock avalanches had occurred by that time (Photographs 288 and 289, Bailey Willis, U.S. Geological Survey, 1901). In 1983 dendrogeomorphological data were collected along the flanks of the Slide Lake (Figure 8) and Slide Pond (Figure 6) sturzstrom lobes where they dam Otatso Creek and impound it. These data were reported elsewhere (Butler *et al.*, 1986, 1991), and revealed that a date of 1910 was the most probable date for deposition of the Slide Lake deposit, and 1946 for the Slide Pond deposit. We recently unearthed historical photographs of Slide Lake and the sturzstrom deposit in the Photographic Archives Library of the U.S. Geological Survey in Denver, Colorado, dating to the year 1911 (Photographs 891 and 892, M.R. Campbell, U.S. Geological Survey), corroborating our interpretation of the dendrogeomorphic data. Additionally, a recently published collection of diaries from a GNP Ranger (Lee, 1994, p. 38) contains the following unedited log entry for 8 August 1910:

"It did not seem far down & we came upon a lake at the bottom on the North fork of Kennedy Creek that is not on the map. The mt. has slid down since the map was made & consequently the lake. We camped on the shore about 5 ft from the H2O & we sure have an abundance of wood. Wood wood everywhere millions of pieces brot down by the slides."

It should be noted that the North Fork of Kennedy Creek has subsequently been renamed Otatso Creek. It is clear, then, that the Slide Lake sturzstrom occurred before 8 August 1910, and that the Slide Pond sturzstrom occurred in 1946 (recall that neither slide appears on B. Willis' photographs from 1901).

Because the dates of these two sturzstrom occurrences are known to the year, lichen data were collected on rocks comprising the well-drained surface portions of each deposit. The orange-colored lichen *Xanthoria elegans* is one of the few calcium-tclerant lichens found on rocks in this area (also see McCarthy and Smith, 1995, for details on application of lichenometry in nearby portions of the Canadian Rocky Mountains). In 1983, data were collected from the largest *X. elegans* thalli found on the Slide Lake sturzstrom (Table 2) (Oelfke and Butler, 1985b). No lichens were located on the Slide Pond deposit in that year. In 1995 we collected maximum lichen thalli diameters for 100 *X. elegans* lichens on both the Slide Lake and Slide Pond deposits, noting with interest the relatively rapid colonization by lichens of the Slide Pond deposit in the intervening 12 years (Table 3). The data from the Slide Pond versus the Slide Lake lobes are statistically significantly different (T test, 0.05 level), and thus the data from these two periods of collection (Tables 1 and 2) provide a well-controlled chronology for use in the future on deposits of unknown age such as the Gable Mountain and Curly Bear sturzstrom deposits.

Figure 5. - U.S. Geological Survey aerial photograph, taken 1966, showing Gable Mountain rockslide (open arrow 1), Slide Lake sturzstrom (open arrow 2), and dammed Slide Lake (left solid arrow), Slide Pond sturzstrom (open arrow 3) and dammed Slide Pond (right solid arrow), Slide Pond adjacent sturzstrom (open arrow 4), and the source material (open arrow 5) for the 1992 Otatso Creek sturzstrom.

An additional sturzstrom (the "Slide Pond adjacent") deposit (Figures 5, 6) also post-dates the 1901 photographs of Bailey Willis, and appears on the 1966 aerial photograph. This sturzstrom therefore occurred between those dates. Measurement of lichen thalli on the surface of this deposit should in future provide a more accurate estimate of the year of deposition. It should also be noted that this sturzstrom traveled a linear distance of nearly 1.5 km from the base of Yellow Mountain, and came within *ca.* 200 m of producing a third impoundment of Otatso Creek. Very heavy rains during June of 1995 heavily scoured the surface of the upper portion of the sturzstrom lobe (Figure 6), suggesting that remobilization of the lower half of the deposit could occur. In such a case, subsequent damming of Otatso Creek would be very likely.

3.3.3 The Seismically Triggered Otatso Creek Deposit

On 2 July 1992, an earthquake occurred approximately 50 km southeast of the Chief Mountain/Otatso Creek area. The earthquake triggered two sturzstrom in our study area, one east of the "Slide Pond Adjacent" lobe in the Otatso Creek drainage, and one on the north face of Chief Mountain (Figure 1) (Anonymous, 1992a, 1992b, 1992c).

TABLE 1. Lithologic Cross-section of Stratigraphic Units
Associated with the Lewis Overthrust Fault in GNP

Unit	Characteristics	Thickness (m)
Helena (Siyeh) Formation	Tan & grey limestone, with abundant fossil algal stromatolites	760-1,000
Grinnell Formation	Red mudstones and argillites, with some layers of green mudstones and white sandstones	760
Appekunny Formation	Green mudstones and argillites, with beds of white sandstone	760-1,000
Altyn Formation	White limestone that weathers to shades of tan, with numerous thin layers of quartz sand	670
LEWIS	OVERTHRUST	FAULT
Cretaceous Sediments	Black, marine mudstones, rich in organic matter, interbedded with sandstone	Largely buried

The starting zone of the Otatso Creek sturzstrom is just visible at upper left in Figure 6. Because of instability associated with the earthquake, the Otatso Creek drainage was closed to visitors for about two weeks. A closer examination of the resulting deposit (Figure 9a) reveals that movement out of a bifurcated starting area merged about 100 m downslope, and continued as a rockflow downslope to within *ca.* 30 m of Otatso Creek (just out of sight at the base of Figure 9a). The lobate deposit carried a mature forest along its surface, and a watery surge probably produced the narrow tongue extension reaching beyond the main lobe towards Otatso Creek. Because we know essentially to the day when this sturzstrom was deposited, it will provide excellent lichenometric control data in the future as lichens colonize the surface upon stabilization.

Stabilization of the Otatso Creek deposit is not yet complete, however. As is illustrated in Figure 9b, the heavy rains of June 1995 heavily scoured upper portions of the deposit and remobilized portion of the central lobe that bears the "floating forest". Continued movement may be expected, and an additional seismic trigger could sufficiently destabilize the deposit so that it would reach and impound Otatso Creek, leading to the production of yet another potentially dangerous landslide-dammed lake.

3.4. ROCKSLIDES AT CHIEF MOUNTAIN

Chief Mountain (Figure 10) is a klippe separated by erosion from the bulk of the Lewis Overthrust Fault to the west. The Chief Mountain sturzstrom was believed to "be caused by a combination of freezing and thawing and a July 2 earthquake near Browning" (Anonymous, 1992b). The rockslide extended nearly two kilometers from

TABLE 2. Lichen Thalli Data Collected in 1983,
Slide Lake Sturzstrom

Minimum Diameter (In Millimeters) of Largest *Xanthoria elegans* Thalli

21	13	10	9
17	12	10	9
16	12	10	8
16	12	10	8
15	11	10	8
15	11	10	8
14	11	9	8
14	10	9	7
13	10	9	7
13	10	9	7
13	10	9	7

Mean Minimum Diameter of Largest Thalli = 10.9 Millimeters

the north face of Chief Mountain (the source area of the movement is visible on the left side of Chief Mountain in Figure 10). The landslide remained unstable over most of the summer of 1992. Local eyewitnesses were quoted as saying:

"'You can walk up there and hear trees cracking and popping - trees are actually being uprooted and turned around', said Bruce Connell, Bureau of Indian Affairs forest manager. 'As it was going down it was pulverizing that rock like powder. It was almost a liquidy-type substance' Connell said." (Anonymous, 1992b, p. 1).

Officials also warned that although the area was stabilizing as summer progressed, continued rainfall or other factors could send debris moving again (Anonymous, 1992c). Some debris from the rock-avalanche actually reached the Chief Mountain Highway, a major tourist route connecting GNP with Waterton Lakes National Park, Alberta. This debris had to travel > 4 km to reach the highway. Future lichenometric work will also be valuable at this site, in concert with the Otatso Creek sturzstrom deposit, to provide an accurate assessment of the period of time necessary for lichen colonization following sturzstrom deposition and stabilization.

TABLE 3. *Xanthoria elegans* Thalli Data Collected in 1995,
Slide Pond and Slide Lake Sturzstroms

	Slide Lake	Slide Pond
Mean Diameter of 100 Largest Thalli	32.57 mm	25.91 mm
Minimum Diameter	15.00 mm	5.50 mm
Maximum Diameter	70.00 mm	51.00 mm

4. Discussion and conclusions

A glance at Figure 10, showing the western and southern base of Chief Mountain, illustrates that it has been subjected to additional rock avalanches in the recent past. Several of these have been observed by eyewitnesses during the 20th Century (Butler *et al.*, 1991), whereas others predate European colonization of the area. Because of the inherently unstable geologic and topographic setting at Chief Mountain, additional rock avalanches should be expected periodically in the coming decades.

Figure 6 - Helicopter view of Yellow Mountain; the Slide Lake, Slide Pond, Slide Pond adjacent sturzstroms
(right to left), and Slide Lake (right) and Slide Pond (left). Photo by FDW, 1995.

Every major sturzstrom deposit described herein, with the exception of the lithologically extremely unstable Gable Mountain case, occurred on aspects with a north-facing component. During a helicopter overflight in 1995, we observed ponded water on the surface of the Yellow Mountain slide complex, near the heads of the "Slide Pond Adjacent" and Slide Pond lobes. The extensive shading of the source areas, provided by the precipitous north face of Yellow Mountain (Figures 5, 6), provides a cool, moist microclimate. The underlying Cretaceous mudstones tend to be impervious to drainage from above. Continued lubrication of the interface of the overlying Precambrian bedrock and talus with the underlying Cretaceous bedrock apparently exists. There is no reason to doubt that such conditions also exist on the northward-facing aspects of the Curly Bear and Napi Point deposits. The obvious conclusion is that danger exists throughout the study area, either from the possibilities for new sturzstrom or from remobilization of existing deposits. Our current plans are to statistically analyze the site conditions associated with the deposits described and illustrated herein, in concert with digital elevation models of the sites. Such analyses should allow us to develop a predictive model that will assist in identifying the potential of sites for future sturzstrom deposition. Integrating these data with an existing geographic information system of GNP will enable identification of locations where the potential for sturzstrom deposition most significantly interacts with transportation corridors and tourist facilities. Sheer good fortune has prevented injuries and death from sturzstroms in GNP during the 20th Century. It is our hope that resulting hazard-zone maps will assist in avoiding casualties in the future.

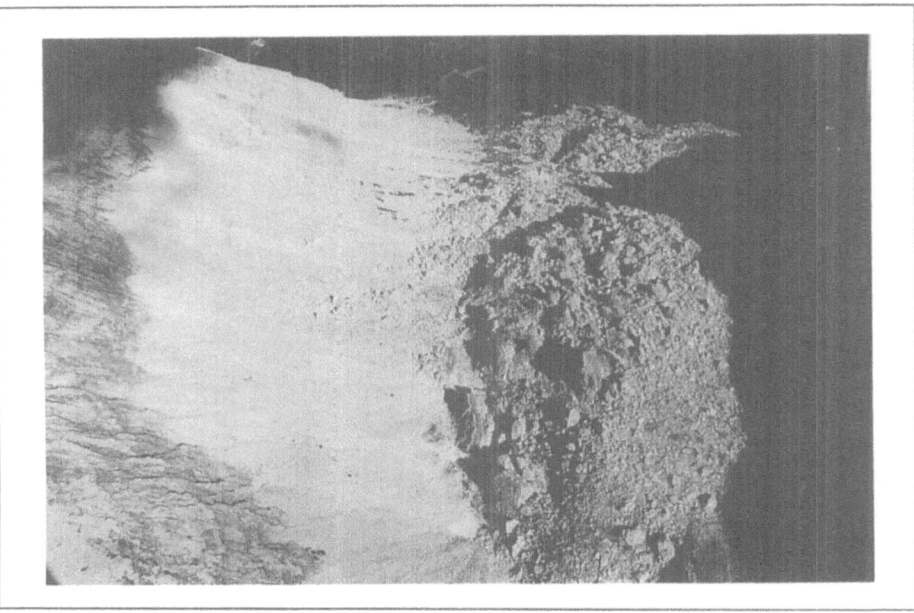

Figure 7 - Gable Mountain rockslide deposit. 7a (above), view emphasizing the sliding plane down which large blocks moved. Photo by FDW, 1995. 7b (below), view of the surface of the Gable Mountain deposit. Photo by DRB, 1995.

Figure 8 - Surface view of the Slide Lake sturzstrom deposit where it impounds Slide Lake. Note co-author GPM for scale. Photo by DRB, 1987.

Figure 9 - a (left), The 1992 Otatso Creek deposit east of Slide Pond; photo by DRB, 1994. b (right), The deposit in 1995. Note the degree of incision near the headscarps, and through the "floating island" of trees.

Figure 10 - View of Chief Mountain from overtop Gable Mountain. Note the numerous rockfall deposits around the base of Chief Mountain, and the prominent scar on lower right. Photo by DRB, 1995.

Acknowledgments

Funding for fieldwork and analyses was provided by UNC University Research Council Grants to DRB in 1994 and 1995. Fieldwork in 1983 was also carried out by J.G. and L.A. Oelfke, in association with DRB. Housing and logistical assistance were provided by officials of Glacier National Park and the National Biological Service. The unique and masterful flying of helicopter pilot J. Kruger provided aerial views of the sturzstrom sites described herein.

References

1. Anonymous (1992a) Major slide on Chief Mountain threatens to close area to hikers, *Waterton and Glacier Park Views* 1, 1. Issue of 15 July 1992, Hill Spring, Alberta, Canada.

2. Anonymous (1992b) Big landslide rips Chief's north face, *Hungry Horse News* 46, 1, 9. Issue of 16 July 1992, Columbia Falls, Montana, U.S.A.

3. Anonymous (1992c) Chief Mountain area still closed to hikers, *Waterton and Glacier Park Views* 1, 1. Issue of 22 July 1992, Hill Spring, Alberta, Canada.

4. Butler, D.R. (1979) Snow avalanche path terrain and vegetation, Glacier National Park, Montana, *Arctic and Alpine Research* 11, 17-32.

5. Butler, D.R. (1983) Observations of historic high magnitude mass movements, Glacier National Park, Montana, *The Mountain Geologist* 20, 59-62.

6. Butler, D.R. (1989) Glacial hazards in Glacier National Park, Montana, *Physical Geography* 10, 53-71.

7. Butler, D.R. and Malanson, G.P. (1990) Non-equilibrium geomorphic processes and patterns on avalanche paths in the northern Rocky Mountains, U.S.A., *Zeitschrift für Geomorphologie* 34, 257-270.

8. Butler, D.R. and Malanson, G.P. (1992) Effects of terrain on excessive travel distance by snow avalanches, *Northwest Science* 66, 77-85.

9. Butler, D.R. and Malanson, G.P. (1993) Characteristics of two landslide-dammed lakes in a glaciated alpine environment, *Limnology and Oceanography* 38, 441-445.

10. Butler, D.R. and Malanson, G.P. (1996) A major sediment pulse in a subalpine river caused by debris flows, *Zeitschrift für Geomorphologie* 40, in press.

11. Butler, D.R. and Schipke, K.A. (1992) The strange case of the appearing (and disappearing) lakes: the use of sequential topographic maps of Glacier National Park, Montana, *Surveying and Land Information Systems* 52, 150-154.

12. Butler, D.R. and Walsh, S.J. (1990) Lithologic, structural, and topographic influences on snow-avalanche path location, eastern Glacier National Park, Montana, *Annals of the Association of American Geographers* 80, 362-378.

13. Butler, D.R. and Walsh, S.J. (1994) Site characteristics of debris flows and their relationship to alpine treeline, *Physical Geography* 15, 181-199.

14. Butler, D.R., Malanson, G.P. and Cairns, D.M. (1994) Stability of alpine treeline in Glacier National Park, Montana, U.S.A., *Phytocoenologia* 22, 485-500.

15. Butler, D.R., Malanson, G.P. and Oelfke, J.G. (1991) Potential catastrophic flooding from landslide-dammed lakes, Glacier National Park, Montana, USA, *Zeitschrift für Geomorphologie* Supplementband 83, 195-209.

16. Butler, D.R., Malanson, G.P. and Walsh, S.J. (1992) Snow-avalanche paths: conduits from the periglacial-alpine to the subalpine-depositional zone, in Dixon, J.C. and Abrahams, A. (eds.) *Periglacial Geomorphology*, John Wiley and Sons Ltd., London, 185-202.

17. Butler, D.R., Oelfke, J.G. and Oelfke, L.A. (1986) Historic rockfall avalanches, northeastern Glacier National Park, Montana, U.S.A., *Mountain Research and Development* 6, 261-271.

18. Carrara, P.E. (1990) Surficial geologic map of Glacier National Park, Montana, *U.S. Geological Survey Miscellaneous Investigations Series Map* I-1508-D.

19. Clague, J.J. and Evans, S.G. (1987) Canadian landform examples: rock avalanches, *The Canadian Geographer* 31, 278-282.

20. Cruden, D.M. (1982) The Brazeau Lake slide, Jasper National Park, Alberta, *Canadian Journal of Earth Sciences* 19, 975-981.

21. Cruden, D.M. (1985) Rock slope movements in the Canadian Cordillera, *Canadian Geotechnical Journal* 22, 528-540.

22. Cruden, D.M. and Hungr, O. (1986) The debris of the Frank Slide and theories of rockslide-avalanche mobility, *Canadian Journal of Earth Sciences* 23, 425-432.

23. Eisbacher, G.H. (1978) Observations on the streaming mechanism of large rock slides, northern Cordillera, *Geological Survey of Canada Paper* 78-1A, 49-52.

24. Eisbacher, G.H. (1979) Cliff collapse and rock avalanches (sturzstroms) in the Mackenzie Mountains, northwestern Canada, *Canadian Geotechnical Journal* 16, 309-334.

25. Eisbacher, G.H. and Clague, J.J. (1984) Destructive mass movements in high mountain terrain, *Geological Survey of Canada Paper* 84-16, 1-230.

26. Eppler, D.B., Fink, J. and Fletcher, R. (1987) Rheologic properties and kinematics of emplacement of the Chaos Jumbles rockfall avalanches, Lassen Volcanic National Park, California, *Journal of Geophysical Research* 92, 3623-3633.

27. Evans, S.G. (1987) A rock avalanche from the peak of Mount Meager, British Columbia, *Geological Survey of Canada Paper* 87-1A, 929-934.

28. Evans, S.G. (1989a) Rock avalanche run-up record, *Nature* 340, 271.

29. Evans, S.G. (1989b) The 1946 Mount Colonel Foster rock avalanche and associated displacement wave, Vancouver Island, British Columbia, *Canadian Geotechnical Journal* 26, 447-452.

30. Evans, S.G. and Clague, J.J. (1990) Reconnaissance observations on the Tim Williams Glacier rock avalanche, near Stewart, British Columbia. *Geological Survey of Canada Paper* 90-1E, 351-354.

31. Evans, S.G. and Clague, J.J. (1994) Recent climatic change and catastrophic geomorphic processes in mountain environments, *Geomorphology* 10, 107-128.

32. Evans, S.G., Aitken, J.D., Wetmiller, R.J. and Horner, R.B. (1987) A rock avalanche triggered by the October 1985 North Nahanni earthquake, District of Mackenzie, N.W.T., *Canadian Journal of Earth Sciences* 24. 176-184.

33. Evans, S.G., Clague, J.J., Woodsworth, G.J. and Hungr, O. (1989) The Pandemonium Creek rock avalanche, British Columbia. *Canadian Geotechnical Journal* 26, 427-446.

34. Griggs, R.F. (1920) The great Mageik landslide, *The Ohio Journal of Science* 20, 325-354.

35. Hadley, J.B. (1964) Landslides and related phenomena accompanying the Hebgen Lake earthquake of August 17, 1959, *U.S. Geological Survey Professional Paper* 435, 107-138.

36. Hungr, O. (1995) A model for the runout analysis of rapid flow slides, debris flows, and avalanches, *Canadian Geotechnical Journal* 32, 610-623.

37. Jackson, L.E. and Isobe, J.S. (1990) Rock avalanches in the Pelly Mountains, Yukon Territory, *Geological Survey of Canada Paper* 90-1E, 263-269.

38. Kaiser, P.K. and Simmons, J.V. (1990) A reassessment of transport mechanisms of some rock avalanches in the Mackenzie Mountains, Yukon and Northwest Territories, Canada, *Canadian Geotechnical Journal* 27, 129-144.

39. Lee, L. (ed.) (1994) *Backcountry Ranger - In Glacier National Park 1910-1913*. Leslie Lee, Publisher, Elk Rapids, Michigan, 264 pp.

40. McCarthy, D.P. and Smith, D.J. (1995) Growth curves for calcium-tolerant lichens in the Canadian Rocky Mountains, *Arctic and Alpine Research* 27, 290-297.

41. Mudge, M.R. (1965) Rockfall-avalanche and rockslide-avalanche deposits at Sawtooth Ridge, Montana, *Geological Society of America Bulletin* 76, 1003-1014.

42. Oelfke, J.G. and Butler, D.R. (1985a) Landslides along the Lewis Overthrust Fault, Glacier National Park, Montana, *The Geographical Bulletin* 27, 7-15.

43. Oelfke, J.G. and Butler, D.R. (1985b) Lichenometric dating of calcareous landslide deposits, Glacier National Park, Montana, *Northwest Geology* 14, 7-10.

44. Ryder, J.M., Bovis, M.J. and Church, M. (1990) Rock avalanches at Texas Creek, British Columbia. *Canadian Journal of Earth Sciences* 27, 1316-1329.

45. Shreve, R.L. (1966) Sherman landslide, Alaska, *Science* 154, 1639-1643.

46. Shreve, R.L. (1968a) Leakage and fluidization in air-layer lubricated landslides, *Geological Society of America Bulletin* 79, 653-658.

47. Shreve, R.L. (1968b) The Blackhawk landslide, *Geological Society of America Special Paper* 108, 1-47.

48. Van Gassen, W. and Cruden, D.M. (1989) Momentum transfer and friction in the debris of rock avalanches, *Canadian Geotechnical Journal* 26, 623-628.

49. Voight, B. (1978) Lower Gros Ventre slide, Wyoming, USA, in Voight, B. (ed.) *Rockslides and Avalanches*, Elsevier, Amsterdam, 113-160.

50. Walsh, S.J., Butler, D.R., Allen, T.R. and Malanson, G.P. (1994) Influence of snow patterns and snow avalanches on the alpine treeline ecotone, *Journal of Vegetation Science* 5, 657-672.

51. Walsh, S.J., Butler, D.R., Brown, D.G. and Bian, L. (1990) Cartographic modeling of snow avalanche path location within Glacier National Park, Montana, *Photogrammetric Engineering and Remote Sensing* 56, 615-621.

52. Walsh, S.J., Malanson, G.P. and Butler, D.R. (1992) Alpine treeline in Glacier National Park, Montana, in Janelle, D.R. (ed.) *Geographical Snapshots of North America*, Guilford Press, New York, 167-171.

Authors

David R. Butler
Department of Geography and Planning
Southwest Texas State University
San Marcos, TX 78666-4616
U.S.A.

George P. Malanson
Department of Geography
University of Iowa
Iowa City, Iowa 52242
U.S.A.

Forrest D. Wilkerson
Department of Geography
University of North Carolina - Chapel Hill
Chapel Hill, North Carolina 27599-3220
U.S.A.

Ginger L. Schmid
Department of Geography
North Carolina Central University
Durham, North Carolina 27707
U.S.A.

STORM INDUCED MASS-WASTING IN THE OREGON COAST RANGE, U.S.A.

CHARLES L. ROSENFELD

1. Introduction

The rugged slopes of the Oregon Coast Range rise abruptly from the Pacific Ocean, where their folded sediments and volcanic intrusives interrupt the flow of the moisture laden subtropical jet stream yielding orographic precipitation exceeding 2000 milli-meters annually. The dense conifer forest cover and deep soils allow copious rates of infiltration and throughflow, leaving sheetwash and rainsplash as relatively ineffective sediment transport processes. The steep slopes show ample evidence of deep-seated earthflow in areas of weaker bedrock, and frequent debris avalanche activity stripping soils from the surfaces covering the more resistant volcanics. The stream channels impacted by this debris quickly winnow away the fines, leaving behind gravels and large organic debris, the materials which form the gravel spawning beds and rearing pools for the region's anadromous fish.

Between February 6 - 8, 1996 a severe winter storm, brought more than 500 mm of subtropical moisture to the northern Oregon Coast Range. Doppler radar located especially dense precipitation bands over the crest of the mountains west of Portland, Oregon on February 7. Several local streams crested at levels above the previous flood of record (December 1964), and landslides blocked roads throughout the region.

The immediate impact of floods and landslide events focused the attention of both the news media and government agencies on the magnitude of the effects, and their possible causes. As reports of landslides tallied into the hundreds, reconnaissance surveys were hastily organized by resource management agencies and environmental groups alike. Much of the interest is prompted by concerns that the cumulative impacts upon watersheds may seriously diminish efforts to protect and enhance the spawning habitat of anadromous salmon and steelhead runs at a time when fish counts in certain streams were at record lows, and approaching endangered status. Research linking fish population dynamics to land management activities resulted in changes in manage-ment practices on both public and private lands, designed to protect watershed values and remediate declining channel and riparian habitat conditions.

Initial damage reconnaissance efforts were compiled by either field reports or aerial observations conducted immediately following the storm to capture the storm induced events. These hastily implemented surveys, while responding to the immediacy of the situation, often reflect either the observational bias of the technique or the observers. As many secondary roads were blocked by slides, field observations tended to over emphasize mass wasting activity along major highway corridors, while visual surveys by aircraft tended to concentrate on the most visible effects of the storm, mainly on managed forest lands.

J. Kalvoda and C.L. Rosenfeld (eds.), Geomorphological Hazards in High Mountain Areas, 167-176.
© 1998 *Kluwer Academic Publishers.*

An aerial videography mapping project was flown along the flooded banks of the Willamette and Columbia rivers on February 9 and 10, 1996, to map the extent of inundation for the Portland District, Army Corps of Engineers. A segment of this aerial survey flew diagonally across the northern Coast Range from Gaston to Astoria, Oregon. While this vertical imagery was not intended to sample mass wasting in this heavily impacted region, it captured a random transect which depicts 71 large-scale features related to the storm. This imagery was interpreted and compiled to explore associations between mass wasting mechanics, morphometry, geology, and site conditions. In addition, comparisons with other contemporary surveys allows us to evaluate the potential for a more formal application of aerial videography for sampling landscape characteristics.

2. Mass wasting regimes in the Coast Range

Mass wasting is a dominant geomorphic process in the Oregon Coast Range, with significant mechanisms ranging from soil creep due to frost heave (pipkrake) to deep-seated rotational slides. Activity is concentrated in the winter months of December through March, and is usually associated with significant precipitation or snow-melt events. The micro-morphology of most slopes in the Coast Range is a complex of overlapping landslide hollows, of varying age, interspersed by debris fans and hummocky toe slopes.

Shively (1989)[1] carefully examined such a hillslope complex and determined the magnitude and frequency of mass wasting events over the past 250 years using dendrochronology. By delineating 25 of the dated mass wasting features, and applying Crozier's (1973)[2] morphometric indicies, the 'classification index' (D/L) used by Skempton[3] was clearly shown to be a reliable indicator of mass wasting type. Additionally, clustered dates indicate a 'peak periodicity' of 20 to 30 years in mass wasting frequency. Dietrich et.al. (1982)[4] recognize mass wasting as a primary component of sediment budgets in northwest watersheds, and illustrate that progressively older landslide scars fill with increasingly finer materials. They also point out at debris flow materials are frequently deposited in active channels, where fines are removed by channel flows, leaving behind cobbles and gravel in the channel bed.

Wong (1991)[5] monitored the movement characteristics of landslides in the Coast Range over a ten-year period. In his observations, all the landslides exhibited brief movement in response to significant precipitation (>160mm threshold) events, with a lag time of less than 3 days. All major movements (>10mm in 4-10 days) were precipitation induced.

[1] Shively, David D., 1989, Landsliding Processes Occurring on a McDonald-Dunn Forest Hillslope, Unpubl. MS thesis, Department of Geosciences, Oregon State University, 35pp.
[2] Crozier, Michael J., 1973, 'Techniques for the Morphometric Analysis of Landslips', Zeit. fur Geomorphologie, N.F. Suppl. Band 46:67-77.
[3] Skempton, A.W., 1953, 'Soil Mechanics in Relation to Geology', Proc. of the Yorkshire Geol. Soc., 29:33-62.
[4] Dietrich, William E., Thomas Dunne, Neil Humphrey and Leslie Reid, 1982, 'Construction of Sediment Budgets for Drainage Basins', Sediment budgets and routing in forested drainage basins, U.S. Forest Service, Tech. Rept. PNW-141, pp. 5-23.
[5] Wong, Bernard Bong-Lap, 1991, Controls on movement of selected landslides in the Coast Range and western Cascades, Oregon, Unpubl. M.S. thesis, Department of Geosciences, Oregon State University, 193 pp.

Figure 1. Debris slide, originating from sandstone outcrop at center of photo, and debris torrents affecting ephemerial channels, on the left, are typical of the numerous masswasting events throughout this landscape.

Three large-scale mass wasting regimes in the Coast Range landscape were observed in our study: (1) shallow debris slides/torrents, (2) deep-seated slump/earthflow,
and (3) saturated streambank failures. Each of these regimes occur in a variety of settings and often involve a variety of processes and potential influencing factors.

Over half of the events observed were shallow debris slides or torrents. Typically
these events occurred on steep slopes (>30 degrees) and were predominantly associated
with sedimentary bedrock. Depth estimates range from 2 to 5 meters, with maximum
length exceeding 3,000 meters, Figure 1. Topographically, 71% of the debris slides
originated in swales along steep slopes, with 32% of the events delivering sediment to
ephemeral channels with over 50% directly impacting perennial channels. These impacts include both channel scour (sometimes exposing bedrock) and deposition of
landslide and large organic debris. Since these events were precipitation induced, the
flow in the channels was sufficient to 'wash' the landslide debris and now channel
impoundments were observed. Although these events resulted in significant turbidity
along the affected streams, most perennial channels quickly developed gravel bars and
pools along their affected reaches. In some cases however, ephemeral channels were
scoured into bedrock sluices, figure 2.

A combination of high water and significant through-flow from the valley walls
created favorable conditions for saturated earth-flow along the banks of perennial streams. These situations were characterized by multiple 'retrograde' tension cracks
parallel to the stream bank, toppling trees in the riparian zone, and resulting bank-
attached bar development immediately downstream, figure 3. These 'saturation failures' along the banks were frequently obscured from direct observation in the video

Figure 2. Ephemerial channel scoured to bedrock by a debris torrent which entered into a perenial stream and
caused significant downstream deposition.

imagery by riparian tree canopy, and their presence was signaled by multiple downed
trees, and fresh debris in the channels.

While this mass-wasting type comprised 38% of the events, it was difficult to estimate the volume or measure the morphometric indices for individual events. All of
these events (25) occurred in either colluvium from sedimentary slopes (7) or in valley
alluvium (18).

Deep-seated slump/earthflow failures occurred in only 9% of the observed events,
and all but one was along a slope underlain by soft sedimentary rocks. Whether these
were new failures or re-activations of earlier events is unclear, although each exhibited
signs of recent deformation and sediment contribution along its flanks. Undoubtedly,
movements along many such previous failures probably went undetected in our survey.

3. Storm-induced mass wasting implications

Intensive commercial forestry in the Coast Range has altered the natural geomorphic system. Access roads which traverse the steep slopes often interrupt the throughflow, concentrating water as runoff in the ditches and culverts. Unless drainage is carefully controlled, saturated roadbeds and culvert outlets often become source areas of debris avalanches. Clearcuts and thinned forest canopy expose remaining trees to wind stress increasing the potential of blow-down, which in turn exposes bare soil and encourages saturation failures. While modern forest management practices have increased the frequency and size of road culverts, and timber harvest using cable suspension systems has reduced the effects of soil compaction, the large magnitude storms often trigger geomorphic responses from older management features- especially when the event exceeds the 50 year return interval. As a result, the anticipated benefits of modern forest practices are sometimes obscured by disturbances initiated as a result of older management practices or the lack of maintenance to remaining road and culvert systems.

In an effort to evaluate the effects of the February 1996 storm event upon the present characteristics of the Oregon Coast Range, the individual landslide events observed on our videography transect were interpreted to determine several relevant

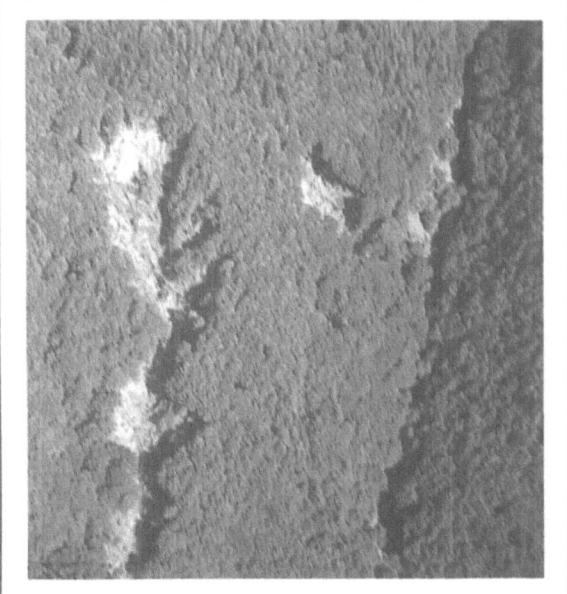

Figure 3. Saturation failures along channel banks were frequently observed phenomena Steeper channels. such as the one on the left, were often characterized by the obvious scars of debris slides. Saturated bank failures occured along lower gradient channls and were most often identified by disturbances to riparian vegetations, as seen at the upper right.

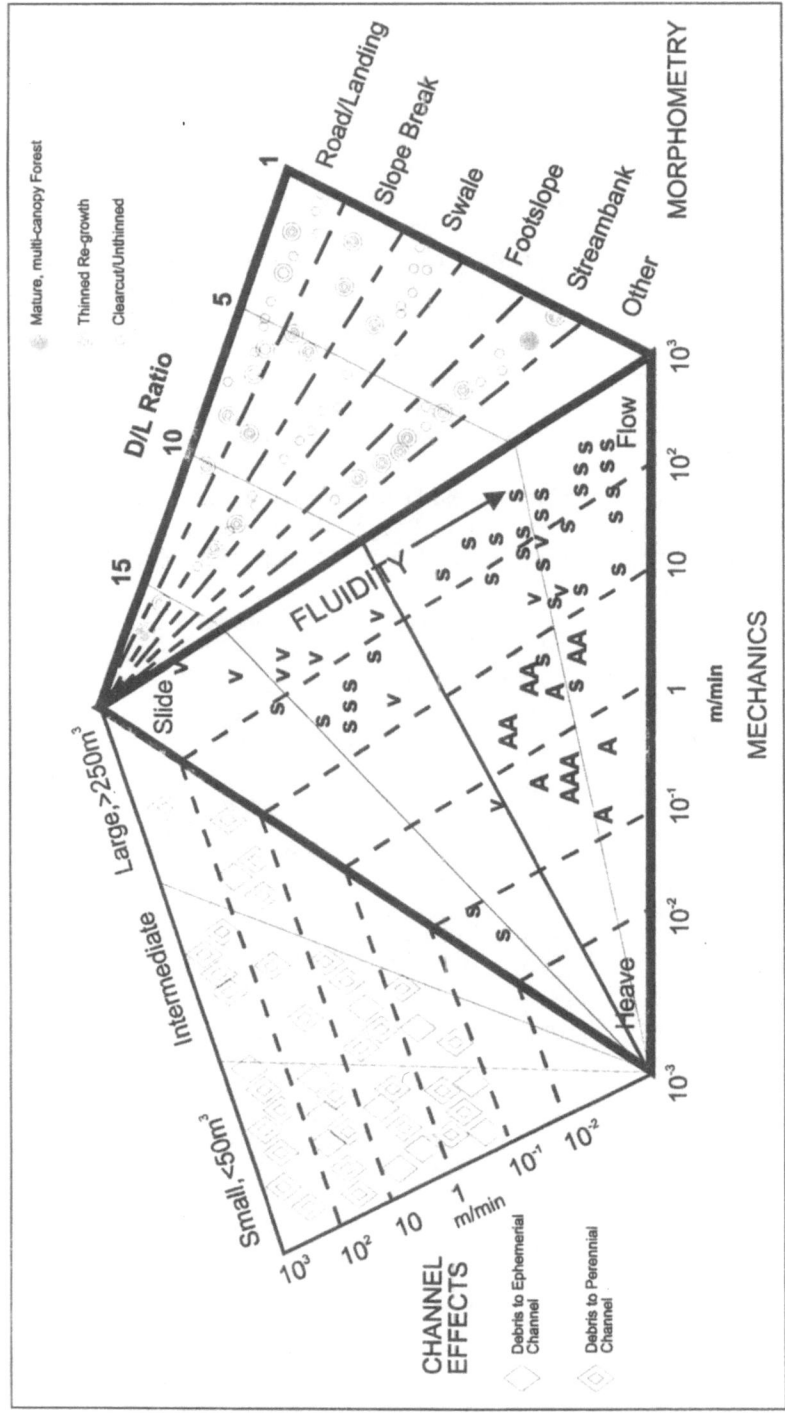

Figure 4. The "Faces" of this pyramid illustrate the relationship between the mechanics, morphometry and channel effects of the slide events observed in our survey. See text for explanation.

landscape associations. The mass wasting mechanics factors included a subjective assessment of the character of the event and its rate of material displacement[6].

When possible, the event was associated with the mapped bedrock type, which was grouped into three classes: sedimentary rocks (s= principally sandstone and mudstone), volcanic (v= basalt and breccia), and alluvium (A= valley fill and channel bank deposits). Some very clear relationships are seen in Figure 4 related to the mechanics of the individual events. All occurrences of alluvial material failures were saturated earthflow and bank failures of slow to moderate motion rates. Most rapid debris torrents occurred in sedimentary bedrock areas, while rockfalls and debris avalanches were evenly distribute among volcanic and sedimentary bedrock types (accounting for 65% of the total slides observed).

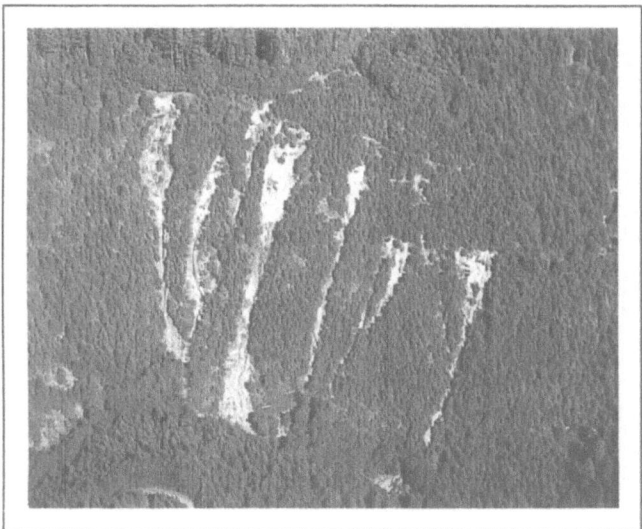

Figure 5. Rockfall and debris avalanches on very steep slopes in volcanic bedrock areas were frequent.

These data demonstrate the significance of topographic position and landform slope as controlling factors in the type and rate of mass wasting events. Alluvial materials, confined to generally low gradient valley bottom areas, dominated the moderate rate events. The actual number of such events observed in our aerial survey probably underestimates the actual number of such events significantly, as our observations were limited to areas where obvious disturbance of trees in the riparian corridor and 'in stream' deposition were seen in the imagery. Many other small alluvial bank failures, without major vegetation disturbance, were hidden under the canopy of riparian trees and were not included in our tallies. The debris avalanches and torrents were the most obvious events observed in our aerial survey- and they were the events which visually caught our attention during our over-flight. These events dominated the valley-side slopes of the mountains, and were most apparent when they originated in young managed forest areas.

---

[6] The technique used follows the classification of Carson and Kirkby (1972),Hillslope form and Process, London:Cambridge Press.

Topographically, the sedimentary bedrock underlies moderate to steep (15 to 60 degree) slopes, with volcanic intrusives boldly forming ridges and peaks at very steep (50 to 80 degree) angles. As a result, most debris torrents (mostly water) occurred in sedimentary terranes, with rockfall and debris avalanches on very steep slopes originating from mainly volcanic areas, Figure 5.

Deep seated slump/earthflow events were observed in sedimentary areas, however the number of events was not significant. Additionally, it was almost impossible if these events were initiated by the storm event in question, or if a re-activation of a pre-existing failure had occurred, Figure 6.

Figure 6. Large slump/earthflow features are present within the study area, although most pre-date this storm event. Re-activation of movement was very difficult to detect by image interpretation.

Crozier's Classification Index[7] was used to further explore the relationship between the morphometry of the landslide events and the topographic position and forest cover type of the initial failure. The classification index is based on length measurements and depth of failure estimates- although the videography may be viewed in 'stereo', the depth estimates are rough at best. We estimated shallow failures at <2 meters, intermediate depths at 5 meters, and deep seated failures at >10 meters. The right flank of the 'pyramid' diagram illustrated in figure 4 divides topographic position into six classes: roads and timber harvest 'landing' areas, significant 'breaks in slope' of greater than 15 degrees, 'swales' along valley-side slopes, footslope areas with colluvial transitions to valley bottoms, lower banks adjacent to stream channels, and other locations such as divides or debris fans. Drainage swales and bedrock controlled 'slope breaks' domina-

---

[7] Crozier (1973), ibid.

ted the failure areas on 'natural' slopes, with man-made sites (mainly roads) contributing 42 % of the total slope failures.

When we examine the forested sites, classified as mature (multi-canopy) forest, thinned (even canopy) forest, or clearcut and unthinned plantations, the relationships become even more interesting. Road and 'landing' associated failure sites all occurred in managed forest areas, with the larger debris torrents mainly starting in thinned forest stands. It is unclear form our study whether these observations result from smaller failures being 'hidden' by taller forest canopy, or perhaps that harvest activity in thinned stands occurred 25-40 years ago under less demanding forest management practices, which has yielded more frequent failures. Another curious association was that although mature multi-canopy forest occupied only about 2% of the area covered by our aerial imagery, this class of landcover yielded 9% of the failures. It is proposed by this observer that the remaining 'old growth' forest cover is frequently found in very steep or unstable areas, where the economics of forest management are impractical, but where the conditions for mass wasting may be pronounced.

Failures along drainage swales tended to occur along the upper slope, just below Horton's zone of 'no erosion'[8]. All of these occurrences were observed in areas of managed forest, with the larger volume slides associated mainly with thinned 'middle aged' stands. It is unlikely that observational error would account for this relationship, and the author suggests that falling trees exposed deeper 'root throw' on the older stands, resulting in larger volume debris torrents and avalanches.

To understand the implications of these 'storm-induced' failures upon watershed values, the contribution of debris directly to ephemeral and perennial channels was surveyed. Estimates of the total volume of the contributing slide event was also recorded along with an estimate of delivery velocity. The left flank of our diagram (figure 4) reflects direct channel contribution for 67% of the total number of events in our survey. Given the precipitation intensity at the time of the storm, it may be assumed that active flow was occurring in all of the channels surveyed. However, channel flow normally does not occur in the small first and second order channels during the drier summer months, and such channels were noted in our observations. Neither size of the event, nor the velocity which would extend its effects along a channel had any apparent relationship with the types of channel affected. In other words, channels of all orders were apparently directly affected by mass wasting deposition by more than 2/3's of the events. Mass wasting debris deposited by a storm in February will be washed by higher flow discharges which will wash away fine sediment, leaving the gravel and cobble fraction as bars and riffles. Although the initial turbidity resulting from this 'sorting' process will inhibit anadromous fish migration up the streams, the resulting gravel beds and riffle/pool sequences will provide future spawning reaches.

4. Conclusions

Overall implications of these initial observations would tend to confirm the significance of 'storm induced' mass wasting 'triggers' in the Oregon Coast Range. The

---

[8] Horton, R.E., (1945), 'Erosional development of streams and their drainage basins', Bulletin of the Geological Society of America, (56), pp. 275-370

impact of forest management activities is somewhat less clear from our observations, although it appears that large storms may actually initiate failures resulting from older management activities. Furthermore, it appears that larger trees (i.e. deeper roots) may be associated with deeper and larger debris torrents and avalanches. The higher than expected frequency of failures originating in mature forest stands may be an indication of the geomorphic character of the remaining 'old growth' forest stands in the Coast Range.

Watershed managers can gain some insight from these initial findings, as direct channel deposition of mass wasting debris occurred from most events observed- regardless of the origin or mechanics of the slide. Equally important was the fact at all classes of channels were involved. This may indicate that efforts to enhance anadromous fish habitat should take into consideration the magnitude and frequency of debris contributions, and the positive aspects of mass wasting as a source of spawning gravel and large woody debris.

Author

Charles L. Rosenfeld
Oregon State University
Corvallis, Oregon, 97331-5506
U.S.A.

NATURAL HAZARDS IN RELATION TO PRESENT STRATOVOLCANO DEGLACIATION: POPOCATEPETL AND CITLALTEPETL, MEXICO.

DAVID PALACIOS

1. Introduction

Popocatepetl (5,450m) is located in the center of the Trans-Mexican Volcanic Belt (19°03'N, 98°35'W), and Citlaltepetl or Pico de Orizaba (5,700m) is located on the eastern edge of the Belt (19°01'N, 97°16'W) (Fig.1). They are the two highest stratovolcanoes in Mexico and the only ones that are active and have glaciers on their slopes. These characteristics make them a high risk area for lahar formation and other geomorphologic processes, so it is very important that the glaciers be monitored.

Popocatepetl and Citlatepetl have similar morphology, including the remains of old volcanic structures that were destroyed as calderas were formed on the northwestern slopes of the great upper volcanic cones. The remains appear on both volcanoes as rocky spurs that act as an orographical barrier against winds and insolation. This protection allows snow to accumulate where the eastern side of the spurs and the northern slopes of the cones meet.

The largest glaciers on the volcanoes were formed after the last great eruptions that took place in the 15th and 17th centuries and evolved as a result of the sheltering effect of the rocky spurs. Both glaciers advanced to their lowest altitudes at the height of the Little Ice Age, at the end of the 19th century. With the exception of a slight advance or stabilization of the terminus in the early 1970's (White 1981), the two glaciers have rapidly receded during the 20th century.

During the L.I.A., the glacier termini advanced over the slopes of the volcanic cones that were covered with ash. The glacier snout pushed the volcanic ash and, depending on the shape of the ice flow itself, caused the ash to accumulate at many different altitudes. Because ash is not dense, it was also transported by meltwater that flowed from the front edge of the glacier. Ramps consisting mainly of ash and blocks were formed by these processes and encircle the entire foot of the glaciated slopes of the stratovolcanoes. Debris flows often occur when the ramps are forming and when the period of glacial recession begins, because glacier meltwater saturates the deposits that form the ramps (Palacios 1996 and Palacios and Vázquez 1996).

When the glaciers retreated, a network of depressions was formed between the proglacial ramps and the foot of the upper cones. These depressions channeled the water thus forming a network of gorges that incised the proglacial ramp. Such activity is especially evident on Popocatepetl. The appearance of major streams in these landforms is testimony to the speed at which the glaciers melted.

If the basins had undeveloped drainage systems, as in the case of Citlaltepet, then the meltwater would accumulate in the depressions left by the receding glacier. The

J. Kalvoda and C.L. Rosenfeld (eds.), Geomorphological Hazards in High Mountain Areas, 177-209.

water filtered through the morainic material and saturated it, thus generating great debris flows on the proglacial ramp (Palacios, Parrilla and Zamorano 1996).

As the glaciers retreated, the durable levels of the upper cones were gradually exposed, while on the other slopes they remained covered by ash. The surfaces beneath the glaciers are composed mainly of levels of nuées ardentes that are compact and resistant (Popocatepetl) and compact andesitic lava flows (Citlaltepetl). The glacier merely smoothed them lightly.

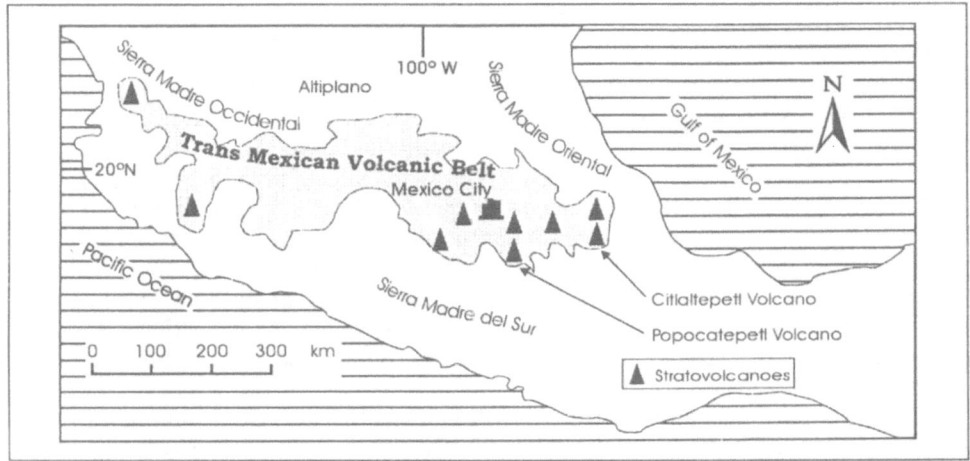

Figure 1. Location map of the study area

At the beginning of this century, there still was fluvial activity in the interior of the gorges. The continuing retreat of the glaciers caused streams to disappear, and ash fell from the slope and covered the bottom of the proglacial depressions, proglacial gorges and debris flow channels. These areas provide a suitable environment for snow accumulation from snowfall and snow avalanches. Ash fell from the slopes and covered the snow that collected, preventing it from melting. The snow serves as a lubricating layer for blockslide tongues. The scree slopes composed basically of ash, was easily eroded, and the ash quickly covered the snow and rockslide deposits. For a long period of time, the mantle of ash kept the snow trapped inside from melting and transformed it into ice lenses. When snowfall was scarce, the deposits lost part of the ice. As melting occurred, numerous thermokarst landforms were formed (Palacios 1995).

Frequently, the ash covered snow forms true permafrost in the basins and lasts for many years. When there is an abundance of water, which is often the result of snowmelt caused by rain or heavy glacier melting provoked by volcanic activity. When the water saturates the suface deposits on the same level of the permafrost, this acts as an impermeable layer, and generates many debris flows (Palacios, Zamorano and Parrila 1996 and Palacios, Parrilla and Zamorano 1996)

Deglaciation exposed the eastern walls of the rocky spurs that appear on both stratovolcanoes. Simultaneously, they set off erosion processes associated with gelifraction and decompression that caused rocks to fall from the walls. As the ice retreated and erosion of the sidewalls became more severe, part of the mass of ice was trapped beneath gravity deposits that protected it from insolation and melting. Eventually, this ice was isolated from the rest of the glacier. Thus, a rockfall cover ice talus was formed where slumping is very common (Palacios and Vázquez 1996 and Palacios and Marcos 1996).

All of these processes illustrate that, as far as volcanic activity and geomorphology are concerned, large glaciated volcanoes have a higher risk factor during the deglaciation process. Projections suggest that risks associated with deglaciation reach a threshold and then decline when the glacier becomes so small that it is ultimately confined to the upper cone of the volcanoes, above the rocky spurs. The terminus is no longer steep and avalanches do not occur. Also, meltwater filters into the ground before it can reach the proglacial depressions.

2. Volcanic and glacial evolution of Popocatepetl and Citlaltepetl prior to this century

The volcanic characteristics of Popocatepetl began with two volcanic periods: a volcanic-constructive phase that began approximately 1 My ago (Demant 1981) and formed an old volcano known as Nexpayantla (Fig. 2). The remains of this volcano can be seen today in the northwest sector of Popocatepetl (Ventorrillo-Nexpayantla area). The old volcano was eliminated by a Bezymianny-type debris avalanche (Robin and Boudal 1987), approximately 100,000-50,000 yr BP. The present cone, which is the subject of this study, was formed during a later phase that began at the end of the Pleistocene and continued through the early Holocene. Initially, it developed 1.5km to the north of the present crater. This first stratovolcano was partially destroyed by a series of explosive phases, the last of which occurred 5,000-3,500 yr BP (Robin 1984; Robin and Boudal 1987). The remains of the cone are found today on Prico del Fraile (Fig.2). Later, a prolonged period of eruptions between 3,800-1,200 yr BP, formed a new cone south of the earlier one (Robin 1984; Robin and Boudal 1987). New St Vincent-type volcanic-destructive processes partially destroyed the volcano in 1,200 y BP, but subsequent activity formed the modern upper-cone. On the northtern side it is 1,200m thick and on the southern side it reaches a thickness of 2,500m. The building process was interrupted by various destructive episodes. The great accumulation of volcanic materials indicates a period of extraordinary eruptive activity that occurred after 1,200 y BP. The upper cone is formed from the products of dacitic and andesitic lavas and pyroclastic deposits.

Historical sources frequently refer to volcanic activity on Popocatepetl. The pre-Hispanic manuscript „Tellerino Remense" makes reference to major volcanic activity there in 1350, possibly of a very destructive nature. Other references are made to eruptions in 1504 and 1512 (compiled by Guzmán 1968). Hernan Cortés relates the occurrence of major emissions of ash in 1519 (Cortés edit. 1942). The historian Bernal Díaz del Castillo, mentions the eruption of 1519 and another great emission of ash in 1539 (Díaz 1519, edit. 1980). Other sources (Guzmán's compilation of 1968) indicate that this period of eruptive activity came to a close in 1594. Activity suddenly started again on October 13, 1663, with emissions of ash that continued until 1697. Since then, the volcano has experienced a period of calm, interrupted only briefly in 1720. From 1802-04, Humboldt made references to emissions of ash (Humboldt von A.

1807/1811), that were still evident even in 1827 (Ward 1985). Another period of inactivity followed (documented by Alzate 1831; anonymous 1836; Payno 1849; Rios 1851; Monnier 1884; Orozco 1887; Ordoñez 1895; Packard 1896; Sánchez 1902) but was interrupted in 1919, when dynamite explosions from a sulphur mine in the interior of the crater provoked new volcanic action (Waitz 1921; Murillo 1939). Activity continued in 1925 (Camacho 1925), and in 1927, a great pyroclastic emission occurred (Hernández-Sosa 1948). From then on, Popocatepetl did not emit solid materials until December 21, 1994, when a new phase of ash emission started and continues today (June 1996).

Glaciation at the end of the Pleistocene must have brought about important developments in the Popocatapetl area (White 1987). The glacial action that left indisputable geomorphologic manifestations, began after the last violent eruptions of 950 yr BP, which were characterized by great emissions of nuée ardentes.

A glacier on the north side of the volcano also extended along the west side until recent years. The lack of references about the glacier in historical sources, however, makes it very difficult to document its evolution. Mention is made of the glacier in accounts dating from the period of the Spanish conquest, but no data are provided regarding its dimensions. Frequent remarks are found in the old documents about the difficulties involved in crossing the levels of permanent snow before ascending to the crater from the south side of the glacier (e.g. Cortés edit. 1942 or Díaz del Castillo edit. 1980).

The earliest precise information about the glacier dates from the 19th century, and is found in descriptions of ascents of the volcano. These ascents were always attempted from the eastern border of the glacier, where the traditional route to the crater had been established. Thus, it can be deduced that the eastern part of the glacier descended to an altitude of approximately 4,600m. Other interesting information reveals that there was no depression between the glacial terminus and the terminal deposits, which would indicate that the glacier was still advancing at that time (Hernández-Sosa 1948 in his compilation of the ascent of the Glennie brothers in 1827; Pérez 1857; Mentz 1980 - compilation of the ascent of Von Muller 1857; Lavarriere 1858; Dollfus 1870) (Fig. 2).

Citlaltepetl volcano forms the southern part of a north-south volcanic range of ca. 35 km long with steep, asymmetrical slopes (Fig. 2). The lithology consists mainly of andesites and dacites (Rodríguez-Elizarrarás and Lozano 1991). Cofre de Perote (4,282 m), the northern point of the range, is an inactive stratovolcano that displays extensive glaciation, probably dating from the Late Pleistocene. Xausta Dome (3,880 m), Las Cumbres Caldera (3,940 m) and several other mountains with altitudes over 3,800 m located between Cofre de Perote and Pico de Orizaba show similar morphologic evidence of glaciation.

The eruptive history of Citlaltpetl has three stages (Robin and Cantagrel 1982). During the initial phase, which lasted 1 million years, the entire base of the stratovolcano developed. The second stage (100,000 to ca. 33,000 yr BP) was marked by the formation of a large caldera -the remains of which is today's, Pico del Sarcófago- and the subsequent appearance of large andesitic and dacitic domes and numerous blocks and ash flows (Robin and Cantagrel 1982; Robin et al. 1983; Hoskuldsson 1990; Hoskuldsson and Robin 1993, Hoskuldsson 1994) (Fig.2). The third phase began 19,000 yr BP with the formation of the new cone, which covers most of the wall of the caldera and the inner domes. It consisted of several alternated phases of andesitic lava flows and pyroclastic eruptions (Hoskuldsson and Robin 1993, Hoskuldsson 1994). An episode of debris avalanche that was transformed into lahar, seemingly due to the presence of glacier ice, took place at *ca.* 13,000 yr BP on the eastern slope (Carrasco-Núñez *et al.* 1993). A series of pyroclastic flows were channeled through a glacier cirque and formed a wide fan on the western slope between 4,660 and 4,040 yr BP (Siebe *et al.* 1993). The last eruptive phase (700 A.D. to 1687 A.D.) included seven separately identifiable eruptions. They were effusive, with the exception of one plinian event (Mooser *et al.* 1958; Simkin *et al.* 1981; Robin and Cantagrel 1982; Robin et al. 1983; Cantagrel *et al.* 1984; and Hoskuldsson and Robin 1993, Hoskuldsson 1994). The historic lavas were emitted from the upper crater and flowed down in all directions (Fig. 2). They advanced through the ravines and glacier valleys and sometimes filled them completely. In 1537 the lava flowed on the north and northeastern sides, while in 1545, 1566 and 1613 it flowed on the south side. The activity continued until 1687 with weak explosive eruptions. Since then, there has only been mixed fumarole activity (Hoskuldsson and Robin 1993, Hoskuldsson 1994).

According to Heine (1975a, 1983, 1988), the maximum advance of the glaciers, prior to the advent of Holocene conditions, took place between 10,000 and 3,500 yr BP on Citlaltepetl and, in general, in Mexico. He also identified signs of only two Neoglacial advances. The first one is evidenced by morainic loops around the base of the terminal cone at 4,000-4,400 m, which were deposited prior to 1730 ± 85 yr BP and probably between 3,000 and 2,000 yr BP (Heine 1983, 1988). The second Neoglacial advance is from the L.I.A. (Little Ice Age), which reached its limit of influence in Mexico by the middle of the 19th century and formed a series of moraines between 4,400 and 4,800 m on Citlalteptl (Heine 1975a, 1983, 1988). In both Neoglacial episodes, the glaciers extended mainly on the northern and western slopes (Fig. 2).

It is interesting to point out that the end of the last eruptive episode coincided with the beginning of the L.I.A., so the eruptions would have destroyed the existing glaciers. Subsequently the glaciers of the L.I.A. formed on top of the new lava.

Evidence on all of the large stratovolcanos in Mexico reveals glacial advances during the second half of the Holocene. Authors who have studied this phase call it the „Neoglacial" period. Up to two or three advances took place during the Holocene and left traces on Iztaccihuatl (5,286m); the most recent advance undoubtedly occurred in the LIA (White 1962 a, b and 1981; Heine 1973; and Nixon 1986 are notable discrepancies among these authors). One advance on Téyotl (4,660m) is considered very recent, although it has not been dated (Vázquez 1991). On Ajusco, traces of two advances exist (more recent than 2000 yr BP) (White 1981; White and Valastro 1984). Evidence of various Holocene advances on La Malinche (4,461m), ended with the formation of rock glaciers in the L.I.A. (Heine 1973; Heine 1984). Two generations of Holocene rock glaciers were identified on Nevado de Toluca, the last of which dates from the L.I.A. (Heine 1976a, b).

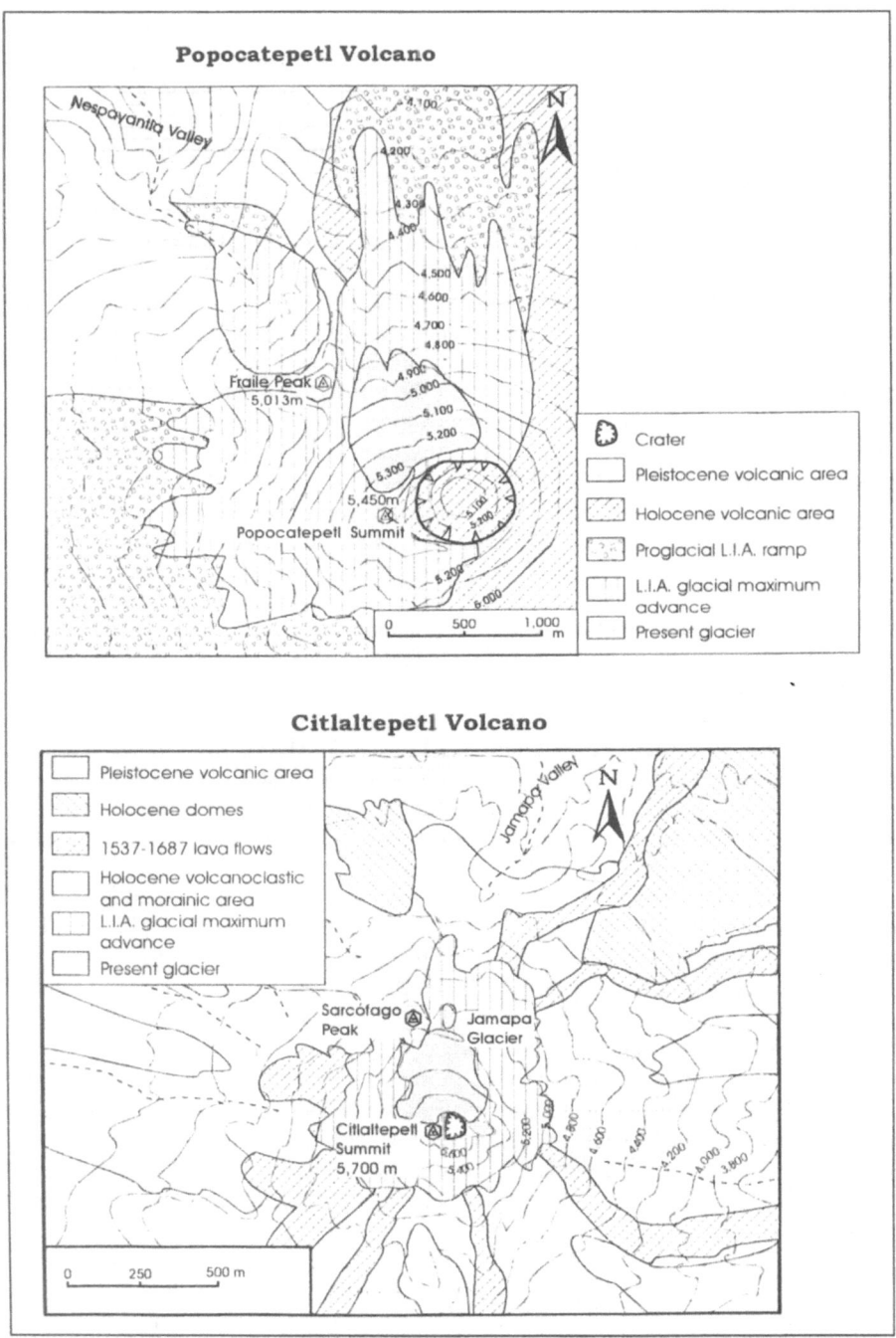

Figure 2. Geomorphological Units of Popocatepetl and Citlaltepetl Volcanoes

3. Deglaciation of Popocateptl and Citlatepetl during the present century

Data available from the beginning of this century is much more informative about the evolution of the glaciers on Popocatepetl and Citlalteptl.

In 1906, the largest tongue of the Popocatepetl glacier descended to an altitude of 4,250m (Anderson's photograph, 1917). Ahead of the glacier, a separate smooth area extends as far as the accumulation of ash. This indicates the occurrence of an earlier advance that reached about 4,150m. By 1910, the tongue had retreated to approximately 4,330m (Melgarejos's photograph, 1910). In 1920, it ended at approximately 4,435m (Weitberg 1923). The eruptions of 1920-27, appear to have contributed further to the retreat of the glacier (Waitz 1921).

Similar characteristics prevailed during the next eruptive phase of the volcano, until 1927 (Camacho 1925; Priester 1927). An aerial photograph from 1945 shows the glacier front at 4,650m (WDC file, Boulder, CO, USA) (Fig.3). In 1950, the snout of the glacier had advanced to 4,573m (White 1954). A 1957 photograph (WDC file) shows the glacier front at 4,670m. In 1958, Lorenzo (Lorenzo 1964) estimated that the minimum altitude of the glacier was 4,690m. Because the exact limit was concealed by snow, he allowed a margin of error of 20m elevation (Lorenzo 1964, page 17). Judging from these photographs, the glacier appeared to have reached a minimum altitude of approximately 4,670m.

From 1960 to 1968, the lower part of the glacier remained at an altitude of approximately 4,700m. During the following decade, it advanced again and descended to an altitude of 4,600m. Also at that time, a great cliff, roughly 100m high, formed the front of the glacier (White 1981 and various collections of aerial photographs). Since then, the glacier has retreated. In 1983, it was at an altitude of 4,630m (aerial photograph from the State of Mexico); in 1987, at 4,680m (various collections of aerial photographs from the State of Mexico); in April of 1992, at 4,694m and in August of 1993, at 4,702m; in February of 1994, at 4,717m (author's observations) (Palacios 1996). Furthermore, the frontal cliff of the glacier was replaced by gentle inclines on all of its borders (author's direct observations). At the beginning of this century, the glacier covered the whole western side of the volcano to an altitude of approximately 4,600m, but between 1950-58, it totally retreated from this side (White 1954; Lorenzo 1964). The amount of ice that had accumulated on northeast face of Pico del Fraile and had been previously recorded by researchers (Weitzberg 1923), had also disappeared.

The glaciers on Citlaltepetl have steadily retreated during this century (Fig. 3). Heine considers that they reached their maximum L.I.A. advance in 1850. The Jamapa Valley moraine associated with the main L.I.A. advance is found at 4,395 m, with recessional positions at 4.460 and 4.485 m.

The extent of the glacier area in 1945 was determined by using a collection of aerial photographs (World Data Center A, Boulder, CO, USA), supplemented with other photos from unknown sources found in different archives and collections. By 1945, the glacier had been reduced to an ice cap whose center lay on the northern side of the new upper cone of Citlaltepetl. A number of tongues extended from the upper cap, the longest of which is Jamapa Glacier, with a length of 6,000 m from the summit. The end of the tongue was divided into three small tongues separated by the levées of the recent andesitic lava flows described in the foregoing section. The western tongue reached an elevation of 4,590 m and was protected by the wall of Pico del Sarcófago. Another tongue located to the east of Jamapa Glacier moved down over the other recent lava flow to an elevation of 4,725 m. It is called Chichimeco Glacier. On the eastern side of the cone Oriental Glacier tongue reached an altitude of 4,900 m and was separated from the ice cap by a narrow corridor. A complex of several glacier tongues on the west side of the cone, was channelled between rocky spurs and even managed to clear steep cliffs. The northwestern tongue descended to nearly 4,600 m. Farther south was a great cascade of séracs that ended at 4,700 m. Two large tongues appeared to the southwest and, despite less favorable aspect conditions, moved down the gentle slope of the recent lavas to 4,900 m.

In 1958 Lorenzo headed a topographic survey of these glaciers (Lorenzo 1964). The changes between 1945 and 1958 are of little significance. Jamapa and Chichimeco tongues had barely receded, although Jamapa was narrower. The glacier which had once covered the youngest lava flows of the north side, only covered its central portion and western edge. Two small tongues at the front were separated by the western levée of the youngest lava flow. The tongue farthest east had advanced to 4,650 m while the western one reached 4,640 m. A space had begun to appear between the western border of the glacier and the wall of Pico del Sarcófago, and a debris talus was forming there. Chichimeco Glacier was also becoming narrower and had retreated to 4,750 m. The front of Oriental Glacier had retreated to 5,070 m. The major differences are more apparent on the western glaciers. The cascades had disappeared and the front of the glaciers were located at the edge of great cliffs. The big blocks of ice that piled up at the base were no longer able to regenerate the glacier. Toro Glacier only reached 4,970 m. The southwestern glaciers had retreated (4,980 m) creating a sharp step in the slope in this sector.

The following information is from 1971 and 1975 aerial photographs (CETENAL, XI 71 and VIII 75) and fieldwork conducted during these years by Heine (1975b) (Fig. 3). The observed changes are not very significant, and they probably illustrate the general tendency of other Mexican glaciers to reach a positive balance at the beginning of the 1970's (White 1981). Jamapa Glacier retreated only slightly, and the tongue that was located in the middle of the lava flow extended down to 4,655 m.

A much more conspicuous space appeared between the glacier and the wall of Pico del Sarcófago, where a great debris talus was forming with glacier ice in its interior (Heine 1975b). The Chichimeco tongue was severely reduced, due in part to the gentle slope of the underlying lava. Its front retreated to 4,930 m. A subglacial moraine appeared that was frozen inside (Heine 1975b). Neither eastern Glacier nor the northwestern glacies had shown significant changes since 1958. The southwestern glaciers, however, had almost completely disappeared.

The next reference is a series of detailed oblique aerial photographs taken in 1987 (Zambrano 1988). Jamapa Glacier had retreated only slightly (4,660 m), while Chichimeco Glacier had disappeared leaving a subglacial moraine. The northwestern and western glaciers had not retreated, but their steep fronts had disappeared. The more than 60 m high cliff formed at the snout had been reduced to a gentle ramp of ice with a front edge that is only 2 to 5 m high. The southwestern glaciers continued to disappear, making it difficult to differentiate between different tongues.

During February 1994, the authors conducted fieldwork at the volcano to determine the extension of Jamapa Glacier with the aid of precision altimeters. Conditions were favorable because there was virtually no snow. According to local climbers, a rocky threshold appeared in 1989 between the ice cap and Jamapa tongue. By 1994 the tongue had been reduced to two patches. The upper edge of the western patch is at 5,013 m (rocky threshold) and splits into two small tongues below the highest visible point of the levée (4,760 m): one tongue is between the levée and the wall of Pico del Sarcófago and reaches a minimum altitude of 4,718 m; the other one, to the east, ends at 4,745 m. The other patch is about 50 meters to the east, between 4,965 m and 4,735 m. Chichimeco Glacier has completely disappeared, so the ice cap on Citlaltepetl has no tongues on its eastern and northern slopes. The ice cap extends to 5,037 m on the north side. Eastern Glacier still exists although it is quite small. Its head is still on the edge of the crater but its minimum altitude is 5,100 m. The northwestern glaciers show less differences in their extension. The thickness of ice is probably, at most, a few meters thick and apparently has lost its capacity to flow (Palacios and Vázquez 1996).

In short, the glaciers of Citlaltepetl have undergone marked transformations in the last decades. At the beginning of the 20th century, they formed a glacial dome with many tongues that extended in all directions, except to the south. Gradually, the tongues disappeared until the dome was transformed into a simple ice cap with no tongues in the early 90's. Some of these changes were no doubt influenced by critical climatic changes, but topography also controls the pace of the retreating glaciers. For example, a sudden retreat may occur when the glacier moves over a rocky escarpment and cannot regenerate anymore at the base of the latter. This happened to the western glaciers and to Eastern Glacier between 1945 and 1958. The opposite might also explain a sudden retreat. When the slope is gentle the glacier does not retreat much over a given time, but its thickness does diminish. When loss is substantial, the glacier retreats very quickly. This was the case of the southwestern glaciers and Chichimeco Glacier between 1971 and 1987. The same process is operating, but in an earlier phase, on the northwestern and western glaciers, which are apparently moving much slower but still have their tongues.

Jamapa Glacier has retreated very little in the last decades in comparison to the rest. Nevertheless, the thickness of Jamapa Glacier has been greatly reduced. The rocky threshold that now separates the tongue from the ice cap has cut off the lower part of the glacier from its source.

4. The retreat of Popocatepetl and Citlatepetl glaciers between February 1994 and October 1995

The head of Tenenepanco Gorge is located between 4,700-4,800m and was exposed during the 1980's and 90's as the glacier retreated (Fig. 4). The spurs of Pico del Fraile form the far western slope where lava and pyroclastics have been uncovered and create a very unstable environment. Rocks fall constantly and form large and very active talus.

The glacier has retreated rapidly during the 90's. In April 1992, it was possible to see tongues that had crept along the slightly inclined lava strata and *nuées ardentes* to advance beyond the rest of the glacial terminus. In February 1994, almost nothing but the very thin, extreme western tongue still existed at an altitude of 4,713m. By October 1995, all of the tongues had completely disappeared, with the exception of the western one that had receded to 4,735m. This process may be associated not only with climatic conditions but with the volcano's increased eruptive activity that partially covered the glacier with volcanic ash on various occasions during the year (Delgado 1996).

The steps that separate the volcanic strata and run the length of the gorge have been exposed. There is also now an abundance of gravity induced rockfall activity on the steps. As the ice disappears, gravity talus forms at the base of the steps. This process is so intense that it has trapped part of the mass of ice under the debris, and beneath each step there is a sedimentary formation consisting of large boulders that covers the remains of the dead ice. The largest formation is located in the extreme western section, and is fed by falling material from the neighbouring rocky spur and the slope of Pico del Fraile (Palacios and Marcos 1996).

Jamapa Glacier has retreated very rapidly during 1995. The reason for this cannot be attributed to eruptive activity (Fig. 4), rather to the glacier's own dynamics and climatic conditions. The tongue of Jamapa Glacier covers a lava flow that dates from 1537 (Hoskuldsson and Robin 1993; Hoskuldsson 1994). This flow surpassed the limits of an earlier volcanic caldera and formed a high ridge in the area, which appeared in the 80's, due to the narrowing of the glacier's width. The ridge was responsible for separating the tongue of Jamapa Glacier from the rest of the cap glacier that extends across the top of the volcanic cone. It was following this separation that Jamapa Glacier began to shrink drastically in volume.

By February 1994, there was still a large remnant of the glacier that was divided in half by a narrow rocky outcrop. The terminus was then located at 4,728m, but by October 1995, the most eastern remnant had all but disappeared. The western section had also lost a great deal of volume, but was still relatively large.

Most of the lava flow and its distinctive levees appeared as the ice retreated. Glacial abrasion has severely eroded the lava surface. No till has been found, which supports the belief that the glacier has a small sediment load.

The slope of Pico del Sarcófago forms the western side of Jamapa Glacier. Boulders break away from the slope constantly, and there are frequent avalanches of snow mixed with debris. The accumulation of this material forms a talus slope that traps the remains of glacial ice (Palacios and Vázquez 1996). This process was active in 1995, and the volume of ice-trapping talus grew as the glacier retreated. Since then, however, the slope of Pico del Sarcófago has been worn down and no longer has the capacity to feed the talus slope. This is occurring near the summit area (Palacios and Marcos 1996).

5. Current climatic characteristics of the deglaciated area

The only high altitude station in Mexico is located on Nevado de Toluca (4,129 m) which is located in the center of the country and has a marked continental climate. Data from the station indicates that there are 220 days of minimum temperatures below 0°C. The average monthly temperature ranges from 0° to -6°C. Average annual precipitation is 850 mm and most of it occurs in summer (79% of the total falls between May and October) in relation to the trade winds. Precipitation in the form of snow is relatively scarce (13% of the total). Snow falls in the summer and also in the winter,

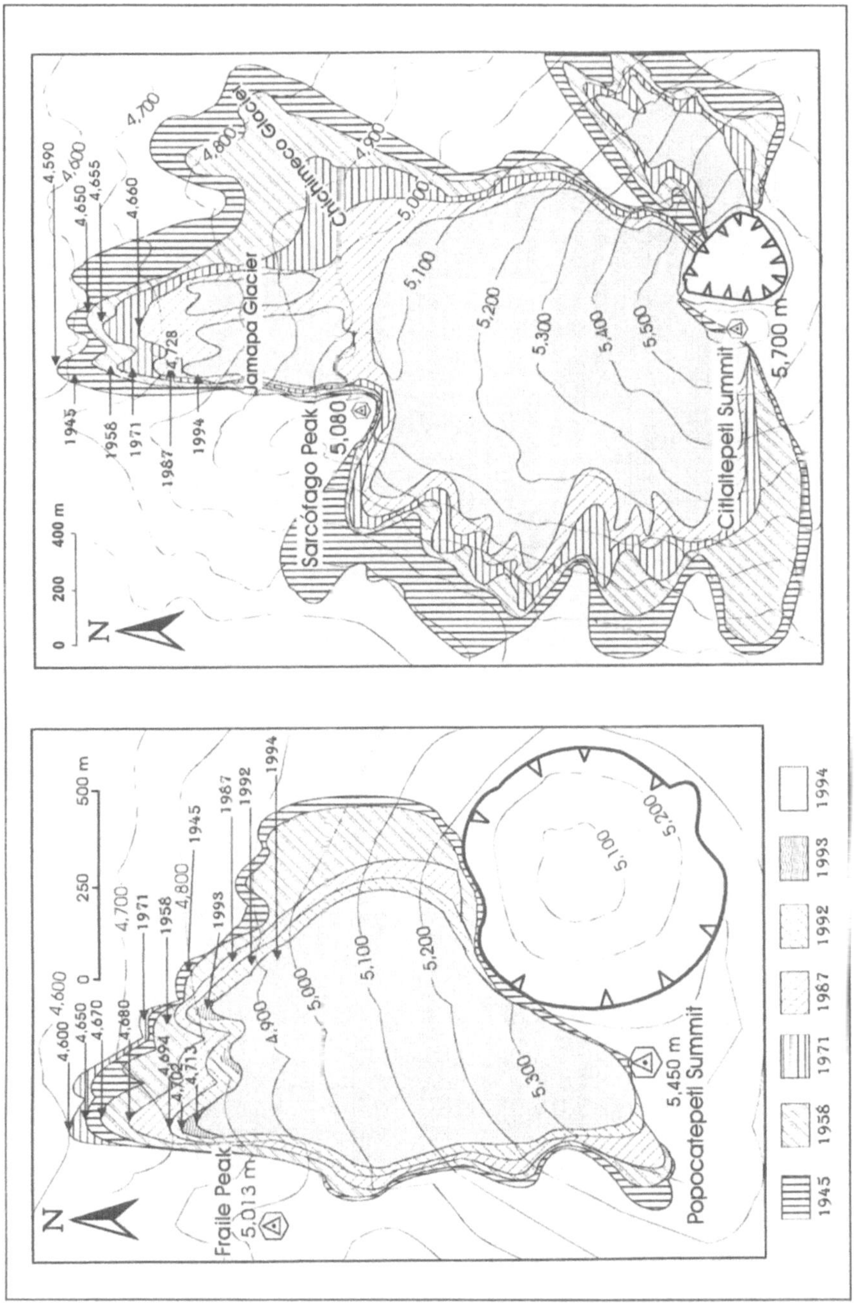

Figure 3. Recent evolution of the glaciers

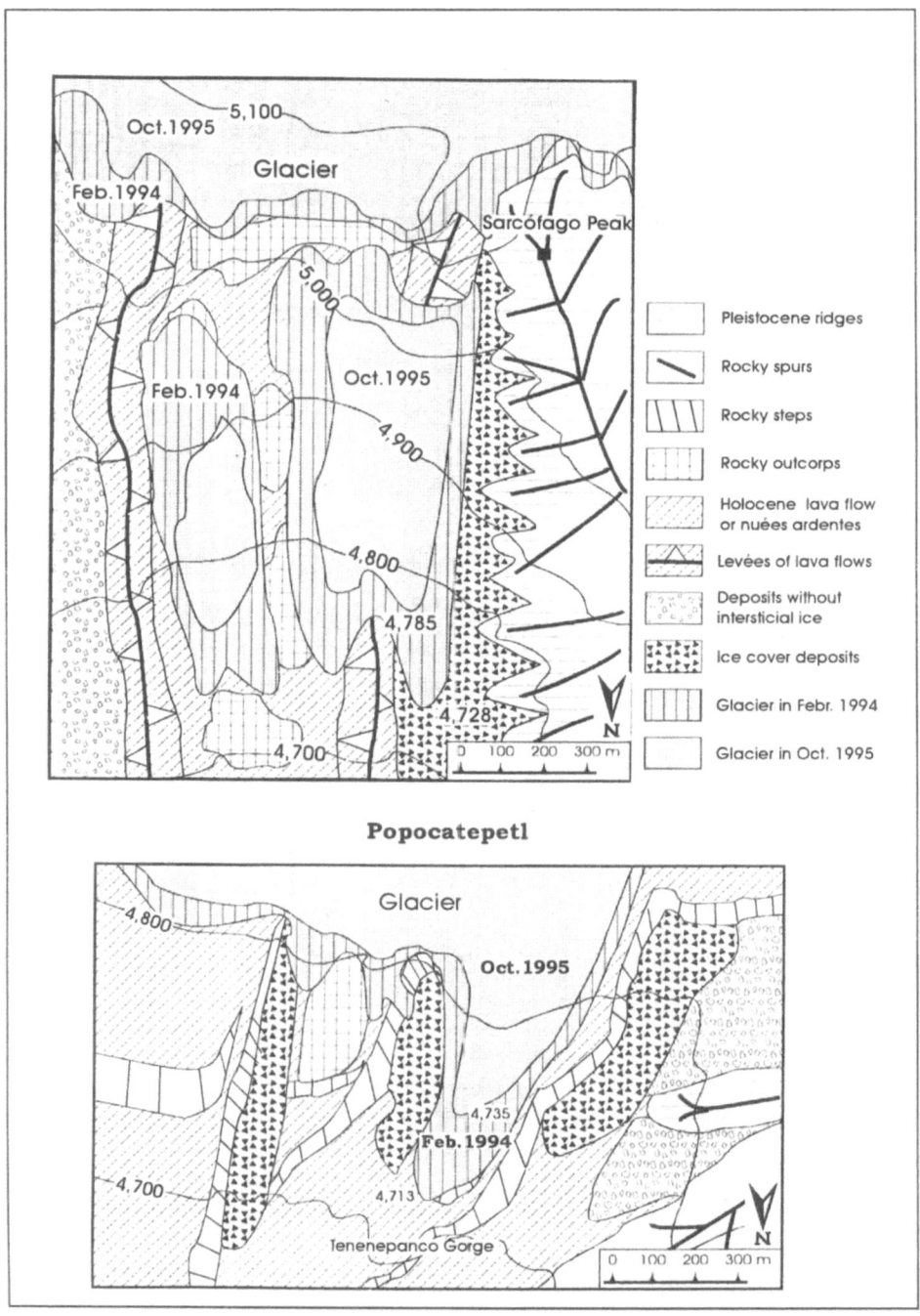

Figure 4. Last year recession of the glaciers

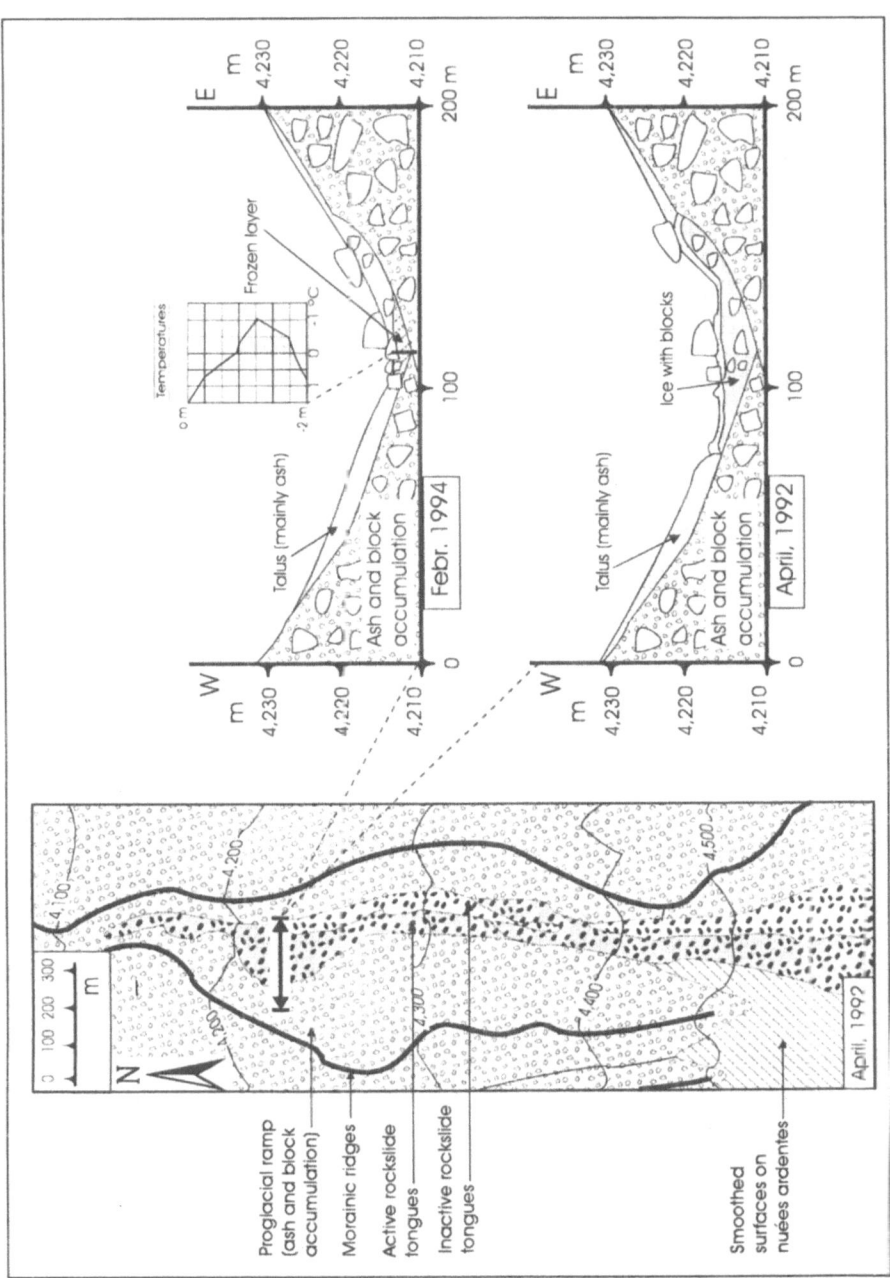

Figure 5. Geomorphologic map and cross-section of northeastern proglacial gorge (Popocatepetl)

when it is associated to the arrival of polar air masses. Hail is the most common type of precipitation (from 21 to 47% of the total) and appears mainly in summer. (Lorenzo 1969)

Data recorded from 1961 to 1968, from the weather station near Paso de Cortés (Cerro Altzomoni, 4,034 m), located between Popocatepetl and Iztaccíhuatl volcanoes, supplemented the readings from Nevado de Toluca (Lorenzo 1969). The data show that the concentration of rainfall in the summer months is the same, but rain rather than hail (17%) is the predominant type of precipitation. This is probably due to the lower elevation. Snow represents only 7% of the precipitation, although this varies greatly depending on winter snowfall. The temperatures also reflect the effects of the 100 meter difference in elevation. There are 180 days with min. temperatures below 0°C.

Studies by Lauer and Klaus (1975) and Lauer (1978) provide significant data on the climatology of Pico de Orizaba. Based on long period records from different parts of central Mexico and on direct monitoring in the area, they estimated annual average values of several climatic parameters at 4,000 m on the western slope of Pico de Oriza-ba. The mean annual temperature was 5°C. The average number of days with minimum temperature below freezing point was 200. The average number of days with maximum temperature below freezing point was 45. The annual precipitation was 900 mm. Lauer and Klaus (1975) discovered that in the area of Jamapa Glacier the changes in temperature affect the rock to a depth of 70 cm.

6. Hazards related to avalanches and rockslide post-deglaciation processes

Today, some of the most geomorphological active and hazardous areas on Popoca-tepetl are the proglacial gorges, where snow and ice has broken off from the glacier and causes major avalanches (Fig.5). Furthermore, the floor of these gorges are cove-red by rockslide deposits that form active tongues, which are sometimes very mobile. The characteristics of the rockslide tongues explains the nature of the processes that occur on the floor of the gorges (see Palacios 1996).

The deposits appear in the proglacial gorges exactly where these and the terminal glacial hollows conjunct. It is very significant that the deposits are found in every gor-ge. The rockslide tongues were probably formed very recently. They had to have originated after the retreat of the glacier and, more importantly, subsequent to the activity of the proglacial waters in the gorges. These factors and the height of the rockslide deposits seem to indicate that the tongues were formed before the end of the 1950's. They are not visible in 1945 aerial photographs (WDC). Although some ton-gues appear in photographs from 1958 (WDC), it is logical to assume that they did not develop until the end of the 1960's.

It is evident that the flows advance rapidly, and their speed can be determined by comparing the minimum altitudes shown in the aerial photographs from December 1983 (DGC DIC-83, ESC 1:37.000, D.F. 151 96 L-337 E 14-B-42) and May 1989 (S.E.C.T.E. -State of Mexico-Volcanoes-scale 1:19,000) and the measurements taken during field work conducted in 1992, 1993 and 1994.

Readings were taken of the maximum distance and minimum altitude of the flow of the deposits located in the Tenenepanco Gorge. This gorge originates to the east of Pico del Fraile and runs northeast. In 1983, the flow reached an altitude of 4,034 m and in 1992/93, 4,009 m, with a total increase in distance of 203 m from 1983 to 1992, or 22,5 m per annum.

The orientation of the flows, however, is not the only factor that plays a role in the deposits' formation and development. There is a more important determinant that is associated with the geomorphologic context. The tongues of the deposits always appear at the foot of the volcanic cone's steepest slope and inside the depressions that form at the glacier's terminus. They increase in size as great masses of rock fall from the glacier's front, or more importantly, from the spurs consisting of volcanic layers.

Also, the tongues advance in relation to a particular geomorphologic context . They always follow the incisions made by the gorges in the ash. The tongues cover the floor and do not rise along the sides, thus their width is closely related to the previous width of the valley floor. As the valley gradually becomes narrower, even to the point of disappearing, so do the tongues.

Many of the factors that determine the generation of rock glaciers also play a role in the formation of rockslide deposits. They occupy the valley floors and are composed of materials of many sizes, although boulders predominate. The tongues appear as the result of the interaction of altitude, radiation, shade and source area, and where flow structures associated with trapped snow and ice are present (Capps 1910; Benedict 1981, Morris 1981; Martin and Whalley 1987). There is a fundamental difference, however, between rockslide tongues and rock glaciers: the former are located below the lower permafrost limit (Haeberli 1983) and therefore, most of deposit's ice and snow melt, and not just seasonally, except when weather conditions are right, such as during 1993/94, when only a thin ice layer remained in the deposits. In this case, geomorphologic structures linked to the melting process, such as spoon shaped depressions, collapse pits, central meandering furrows, etc., are abundant (White 1975). Although there is some geomorphologic evidence of permanent ice, such as transverse crevasses, surface ridges, and shear planes, this is uncommon.

Rock glaciers can originate due to rockslide activity that sometimes the form of catastrophic landslides. When these occur, the creation of the deposits is generally attributed to a single event, although some small movement caused by creep and associated with interstitial ice or snow might occur later. This type of rock glacier formation depends on the relationship between snow-ice input/output and debris supply (Whalley and Martin 1992).

Some authors have characterized this type of accumulation as a feature of lower talus formation (Barsch 1977, 1988 and 1992; Corte 1987), and have pointed out that more snow accumulates as a result of avalanches than of the rock glaciers located below ice glaciers, where ice comes directly from glacial remains. The formations on Popocatepetl can be classified between these types. The fact that Popocatepetl's formations are situated below the glacier is of key importance to the formation of rockslide deposits, although, apart from the avalanche ice blocks, they do not receive any ice from the glacier. Snowfall and snow avalanches are the main sources.

Snow from avalanches is very important in the formation of rock glaciers (Lliboutry 1961 and 1988; Barsch 1977; Johnson 1978, 1984; Haeberli 1985; Corte 1987; Whalley and Martin 1992). There was a period when major snow accumulation after deglatation and rock supply caused by avalanches (Johnson 1987) were responsible for great slope instability on Popocatepetl's north face. Also, the evolution of

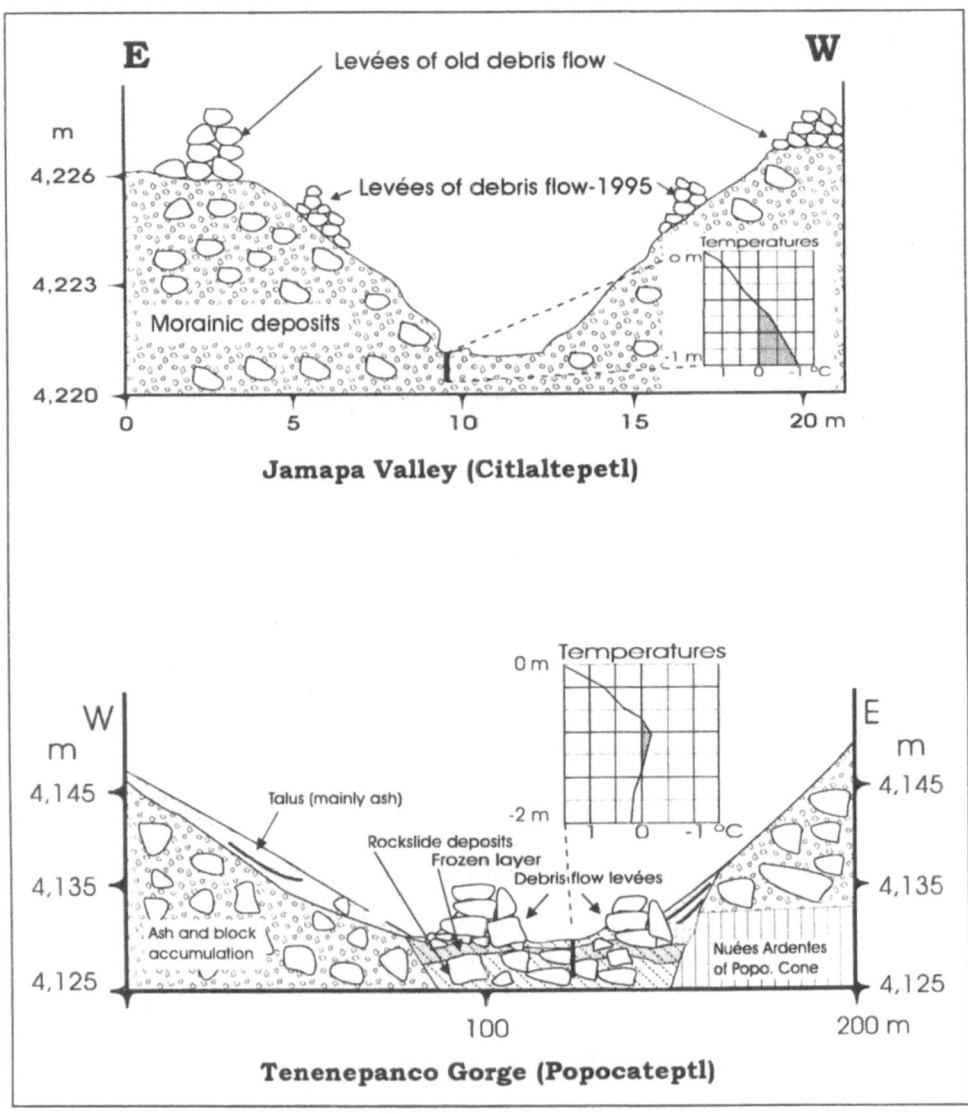

Figure 6. Cross-section of debris flow channels on proglacial ramps

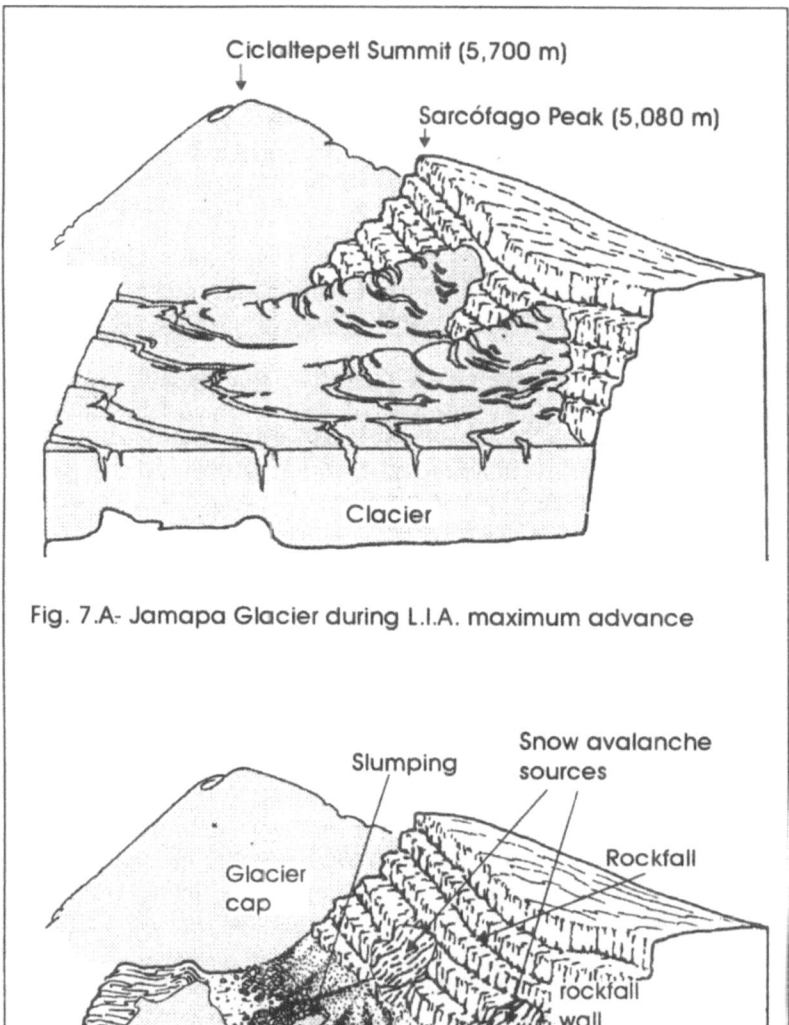

Fig. 7.A- Jamapa Glacier during L.I.A. maximum advance

Fig. 7.B- Present Jamapa Valley, and processes after deglaciation

Figure 7. The origin of rockfall and slumping post-deglaciation processes

certain rock glaciers was related to the quantity of snow generated during a given climatic period, before intensely cold weather set in. These circumstances also provided a suitable environment for the formation of avalanches and the subsequent transformation of the trapped snow into clear ice lenses (André 1992).

The literature on rock glaciers illustrates how important snow patch covering is to the increase in volume of the ice and snow found within the sediments. Trapped snow patches are common in highly contrasting mountain environments and create protalus ramparts or small rock glaciers (Lliboutry 1961 and 1988; Johnson 1973 and 1978; Haeberli 1985; Whalley and Martin 1992). Snow patches form easily thanks to the combined effect of the irregular features in the relief caused by morainic accumulations and their lack of compactness. After the patches have formed and are buried by deposits, ice lenses develop. (Vick 1987).

The formation of rock glaciers on the high volcanoes of Mexico is entirely feasible. In fact, this has already been described in detail for Téyotl and Iztaccihuatl (Vázquez 1991) and Nevado de Toluca (Heine 1976b). The rock glaciers on these volcanoes, however, are actually fossil formations generated during at least two different Holocene periods, prior to the appearance of what is now Popocatepetl's cone. Although the extent of active discontinuous permafrost now occurs at about 4,600 m, in the L.I.A., it appeared 500-600 m down slope (Heine 1994).

The present lack of both perennial interstitial ice and cohesive flow (Haeberli 1985), makes it difficult to classify Popocatepetl's rockslide deposits in the category of rock glaciers. Nevertheless, research conducted on the dynamics of rock glaciers has been helpful in classifying the deposits. It emphasizes characteristics such as the snow's ability to accumulate and be buried by deposits, especially moraine and rock fall talus and to remain in banks. This action, which is very important on Popocatepetl, assists the landslide subsystem (Giardino and Vitek 1988), thus creating talus input, and ice and snow input as part of general slope dynamics.

The Popocatepetl rockslide deposits are also difficult to classify in terms of other slope processes, since they are formed by snow avalanches and are tongue shaped. These characteristics suggest a type of avalanche boulder tongue (Rapp 1959). On Popocatepetl, the geomorphologic context indicates that the avalanche effect is increasing. The glacial terminal hollows where the tongues are formed, have a cone-shaped head and a very narrow channel termination. Because of their shape, the hollows can channel the avalanches that flow from the glacier front that now hangs above an extremely steep slope. The hollows function as avalanche paths, so as they narrow, the avalanches flow with a great deal of energy and are able to move relatively long distances. The hollows behave as conduits and carry the evidence of an upper periglacial environment down to lower altitudes (Butler, Malanson and Walsh 1991).

The materials that are transported by the avalanches on Popocatepetl are from the snout of the glacier itself and the active rockfall talus slopes that cover the sides of the glacial depressions. The erosion at the base of the talus caused by the avalanches reactivates the slope processes and provides materials for future avalanches, thus creating a feedback effect known as „failure of progressive unconsolidated debris" (Johnson 1987).

Other processes provide snow for the formation of Popocateptl's rockslides, such as the accumulation of snow on the floor of the glacial terminal hollows and proglacial gorges during two seasonal snow periods: the middle of the winter and high summer. In fact, the snow from the whole stratovolcano collects most effectively on the floor of the hollows and gorges, where it lasts for the longest periods of time (source: study of the entire collection of aerial photographs and author's observations from 1992-95). The hollows are bordered by longitudinal ridge moraines that jut out over the relief. The ridges descend toward lower levels where glacial terminal hollows are replaced by proglacial gorges. At this point, the ridges meet the remains of the fluvioglacial fans that were incised by the gorges on the central axis.

Alternating ridges and hollows block the northeasterly winds that carry snow. The rest of Popocatepetl's slopes are extremely regular, as is typical of active stratovolcanos. According to the plume model (Föhn and Meister 1983) the regularity of the slopes and the profile of ridges with steep lee slopes, ensures optimum conditions for snow accumulation in the depressions. The flow of deposits that formed over the snow-covered floor of the hollows and gorges after the materials from the avalanches settled is much greater. Spherical clast rolling (Pérez 1988) was observed in this sector and most always occurs when there are mass sliding movements over a lubricating layer of snow (Johnson 1987).

It is a well-known fact that the sediments from avalanches protect underlying snow and even promote the formation of ice lenses below the line of permafrost (Wayne 1981). Equally true is the fact that ash has an even greater capacity to protect snow from melting. The ash comes from glacial and fluvioglacial accumulations that have undergone little widespread reworking. In general, the ash moves freely on scree slopes, and gravity causes frequent down slope rolling and sliding of cohesionless grains (Cas and Wright 1987). The upper levels of the slopes where the rockslides deposits are formed, are free of snow during most of the year, while the lowest levels remain covered. As a result, these slopes are much more unstable. Wind constantly blows ash from the peaks, and both the melting of snow and the lateral erosion of avalanches create a tendency for instability at the base of the slopes.

The ash very effectively protects the snow that accumulates on the floor of the gorge and over the fan avalanches. A thin layer of ash can change a new snow surface albedo from 0.75 to 0.10 and increase melting by 220% on a clear day (Meier 1969). During periods of cloudy weather, when convection and long wave radiation are prominent, the effect of ash on decreased surface reflectivity will be negligible (Tangborn and Lettenmaier 1981). More importantly, however, is the fact that when ash exceeds a thickness of 24 mm it prevents snow from melting (Brugman and Meier 1981; Driedger 1981). In fact, on Mount St. Helens, the thicker mantle of ash effectively insulated the snow, causing a much lower melt rate than in the surrounding area. When the thickness of the ash exceeded 12-15 cm, melt was reduced by 50% (Crook, Davis and Moreland 1981). The same effect was observed on numerous volcanoes in New Zealand (Skinner 1964), Alaska (Trabant and Meyer 1992), the Cascades (Thouret 1989), the Andes (Clapperton 1990) and Kamchatka (Vinogradov 1975, 1981).

Popocatepetl's rockslide deposits are the product of various factors that appear simultaneously: ash accumulation in proglacial areas; the existence of terminal glacial hollows extended by inactive proglacial gorges; the rapid retreat of the glacier whose snout hangs over a very steep slope immediately above the final hollows; the formation of large scale snow avalanches; the massive accumulation of snow on the floor of the hollows and gorges; and finally, the constant precipitation of ash from the scree slopes onto the valley floors. All of these processes are related to the deglaciation of this stratovolcano.

Similar processes and landforms have been observed on many other volcanoes in the Cascades (Thouret 1989) and Kamchatka (Vinogradov 1981). However, the only detailed study that exists is for Redoubt Volcano in Alaska (Trabant and Meyer 1992).

7. Hazards related to debris flow post-deglaciation processes

The activity on Popocatepetl increased at the beginning of 1995. Volcanic emissions occurred early in April 1995, and covered more than 50% of the glacier surface with a fine layer of ash. Since the albedo effect was reduced (Meier 1969, and Driedger 1981), the melting of the glacier's surface was greatly enhanced. The affluence of water in the proglacial gorges that were filled with the deposits described earlier in this article, caused widespread debris flows at the end of April (see Palacios, Zamorano and Parrilla 1996) (Fig.6).

The avalanche deposits facilitated the later formation of debris flows (Nyberg 1985, Osterkamp and Costa 1986, and Luckman 1992). As stated earlier, the layers of subsurface ice form between layers of ash and last for long periods of time. The increase in volcanic activity, which is responsible for partially covering the glacier with a light layer of ash, suddenly accelerated the surface melting process (see similiar cases in Vinogradov 1981, Brugman and Meier 1981, Pierson 1985, Linder and Jordan 1989, Pierson et al. 1990, Trabant and Meyer 1992, and Walder and Driedger 1994). Most of the meltwater from the glacier was not adequately channeled through the avalanche material that covered the floor of the gorges. Approximately 1.5 m of the surface layer of these deposits became saturated, because the underlying level of ice prevented the water from percolating to a greater depth (Larsson 1982, Zimmermann and Haeberli 1989, Haeberli et al. 1990, Nyberg and Lindh 1990, and Harris and Gustafson 1993). Debris flows occurred simultaneously in all of the proglacial gorges described. They appear at the head of the avalanche deposits and spread as far as the elevation at which the underlying layer of ice disappears (approx. 4,100 m). The water no longer saturates the deposits at this low elevation, because it is able to permeate the volcanic material.

The larger debris flow in Tenenepenco Gorge was channeled to the floor of the gorge, forming a single flow (Mizuyama et al. 1993) which reached lower elevations (4,020 m). It was set in motion the same way the debris flows of the other gorges were generated, and moved downward to 4,150 m, where it spread into a massive and chaotic flow. The underlying ice was no longer present and was replaced by bedrock (andesitic lavas), which appeared on the bottom of small fluvial incisions, which channeled the debris and allowed its water load to increase. As a result, the flow became more fluid and some of the larger components settled out, while the mass itself moved another 100 m, although there was very little slope gradient. What began as a debris flow was transformed into a kind of alluvial deposit due to the amount of water it carried (Whipple and Dunne 1992).

Debris flows also occur today on Citlaltepetl on the proglacial ramp that dates from the L.I.A., but their geomorphologic context is somewhat different (see Palacios, Parrilla and Zamorano 1996) (Fig.6). The ramp over which Jamapa Glacier formed is flanked by Holocene lavas to the east, that filled the head of Jamapa Valley, and rocky spurs to the west, that are the prolongation of Pico del Fraile. The ramp narrows at its lower end until it joins the valley floor where it adopts the classic morphology of a glacier, which the Jamapa River has incised. The ramp is formed by large morainic blocks mixed with volcanic ash. Its surface shows the influence of generations of debris flows, many of which have been destroyed by younger flows and other slope processes. The existence of these flows was cited in (Heine 1975b).

Observations from field work carried out in February 1994, revealed three well preserved flows that were more than a kilometer long and 15 m wide. The heads of the flows were located between elevations of 4,350 and 4,370 m. They are small and form depressions that are about 3 m deep, up to 20 m wide and from 50-100 m long. The channel appears abruptly between high levees that parallel the channel's long course. The lobes that extend beyond the main flow are small . The morphology of all of these debris flows is similiar, since each has a very long channel and great lateral deposits or levees.

A new debris flow formed between February 1994 and October 1995, on top of an older one that is located farthest to the east. The recent flow begins midway down the older one and has virtually no head. Levees appear 20 meters from where the flow formed and were deposited on either side of the older flow. This channel and its deposits continue for another 600 m, and then they form a well preserved lobe that is about 15 m wide and 30 m long.

The ground temperature of the floor of the channel drops sharply until it reaches the permafrost layer at 30 cm, which consists of layers of frozen fine particle materal, mainly ash, that is mixed together with ice. The permafrost layer registers tempeatures of about -1°C. In contrast, the temperature of the floor of the channel of the old debris flow drops to 1 or 2°C, but reaches equilibrium at 8-10°C, depending on the altitude. It is interesting that the permafrost appeared at such a low altitude in relation to the permanent snow level (Heine 1975b, 1994, and Palacios and Vázquez 1996). The explanation of this phenomena was discussed in the section on Popocatepetl.

The most abundant precipitation on Pico de Orizaba occurs between 4,000 and 4,500 m (Lauer and Klaus 1975, and Lauer 1978), much of it in the form of snow. Snow accumulates easily in grooves or on the lee side of obstacles that break the wind. As is the case of Popocatepetl, the snow that gathers in the grooves formed on morainic material with a high ash content, is quickly covered by loose ash that acts as an efficient thermal insulator. The buried show melts very slowly. The amount of ash that falls from the walls increases with time until a frozen layer forms at a depth of about 70 cm, and is protected from surface temperature variations (Lauer 1978).

The author believes that the frozen layer that appears on the floor of the channel of the debris flow is from snow that had accumulated on the floor of the old debris flow channels. The ash fell from the walls of the channels as it had done on Popocatepetl and created a frozen layer at a depth of 70-100 cm., creating what was probably the impermeable layer over which the new debris flow formed.

The new debris flow obviously has no lichens, but the old one has thalluses of *Rhizocarpon Geographicum* that are up to 2-3 cm in diameter. These were probably inherited, since they were not destroyed by the flow (Rapp and Nyberg 1981, Innes

1985 and André, 1986). Most of the thalluses are about 6 mm in diameter and rarely reach 1 cm.

All of the data on the new debris flow is similiar to that of the old one. This seems to point to a formation phase in the development of great debris flows, in which the lack of data pertaining to the sector above the growth curve for lichens interferes with any attempt at absolute dating. There is an average difference of 1-1.5 cm among the thalluses associated with the blocks on the ramps that were unaffected by later movements, the levees of other debris flows that were sectioned and partially destroyed and the levees of old debris flows that were analyzed.

This analysis leads to the logical deduction that the debris flows formed at about the same time as the proglacial ramp. As the glacier receded, meltwater accumulated in the depression between the glacier terminus and the frontal morainic deposits. The water filtered through these materials, saturating them and generating frequent debris flows on the proglacial ramp. This processes explains 70% of the debris flows in the Alps (Zimmermann and Haeberli 1989) and much of the same development in the Himalayas (Owen and Derbyshire 1989; Owen 1991 and Owen et al. 1995), and is also associated with the deglaciation of large stratovolcanoes (Crandell 1971). It is therefore reasonable to assume this phenomena is the main cause of debris flows in high mountain areas (Ballantyne et al. 1992). The origin of the flow is always linked to the altitude at which the floor of the inner depression is located (about 4,350-4,370 m). This band marks the area of maximum saturation.

The succession of debris flows was probably constant from the time the ramp formed to the time the glacier retreated to altitudes above these levels. The meltwater filtered through the ground until it reached the finiglacial depression, in much the same way it was observed in the 1950's (observations from aerial photographs). The debris flows moved with great force, transporting oversized material and leaving large channels.

These flows were followed by smaller ones that had a more reduced load capacity. Present climatic conditions that include alternating periods of snow and rain stimulate present debris flow formation, as do the existence of old debris flows and the availability of ash to insulate the snow and later form permafrost layers. Also, the old debris flows provided an ideal base for new flows (Johnson and Rodine 1984 and Owen 1991). A snow patch appeared on the inside of the head of the old debris flow (similar cases were cited by Nyberg and Lindh 1990 and Gottesfeld et al. 1991). Although rain rapidly melted the snow (see, for example, Gardner 1979; Rapp and Strömquist 1976; Wei et al. 1992, and DeGraff 1994), there was still enough to be trapped by the ash that fell from the walls and to form a layer of permafrost on the floor of the channel. The impermeable permafrost layer caused the upper deposits to become saturated, exactly as they had on Popocatepetl.

8. Hazards related to rockfall and post-deglaciation slumping processes

The glaciers that cover the volcanic cones of Popocatepetl and Pico de Orizaba have a reduced load capacity, and the terminal moraines that were deposited during the Little Ice Age, are small. The material they contain is almost exclusively the result of the glacier's bulldozing of the surface materials (Palacios 1995; Palacios and Vázquez

1996). The absence of terminal moraines and subglacial till caused by the recent, quick retreat of the ice seems to confirm this hypothesis. Also true is the fact that the slopes that confine the glaciers contribute no debris, because they are very smooth and composed of highly permeable volcanic material that permits little fluvioglacial erosion activity. There is evidence on both volcanoes that the glaciers have little capacity to dislodge material from the rock walls that run the length of the ice flow. Nonetheless, severe erosion takes place once these walls are exposed by receding ice, thus, it is thought that decompression, gravity, gelifraction and, occasionally, avalanche formation trigger the erosive activity.

The cause and effect relationship between the receding ice and the start of erosion processes is so evident that, normally, the ice that survives at the base of the wall is trapped beneath the deposits. Since the debris serves as an excellent insulator, the ice is kept at low temperatures and lasts (Fig.7).

Meteorological data suggest that there are excellent structural and thermal conditions for the development of frost shattering on the rockwalls (Pico del Fraile and Pico del Sarcófago) in present deglaciated area. Precipitation at this altitude (between 5,000 and 4,600 m) is in form of hail, with the exception of occasional heavy snowfalls. Thanks to the step-like shape of the walls, snow patches form on the ledges. When they melt, they provide the base of the wall with a year round supply of water. The severely fractured structure of the lava is highly permeable, but water remains stagnant in contact with less permeable strata of pyroclastic deposits. The snow patches are also a source of frequent snow avalanches that occur after heavy snowfalls and often carry large amounts of debris. These conditions have fostered the rapid development of talus slopes that cover the entire base of the walls and grow as the glacier retreats. They form so fast that when the glacial ice is no longer able to flow, and they trap and incorporate the ice within the deposit.

The geomorphic processes that develop as the Jamapa Glacier retreats are directly related to the volcanic geomorphology underlying the glacier (see more information in Palacios and Vázquez 1996). The most active processes are found at the base of the eastern wall of Pico del Sarcófago, where the structural conditions are prone to mass movement processes and frost shattering: very steep slopes; alternating layers of permeable lava, loose pyroclastic material and impermeable pyroclastic material; formation of structural steps where snow accumulates year long and provides a constant source of water to the wall; and frequent snow avalanches that originate from the snow that accumulates on the steps.

The climatic conditions make this wall an ideal place for frost shattering processes. Temperatures drop below freezing at least 200 days/year. The eastern exposure and the rocky wall make these maximums and minimums even more extreme (Lauer 1978).

The retreat of the glacier creates a sequence in the formation of rock talus. While the glacier still has the capacity to flow and move along the base of the wall, it drags material with it that has fallen from the wall, which is the major source of the rock load. When the glacier can no longer flow or when its front retreats, the ice remains below the fall deposits and first forms a prolonged surface at the bottom of the wall. Later, when the glacier ceases to flow and the accumulation of material from the wall increases, glacier ice is preserved under thick debris cones and melts on the rest of the footslope.

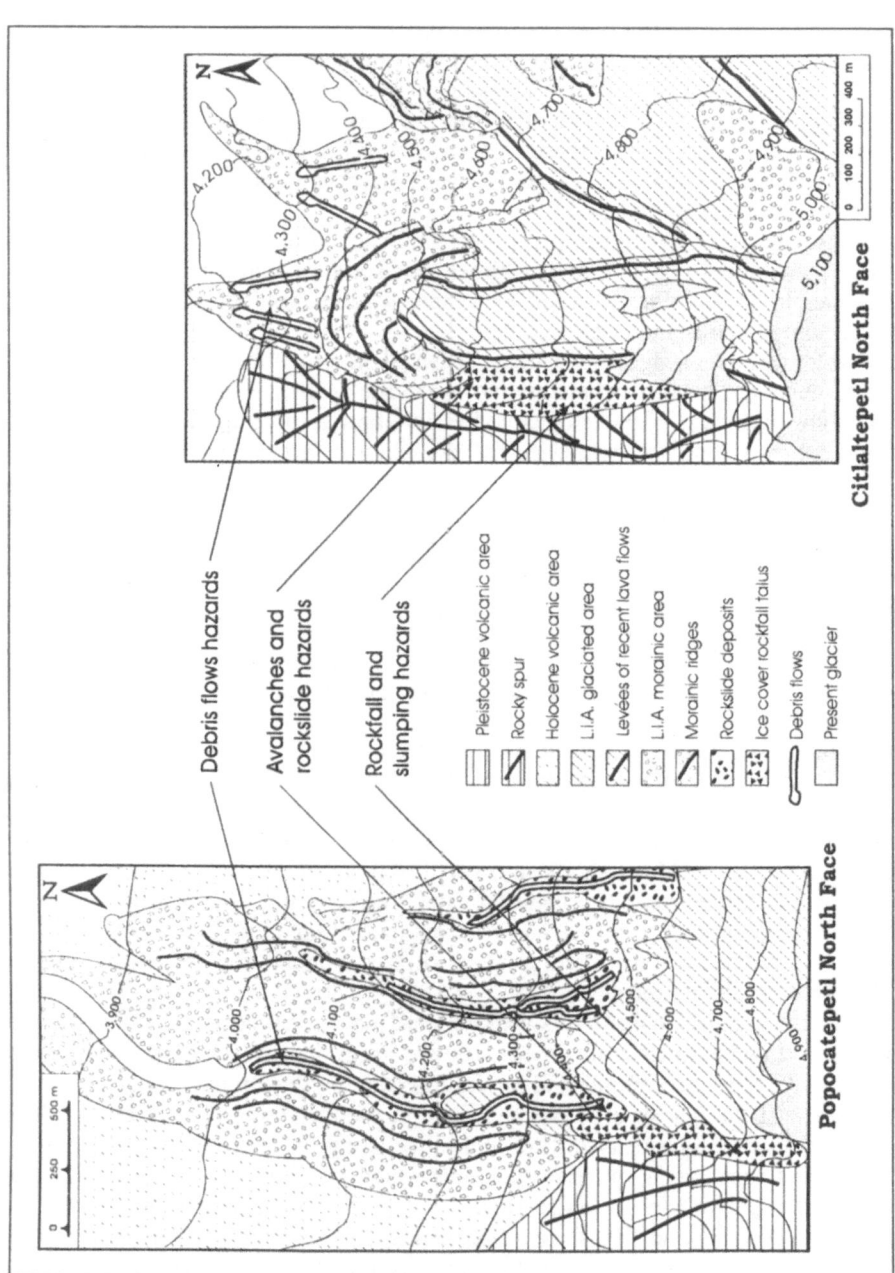

Figure 8. Geomorphologic units and geomorphologic hazard areas in relation to deglaciation

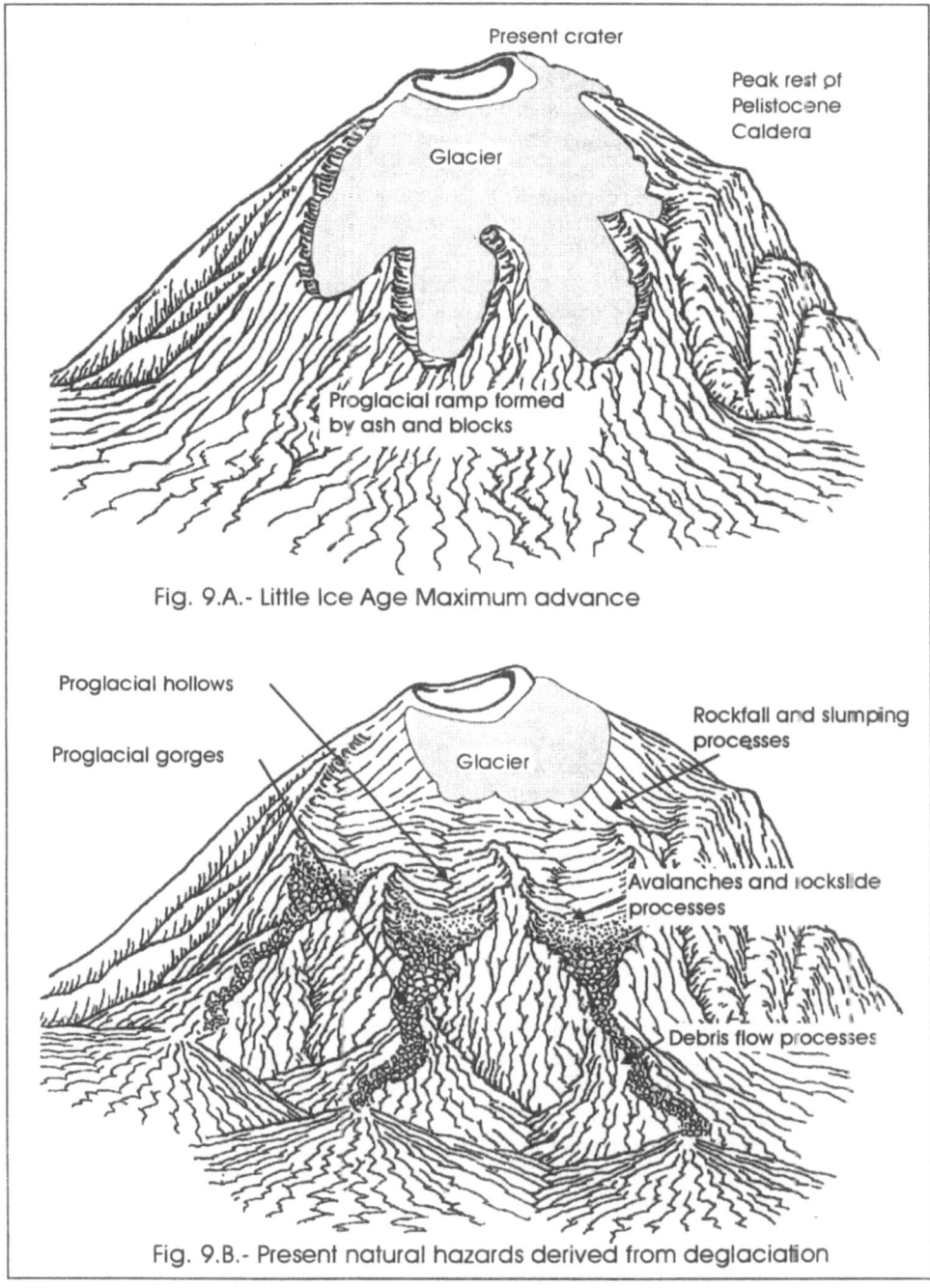

Fig. 9.A.- Little Ice Age Maximum advance

Fig. 9.B.- Present natural hazards derived from deglaciation

Figure 9. Distribution of natural hazards after deglaciation

The presence of buried ice causes frequent ice slumps and debris slides that are also aided by the steepness of the cones. Because of the rock load on the ice, massive flows of ice and rock are produced, forming small rock glaciers at the foot of the cones.

The head of Tenenepanco Gorge at Popocatepetl is located between 4,700-4,800m and was exposed during the 1980's and 90's as the glacier retreated. The spurs of Pico del Fraile form the far western slope where lava and pyroclastics have been uncovered and create a very unstable environment. Rocks fall constantly and form large and very active talus.

The rest of the head of the gorge is formed by lava surfaces and, in particular, very compact *nuées ardentes* that the glacier has severely abrated. Large steps that are 15 to 50m high have formed where the volcanic strata come in contact with each other. In contrast to the upper surfaces, the walls of the steps have been exposed by the ice and are very unstable. Large blocks fall from the walls and either pile up at the base or scatter over the floor of the gorge.

The glacier has retreated rapidly during the 90's. The steps that separate the volcanic strata and run the length of the gorge have been exposed. There is also now an abundance of gravity induced rockfall activity on the steps. As the ice disappears, gravity talus forms at the base of the steps. This process is so intense that it has trapped part of the mass of ice under the debris, and beneath each step there is now a sedimentary formation consisting of large boulders that covers the remains of the dead ice. The largest formation is located in the extreme western section, and is fed by falling material from the neighboring rocky spur and the slope of Pico del Fraile.

9. Conclusions

This work supports Church and Ryder's (1972) theory of deglaciation, according to which paraglacial processes are especially intense after deglaciation. They rapidly transform the landscape, which later stabilizes or is altered by slow erosion

The advance of the L.I.A. generated, in some of younger glaciers, large morainic deposits consisting of boulders, but mostly of volcanic ash. This composition and the presence of volcanic morphology, are responsible for the deposits that take the form of long ramps (Fig. 8 and 9.A).

Since the glacier began its recession at the beginning of this century, meltwater has tended to collect in proglacial depressions. Whenever there is an abundance of water for a long period of time, deep proglacial ravines incise the ramp, as is the situation on Popocatepetl. On Citaltepetl, however, the supply of water was intermittent and the morainic ramp was very steep, so large debris flows formed. In fact, the terminal moraines were often reworked by debris flows (Fig. 8 and 9.B).

The glacier continued to retreat to a point where it hung over the side of the very steep upper volcanic cone. This situation gave rise to frequent avalanches that are channeled by the proglacial depressions and by the proglacial ravines. As the glacier receded, the meltwater filtered through the permeable volcanic materials and never reached the proglacial ravines, so there was no stream water. Instead, the proglacial depressions and the proglacial ravines filled with avalanche deposits, and this gave way to rockslide processes as they were protected by ash that fell from the walls (Fig. 8 and 9.B).

Snow collected in the proglacial ravines and in the channels of the old debris flows and was proteced from melting by the ash that fell from the walls. As time passed, the snow layer was transformed into a layer of frozen ash that lasts for many years. Sudden melt outs associated with volcanic activity or heavy rain on snow patches, made the frozen layer impermeable and furthered the formation of debris flows (Fig. 8 and 9. B).

Deglaciated rockslopes rapidly tend towards equilibrium in response to changes in stress conditions associated with the failure of the rockmass. This is especially important on the walls of old volcanic structures composed of very unstable accumulations of lava and pyroclasts, such as those on Pico del Fraile and Pico del Sarcófago. This is why rockfall processes are particularly intense just after deglaciation. The taluses that are formed trap the glacier ice within which promotes frequent slumping (Fig. 8 and 9.B).

References

1. Alzate, J.A. 1831. Observaciones físicas en la Sierra Nevada. Gaceta literaria, I, 99-107.

2. Anderson,T. 1917. Volcanic studies in many lands. John Murray. London, 213 p.

3. André, M.F. 1986 .Dating slope deposits and estimating rates of rock wall retreat. In André, M-F. 1988 Vitesses d'accumulation des debris rocheux au pied des parois supraglaciaires du nord-ouest du Spitsberg. Annals of Geomorphology Bd. 32(3), 351-373.

4. André, M.F. 1992. Les glaciers rocheux du Spitsberg central et nord-occidental. Revue de géomorphologie dynamique, XII (2), 47-63.

5. Anonymous. 1836. El ascenso al Volcán de Popocatepetl en América, Pfenning Magazin, 196, 421.

6. Ballantyne, C.K., Bennett, M, R.,. Landgridge, A.J.. 1992. Rock slope failure and debris flow, Gleann na Guiserein, Knoydart [discussion and reply].. Scottish Journal of Geology, 28: 77-80.

7. Barsch, D. 1977. Nature and importance of mass wasting by rock glaciers in alpine permafrost environments. Earth Surface Processes, 2, 231-245.

8. Barsch, D. 1988. Rockglaciers. In Clark, M. J. (Edi.) Periglacial Geomorphology. John Wiley & Sons. Chichister, 69-89.

9. Barsch, D. 1992. Permafrost creep and rockglaciers. Permafrost and Periglacial Processes. Vol. 3 , 175-188.

10. Benedict, D. 1981. The Fourth of July Valley; Glacial geology and archeology of timberline ecotone. Centre for Mountain Archeology, Research Paper Report, 2, 1-139.

11. Brugman, M.M. & Meier, M.F. 1981. Response of glaciers to the eruptions of Mount St. Helens, In: Litman, P.W. & Mullineaux, D.R. (Edi.): The 1980 Eruptions of Mount St. Helens, Washington., U.S. Geol. Survey Prof. Paper No 1250, 743-756.

12. Butler, D.R., Malanson, G.P. & Walsh, S.J. 1991. Snow-avalanche paths: conduits from the periglacial-alpine to subalpine depositional zone. In Dixon, J.C. & Abrahams, A.D. (Edi.) Periglacial Geomorphology, Wiley and Son limt. Chichister, 185-200.

13. Camacho, H. 1925. Apuntes acerca de la actividad actual del Popocatéopetl en relación con la sismología. Anales del Instituto de Geología de México, II (1-3), 38-60.

14. Cantagrel, J.M., Gourgaud, A., Robin, C. 1984. Repetitive mixing events and Holocene pyroclastic activity at Pico de Orizaba and Popocatépetl (Mexico). Bulletin of Volcanology, 47-4(1): 735-748.

15. Capps, S.R. 1910. Rock glaciers in Alaska. Journal of Geology, 18, 359-375.

16. Carrasco-Núñez, G., Vallance, W.J. and Rose, W.I. 1993. A voluminous avalanche-induced lahar from Pico de Orizaba volcano, Mexico: Implications for hazard assessment. Journal of Volcanology and Geothermal Research, 59: 35-46.

17. Cas, R.A.F. & Wright, J.V. 1987. Volcanic Succesions. Chapman & Hall, Hampshire, 528 p.

18. Clapperton, C.M. 1990. Glacial and volcanic geomorphology of the Chimborazo-Carihuairazo Massif, Ecuadorian Andes. Transactions of the Royal Society of Edinburgh: Earth Sciences, 81, 91-116.

19. Corte, A.E. 1987. Rock Glacier Taxonomy. In Giardino, J.R., Shroder, J.F, Vitek, J.D. (Edi.): Rock Glaciers. Allen & Unwin, Boston, 27-39.

20. Cortés, H. 1942 (Ed.). Cartas de relación de la conquista de México. Espasa Calpe. Madrid, 136 p.

21. Crandell, D.R. .1971 .Postglacial lahars from Mount Rainier volcano, Washington; .U.S. Geological Survey, Professional Paper 677: 75 p.

22. Crook. A.G., Davis, R.T. & Moreland, R.E 1981. Snow Surveys and Mount Saint Helens. In: 49th Annual Meeting Western Snow Conference. St. George, Utah, 77-84.

23. Church,M. & Ryder, J.M. 1972. Paraglacial sedimentation: consideration of fluvial processes condicionated by glaciation. Bulletin of the Geological Society of America, 83: 3059-72.

24. DeGraff, J.V. 1994. The geomorphology of some debris flows in the southern Sierra Nevada, California. Geomorphology, 10, (1-4), 231-252.

25. Delgado, H. 1996. The glaciers of Popocatepetl Volcano (Mexico): changes and causes. Quaternary (in print).

26. Demant, A. 1981. Láxe néo-volcanique transmexicain: etude volcanologique et pétrographique. Univ. Aix-Marseille. Aix-Marseille. 259 p.

27. Díaz del Castillo, B. 1519 (1980 Edi). Historia verdadera de la Conquista de la Nueva España. Porrua. México. 700 p.

28. Dollfus, A. 1870. Una ascensión al Popocatepetl. La Naturaleza. I, 180-195.

29. Driedger, C.L. 1981. Effect of ash thickness on snow ablation, in Litman, P.W. & Mullineaux, D.R. (Edi.) The 1980 eruptions of Mount St. Helens, Washington. U.S. Geological Survey. Prof. Paper 1250, 757-760.

30. Föhn, P.M.B. and Meister, R. 1983. Distribution of snow drifts on ridge slopes: measurements and theoretical approximations. Annals of Glaciology. N° 4, 52-57.

31. Gardner, J. 1979 ."The movement of material on debris slopes in the Canadian Rocky Mountains" .Zeitschrift für Geomorphologie, 23(1): 47-57.

32. Giardino, J.R. and Vitek, J.D. 1988 Rock glacier rheology: a preliminary assessment. In: Proceedings V International Conference on Permafrost,Tapir Publishers.,Trondheim.,Vol. 1, 744-748.

33. Gottesfeld, A.S., Mathews, R.W. & Gottesfeld, L.M.J. 1991 .Holocene debris flow and environmental history, Hazelton area, British Colombia. .Canadian Journal of Earth Sciences, 28: 1583-93.

34. Guzmán, M. 1968. Las montañas de México, El testimonio de los cronistas. Costa Amic. México. 230 pp.

35. Haeberli, W. 1983. Permafrost-glacier relationship in the Swiss Alps-today and in the past. Fourth International Conference on Permafrost. National Academy Press, Washigton, 415-420.

36. Haeberli, W. 1985. Creep of mountain permafrost: internal structure and flow of alpine rock glaciers. Mitteilungen der Versuchsanstalt für Wasserbau, Hydrologie und Glaziologie N° 77, 142 pp.

37. Haeberli, W., Rickenmann, D., Zimmermann, M. & Rosli, U. .1990 .Investigation of 1987 debris flow in the Swiss Alps: general concept and geophysical soundings. .In: Hydrology in Mountains regions, II- Artificial Reservoirs; Water and Slopes (Proceedings of two Lausanne Syposia, August 1990),; IAHS; 194: 303-310.

38. Harris, S.A., Gustafson, C.A. .1993 .Debris flow characteristics in an area of continuous permafrost, St. Elias Range, Yukon Territory .Z. Geomorph. N.F., 37(1): 41-59.

39. Heine, K. 1973. Zur glazialmorphologie und Präkeramischen archäologie des Mexicanischen Hochlandes Während des Spätglazials (Wisconsin) und Holozäns. Erdkunde. XXVII (3), 161-180.

40. Heine, K. 1975a. Studien zur jungquartären Glazialmorphologie mexikanischer Vulkane. In: Lauer, W. (Ec.) Das Mexiko-Projekt der deutschen Forschungsgemeinschaft, 7: 1-178.

41. Heine, K. 1975b. Permafrost am Pico de Orizaba, Mexiko. Eiszeitalter und Gegenwart, 6: 212-217.

42. Heine, K. 1976a. Schneegrenzdepresion, Klimaentwicklung, Bondenerosion und Mensch im zentralmexikarischen Hocholand im jüngeren Pleistozän und Holozän. Z.Geomorph. N. F. Suppl.-Bd.24, 160-176.

43. Heine K. 1976b. Blockglescher- und Blockzungen- Generationen am Nevado de Toluca, Mexiko. Die Erde, 107, 330-352.

44. Heine, K. 1983. Mesoformen der Periglazialstufe der semihumiden Randtropen, dargestellt an Beispilen der Cordillera Neovolcanica, Mexiko. Abh. Akademie der Wissenschaften in Göttingen, Mathematisch-Physicaische Klasse, Series III, 35: 403-424.

45. Heine, K. 1984. The Classical Late Weichselian Climatic Fluctuations in Mexico. in Morner, A. & Karlen, W. (Ed.) Climatic Changes on a Yearly to Millennian Basts, Reidel P.C. London, .95-115.

46. Heine, K. 1988. Late Quaternary Glacial Chronology of the Mexican Volcanoes. Die Geowissenschafter, 6: 197-205.

47. Heine, K. 1994. Present and past geocryogenic processes in Mexico. Permafrost and Periglacial Processes, 5: 1-12.

48. Hernández-Sosa, P. 1948. Parque Nacional Iztaccihuatl-Popocatepetl. Secretaría de Agricultura y Ganadería. México. 416 pp.

49. Hoskuldsson, A. 1990. Les debris-avalanches du Pico de Orizaba (Mexique). Bulletin de la Section de Volcanologie de la Societé Géologique Française, 19: 5-7.

50. Hoskuldsson, A. 1994. Le complexe volcanique Pico de Orizaba-Sierra Negra-Cerro de las Cumbres (sud-est Mexicain): structure, dynamismes érupyifs et évaluation des aléas. Université Blaise Pascal, Clermont-Ferrand

51. Hoskuldsson, A., and Robin, C. 1993. Late Pleistocene to Holocene eruptive activity of Pico de Orizaba, Eastern Mexico. Bulletin of Volcanology, 55: 571-587.

52. Humboldt, A. 1807/1811. Ensayo político sobre el Reino de la Nueva España. in Gonzalez, V. (traslation in 1966), Ed. Porrua. México. 354 pp.

53. Innes, J. L. 1985. Lichenometry. Progress in Physical Geography, 9(2): 187-254.

54. Johnson, P.G. 1973. Some problems in the study of rock glaciers. In Fahey, B.D. & Thompson, R.H. (Edi.) Research in Polar and Alpine Geomorphology. Geo.Abstracts Ltd. Norwich, 84-95.

55. Johnson, P.G. 1978. Rock glacier types and their drainage systems, Grizzly Creek, Yokon Territory. Canadian Journal of Earth Science, 15, 1496-507.

56. Johnson, P.G. 1984. Rock glacier formation by high-magnitude low-frequency slope processes in the Southwest Yukon. Annals of the Association of American Geographers, 74, 408-419.

57. Johnson, P.G. 1987. Rock glacier: glacier debris sytems or high-magnitude low-frequency flows. in Giardinc, J.R., Shroder, J.F. & Vitek, J.D. (Edi.) Rock Glaciers. Allen & Unwin, Boston, 175-192

58. Johnson, A.M. & Rodine, J.R. 1984 .Debris flow .In: Brunsden, D. and Prior D.B. Slope instability 1984. John Wiley & sons Ltd: 257-360.

59. Larsson, S. 1982 .Geomorphological effects on the slopes of Longyear Valley (Spitsbergen), after a heavy rainstorm in July 1972. .Geogr. Ann. 64A, 3/4: 105-125.

60. Lauer, W. 1978. Timberline studies in Central Mexico. Arctic and Alpine Research, 10: 383-396.

61. Lauer, W. and Klaus, D. 1975. Geoecological investigations on the timberline of Pico de Orizaba, Mexico. Arctic and Alpine Research, 7: 315-330.

62. Lavarriere, J. 1858. Exploración del Valle de México. Bol. Soc. Mexicana de Geografia e Estadística. VI, 191.

63. Linder, W. & Jordan, E..1989.Gelandedeformationen im Rahmen des aktuellen Ruiz-Vulkanismus, Kolumbien (Terrain deformation in the framework of current volcanism, Ruiz, Colombia).Bayreuther Geowissenschaftliche Arbeiten, 14, 135-140.

64. Lorenzo, J.L. 1964. Los Glaciares de México. Monografias del Instituto de Geofísica (UNAM), 1, 123 pp.

65. Lorenzo, J.L. 1969. Condiciones periglaciares de las altas montañas de México. Mexico City: Instituto Nacional de Antropología e Historia.

66. Luckman, B.H .1992 Debris flows and snow avalanche landforms in the Lairig Ghru, Cairngorm Mountains, Scotland. .Geografiska Annaler, 74A(2-3): 109-121.

67. Lliboutry, L. 1961. Les glaciers enterres et leur role morphologique. International Association of Scientific Hydrology, Publ. 54, 272-280.

68. Lliboutry, L. 1988. Rock glaciers in dry Andes. Materalny Glyatsilogy Isseldovany, 58, 139-144.

69. Martin, H.E., and Whalley, W.B. 1987. Rock glaciers. Part 1: Rock glacier morphology: classification and distribution.. Progress in Physical Geography, 11: 260-282.

70. Meier, M.F. 1969. Glaciers and water sypply. Jour. Amer. Water Works Ass., 61(1), 8-12.

71. Melgarejo, A. 1910. The greast volcanoes of Mexico. National Geographic Magacine. 21(9), 741-760.

72. Mentz, B.M. Von 1980. México visto por los alemanes. UNAM. México. 482 pp.

73. Mizuyama, T., Kobashi, S. & Guogiang, Ou..1993. Development of debris flow.In: Sediment problems. Proc. international symposium, Yokohama, 1993, ed R.F. Hadley & T. Mizuyama, (IAHS; Publication, 217), pp 141-145

74. Monnier, M. 1884. Ascensión du Popocatepetl. Anuario du Club Alpin Francais, 40, 38.

75. Mooser, F. Meyer, A.H., and McBirney, A.R. 1958. Catalogue of active volcanoes of the World, part VI, Central America. International Volcanic Association, 6, 36 pp.

76. Morris, S.E. 1981. Topographic factors and the development of rock glaciers facies, Sangre de Cristo Mountains, Southern Colorado. Artic and Alpine Research, 13, 329-338.

77. Murillo, G. 1939. La actividad del Popocatepetl. Polis. México., 30 pp.

78. Nixon, G.T. 1986. The geology of Iztaccihualt Volcano. Geological Society of American, Special Paper, 219. Washington, 58 pp.

79. Nyberg, R. 1985 Debris flows and slush avalanches in northern Swedish Lappland. .Meddelanden frân Lunds Universitets Geografiska Institution. Avhandlingar, XCVII: 5222.

80. Nyberg, R. & Lindh, L. .1990 Geomorphic features as indicators of climatic fluctuations in aperiglacial environment. Geografiska Annaler, 72A: 203-210.

81. Ordoñez, E. 1895. Observaciones relativas a los volcanes de México. Men. Soc. Cien. Antonio Alzate. 8, 183-196.

82. Orozco,J. 1887. Seismología. Efemérides Seísmicas mexicanas. Men. y Rev. Soc. Cien. Antonio Alzate. II, 305-337.

83. Osterkamp, W. R. Costa. J. E.1986.Denudation rates in selected debris-flow basins..Glysson, G. Douglas. Proceedings of the Fourth Federal interagency sedimentation conference; in two volumes. U. S. Geol. Surv., United-States, 4.91-4.99.

84. Owen, L. A. 1991. Mass movement deposits in the Karakoram Mountains:. their sedimentary characteristics, recognition and role in Karakoram landform evolution . Zeitschrift fur Geomorphologie, 35 (4): 401-424.

85. Owen, L.A., Derbyshire, E. .1989 The Karakoram glacial depositional System. .Z.Geomorph. N.F., Suppl. 76: 33-73

86. Owen, L.A., Benn, D.I., Derbyshire, E., Evans, D.J.A., Mitchell, W.A., Thompson, D., Richardson, S., Lloyd,M. & Holden, C. 1995. The geomorphology and lanscape evolution of the Lahul Himalaya, Northen India. Z. Geomorph., 39 (2): 145-174.

87. Packard, A.C. 1896. Ascent of volcano of Popocatepetl. The American Naturalist. II, 109-123.

88. Palacios, D. 1995. Rockslide processes on the north slope of Popocatepetl Volcano, Mexico. Permafrost and Periglacial Processes 6 (4): 345-356.

89. Palacios, D. 1996. The glacio-volcanic evolution of Popocatepetl Volcano: Geomorphologic consequences. Geomorphology (in print).

90. Palacios, D. and Marcos, J. 1996. Deglaciation on Mexico's Stratovolcanoes from 1994-95. Permafrost and Periglacial Processes (in print).

91. Palacios, D. & Vázquez, L. 1996. Geomorphic consequences of a glacial retreat: Pico de Orizaba Volcano Geografiska Annaler (in print).

92. Palacios, D., Zamorano, J.J. and Parrilla. G.(1996) Proglacial debris flows in Popocatepetl North Face and their relation to 1995 eruption. Geomorphology, (in revision).

93. Palacios, D., Parrilla, G. and Zamorano, J J. (1996) Paraglacial and postglacial debris flows on Little Ice Age frontal moraine: Jamapa Glacier, Pico de Orizaba (Mexico). Zeitschrift für Geomorphologie (in revision).

94. Payno, M. 1849. La falda de los volcanes. El Albún Mexicano. I, 79 p.

95. Pérez, L. 1857. Un viaje al Popocatepetl por el segundo capitán de ingenieros. Bol. Soc.Mex. de Geogr. y Esta. V. 338

96. Pérez, F.L. 1988. Debris transport over a snow surface: a field experiment. Revue de Géomorphologie Dynamique, XXXVII, 81-101.

97. Pierson, T.C., 1985 .Initiation and flow behavior of the 1980 Pine Creek and Muddy River lahars, Mount St. Helens, Washington. .Geol. Soc. Am. 96: 1056-1069.

98. Pierson, T.C., Janda, R.J., Thouret, J.C. & Borrero, C.A..1990.Perturbation and melting of snow and ice by the 13 November 1985 eruption of Nevado del Ruiz, Colombia, and consequent mobilization, flow and deposition of lahars.Journal of Volcanology & Geothermal Research, 41(1-4): 17-66.

99. Priester, A. 1927. Notas preliminares sobre vestigios glaciares en el Estado de Hidalgo y en el Valle de México. Men. Soc. Cient. Antonio Alzate, 48 . 1-13.

100. Rapp, A. 1959. Avalanche boulder tongues in Lapland (Descriptions of little-known forms of periglacial debris accumulations), Geografiska Annaler. XXXXI-1, 34-48.

101. Rapp. A, & Nyberg R., 1981. Alpine deoris flows in northern Scandinavia; morphology and dating by lichenometry.Geografiska Annaler. Series A: Physical Geography. 63, (3-4), 183-196.

102. Rapp, A. & Strömquist, L. 1976 .Slope erosion due to extreme rainfall in the Scandinavian Mountains. .Geografiska Annaler. 58A: 193-299.

103. Rios, J.M. 1851. Ascensión al Popocatepetl. La Ilustración Mexicana. V., 423.

104. Robin, C. 1984. Le Volcan popocatepetl (Mexique): structure, evolution pétrologique et risques. Bull. Volcanol. . 47. (1), 1-23.

105. Robin, C. and Cantagrel, J.M. 1982. Le Pico de Orizaba (Mexique): Structure et évolution d'un granc volcan andésitique complexe. Bulletin of Volcanology, 45: 299-315.

106. Robin, C., Cantagrel, J.M., and Vicent, P. 1983. Les nuées ardentes de type Saint-Vincent, épisodes remarquables de l' évolution récente du Pico de Orizaba (Mexique). Bulletin de la Societé Géologique Française, 25: 727-736.

107. Robin,C. & Boudal, CH. 1987. A gigantic Bezymianny-Type event at the beginning of modern Volcan Popocatepetl. Jour. Volc. and Geother. Research. 31, 115-130.

108. Rodríguez-Elizarrarás, S. and Lozano, A. 1991. Las asociaciones alcalina y calcoalcalina en la parte central de la cordillera volcánica Cofre de Peróte- Picc de Orizaba. Convención sobre la evolución geológica de México. Memoria. Pachuca, México, 174-175.

109. Sánchez,G. 1902. Descripción científica del Volcán Popocatepetl. Bol. Soc. Mexi. de Geogr. y Estadística. 3,1,133 p.

110. Siebe, C., Abrams, M., and Sheridan, M.F. 1993. Major Holocene block-and-ash fan at the western slope of ice-capped Pico de Orizaba volcano, México: Implications for future hazards. Journal of Volcanology and Geothermal Research, 59: 1-33.

111. Simkin, T., Siebert, L., McCelland, L., Bridge, D., Newhall, C., and Latter, J.H. 1981. Volcanoes of the World. Smithsonian Institution. Stroudsbourg, Pennsylvania: Hutchinson & Ross.

112. Skinner, B. E. 1964. Measurement of Twentieth Century ice loss on Tasman glacier, New Zeland. J. Geol. and Geophys. 7(4), 37-46.

113. Tangborn, W.V. & Lettenmaier, D.P 1981. The impact of Mount St. Helens ash deposition on snowmelt. 49th Annual Meeting Western Snow Conference. St. George, Utah. 85-94.

114. Thouret, J.C. 1989. Processus éruptifs et glaciaires durant la période „Pollalie" au Mount Hood (Chaine des Cascades, Oregon, USA). Bull. Assoc. Géogr. Franç., 5. 351-360.

115. Trabant, D.C. & Meyer, D.F. 1992. Flood generation and destruction of „drift" glacier by the 1989-90 eruption of Redoubt Volcano, Alaska. Annals of Glaciology 16, 33-38.

116. Vázquez, L. 1991. Glaciaciones del Cuaternario tardío en el volcán del Téyotl, Sierra Nevada. 4Bol. Inst. Geografia 22, 25-45.

117. Vick, S.G. 1987. Significance of landsliding in rock glacier formation and movement. In Giardino,J.R., Shroder,F. & Vitek,J.D. (Edi.) Rock glaciers, Allen & Unwin, Boston , 239-263.

118. Vinogradov, V.N 1975. The peculiarities of accumulation and ablation on glaciers of the volcanic regions of Kamchatka. Snow and Ice. Symposium-Neiges et Glaces (Proceedings of the Moscow Symposium, August 1971): Iahs-Aish Publ. No 104, 129-133.

119. Vinogradov, V.N. 1981. Glacier erosion and sedimentation in volcanic regions of Kamchatka. Annals of Glaciology, 2, 164-169.

120. Waitz, P. 1921. La nueva actividad y el estado actual del Popocatepetl. Men. Soc. Cient. Antonio Alzate. 37, 295-313.

121. Walder, J. S., & Driedger, C. L. 1994. Geomorphic change caused by outburst floods and debris flows at Mount Rainier, Washington with emphasis on Tahoma Creek valley. Water-Resources-Investigations. Geological-Survey, 93 pp.

122. Ward, H.G. 1985. México en 1927. SEP. México, 204p.

123. Wayne, W.J. 1981. Ice segregation as an origin for lenses of non-glacial ice in „ice-cemeted" rock glaciers. Journal of Glaciology, 27, 506-510.

124. Wei W., Gao C. Walling E. 1992.Studies of ice-snow melt debris flows in the western Tian Shan Mountains, China. IAHS-AISH publication 209: 329-336.

125. Weitzberg, F. 1923. El Vestinquero del Popocatepetl. Men. Soc. Cient. Antonio Alzate. 41(2-3), 67-90.

126. Whalley, W.B. & Martin, H.E. 1992. Rock Glacier: Models and Mechanisms. Progress In Physical Geography 16(2), 127-186.

127. Whipple, K. X., and Dunne T. 1992.The influence of debris-flow rheology on fan morphology, Owens Valley, California . Geol. Soc. Am. Bull. 104 (7), 887-900.

128. White, S.E. 1954. The firn field on the Volcano Popocatepetl, Mexico. Journal of Glaciology. 2(16), 389-393.

129. White, S.E. 1962a. El Iztaccihuatl. Ins. Nac. de Antropología e Hist., Serie 6. México. 80 p.

130. White, S.E. 1962b. Late Pleistocene glacial sequence for the West side of Iztaccíhuatl volcano, Mexico. Geological Society of America Bulletin, 73: 935-958.

131. White, S.E. 1976. Rock glaciers and blockfields, review and new data. Quaternary Research 6, 77-97.

132. White, S.E. 1981. Neoglacial to recent fluctuations on volcano Popocatépetl, Mexico. Journal of Glaciology, 27: 356-363.

133. White, S.E. 1987. Quaternary glacial stratigraphy and chronology of Mexico. Quaternary Science Reviews. 5 (1-4), 201-206.

134. White, S.E. & Valastro, S. 1984. Pleistocene Glaciation of volcano Ajusco, Central Mexico, and Comparasicn with the Standard Mexican Glacial Sequence. Quaternary Research. 21, 21-35.

135. Zambrano, S. 1988. Pico de Orizaba. Mexico City: Catalogos Artísticos y Comerciales S.A.

136. Zimmermann, M. & Haeberli, W. 1989 Climatic change and Debris flow activity in High-Mountain Areas .In. Rupke, J. and Boer, M.M. (ed) Landscape ecological Impact of climatic change on alpine regions with enphasis on the Alps: 52-66.

Author

David Palacios
Dept. of Physical Geography
Complutense University
28040 Madrid,
Spain

ANDEAN LANDSLIDE HAZARDS

T. A. BLODGETT, C. BLIZARD, B. L. ISACKS

1. Introduction

1.1. GLOBAL LANDSLIDE OVERVIEW

Worldwide, millions of dollars in property damage and loss of life are attributed to landslides annually. Landslides are frequent hazards in tectonically active mountain belts such as the Himalayas, Southern Alps of New Zealand, New Guinea, Japan, and the Andes, where uplift generates high topographic relief. Steep slopes associated with regions of high relief provide sufficient gravitational potential for landslides to occur. Hazards exist wherever roads, dwellings, or people lie in the path of potential landslides. In many cases, human land use practices such as the cutting of roads and clearing of forest can also serve to increase the potential for landslides, and/or influence where they may happen (e.g., Wright and Mella, 1963; Preciado and Llinas, 1990).

Landslides may occur in the form of rock falls, rotational slumps, debris avalanches, or debris flows. These landslide categories reflect coherence of material moved, presence of water in the material, and morphological characteristics of the slide (Varnes, 1984). Geological agents typically blamed for landslides include earthquakes, heavy and/or sustained rainfall, rapid snowmelt, and flooding. In many cases, several factors contribute to induce a landslide. A synergy of causative agents, such as earthquakes in tandem with heavy rains, may produce a landslide which neither one could cause by itself (Belloni and Morris, 1991). Nevertheless, one of the several agents is commonly singled out as the landslide **trigger**, because it is the agent that acted on the slope immediately preceding the landslide. Other factors which contribute to landslide susceptibility, distribution, and extent include vegetation, lithology, soil type, and topography. The presence and magnitude of all the above factors are influenced directly or indirectly by regional tectonism.

1.2. ANDEAN TECTONICS AND CLIMATE

1.2.1. Earthquakes

The fundamental cause of tectonic and volcanic activity in the Andes is the convergence of two major lithospheric plates, the Nazca and South American Plates, and the subduction of the sub-oceanic Nazca plate beneath the South American plate. The shape of the subducted Nazca plate is indicated in Figure 1 by contours derived from epicenters of teleseismic events (Cahill and Isacks, 1992). The largest earthquakes occur within the plate boundary located beneath the coastal areas of Chile, Peru, Ecuador, and Colombia. However large earthquakes also occur within the subducted Nazca

211

J. Kalvoda and C.L. Rosenfeld (eds.), Geomorphological Hazards in High Mountain Areas, 211-227.
© 1998 *Kluwer Academic Publishers.*

Plate and at shallow depths within the continental crust of the South American Plate (e.g., Chinn et al., 1980; Jordan, et al., 1983). Crustal seismicity, related mainly to horizontal compressional stresses in the upper plate, is most active in the two regions above the nearly flat segments of the subducted Nazca plate located beneath Peru and Argentina (27°S - 33°S).

1.2.2. Volcanoes

Most volcanoes in the Andes are associated with subduction zone tectonism. Large volumes of magma are generated as a result of the dewatering of subducted crust. Magmatic intrusions and eruptions on the surface contribute to the topographic relief, especially in the Western Cordillera of the central Andes, where the volcanoes are built upon the high plateau surface and reach very high elevations. The volcanoes are thus often covered with ice and snow which increases the potential for catastrophic volcanic mudflows, or lahars, even though eruptions are relatively rare.

During the Late Cenozoic, subduction-related volcanism ceased as the subducted Nazca plate became nearly flat in dip (Kay et al., 1987; Isacks, 1988). Presently there are no active or Quaternary age volcanoes located above the two nearly flat segments of the subducting Nazca Plate.

1.2.3. Hydrology

The Andes mountains include some of the most diverse climatic regimes on earth, from lowland deserts and rainforests to alpine tundra and glaciers. One of the most influential controls on moisture is the topographic barrier of the Andes. Some regions within the Andean rain shadow, such as the Atacama desert in Western Chile, experience little or no rainfall annually. In contrast, orographically enhanced moisture conditions on the flanks of the Eastern Cordillera cause the region to receive up to 6 meters of precipitation each year. In areas where the mountain belts receive large amounts of rain and snow, hydrologically triggered landslides are common (Blodgett et al., 1996). However, landslides also occur in more arid regions in response to less-frequent, high-precipitation events. Because vegetation acts to stabilize the soil, arid and semi-arid regions where vegetative cover is largely or entirely absent are particularly susceptible to hydrologically triggered mass movements.

2. Andean landslide trigger mechanisms

2.1. RECENT LANDSLIDES IN THE ANDES

Landslide morphology depends upon geology, vegetation, topography, hydrologic condition of the soil, soil type, and the structure of the subsoil. Deep-seated rotational slides, for example, commonly develop in poorly consolidated sediments, and are often caused mainly by hydrological factors. Shallow-seated debris avalanches, on the other hand, occur frequently on steep slopes underlain by bedrock where the gravitational potential of soil and regolith overcomes the frictional cohesive strength binding soil and rock together.

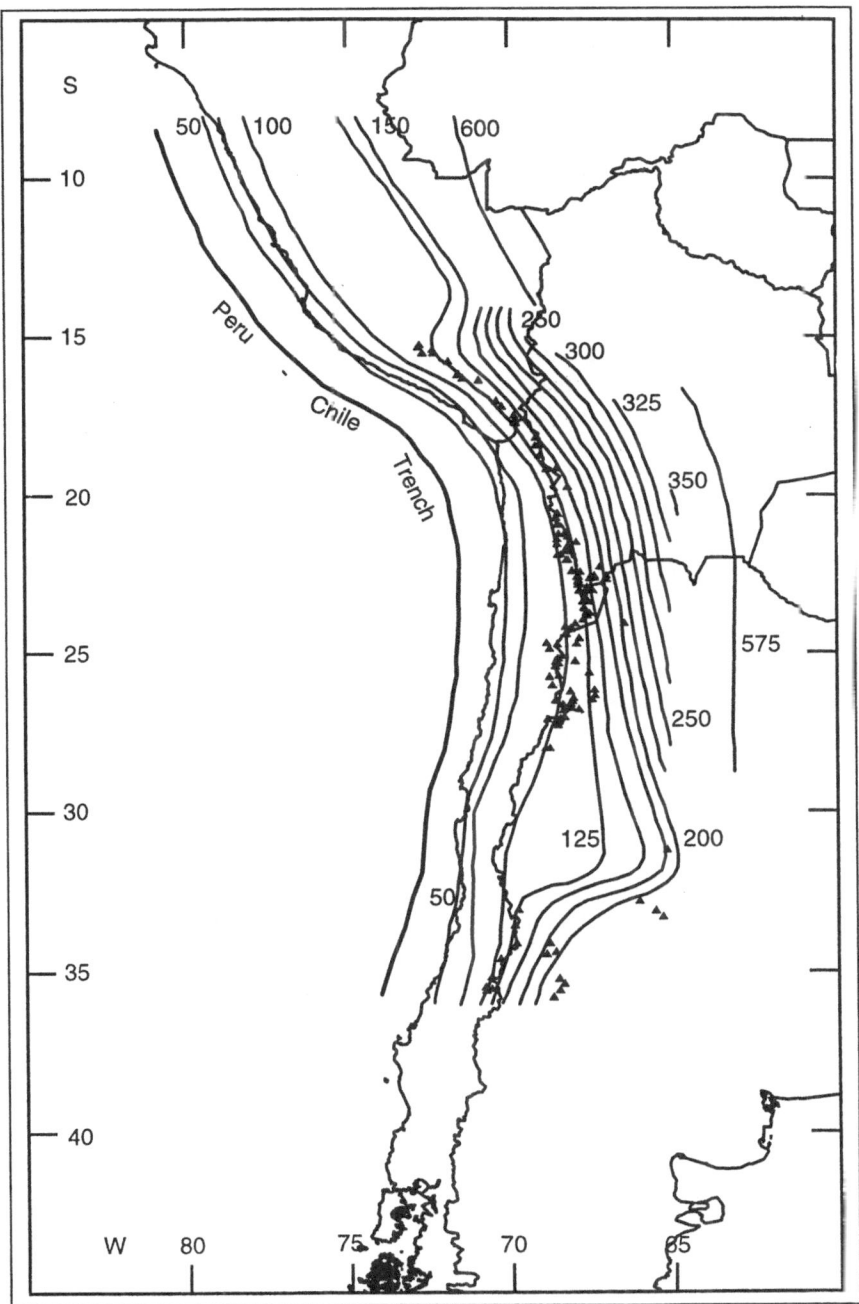

Figure 1. Shape of the subducted Nazca plate as shown via contours determined by epicenters of teleseismic events (Cahill and Isacks, 1992).

In order for shallow slides to occur, regolith and soil must accumulate to a critical thickness. After this condition is reached, a landslide can occur in response to various kinds of triggers, such as a sudden increase in pore pressure and weight of the soil caused by groundwater, the shaking of soils by an earthquake, or the sudden loading by another slide. Regardless of the final trigger, hydrological instability is usually a prerequisite for landslide failure (rockfalls induced by seismic shaking being a notable exception).

Tables 1 and 2 provide a sampling of landslides triggered by earthquakes, rain, snowmelt, or flooding, in the Andes. In addition, a case of a landslides triggered by a volcanic eruption is discussed below, although not included in the tables. Data on landslides with hydrologic triggers have been obtained primarily from news media reports from 1970-1996. Data on landslides triggered by earthquakes has been obtained largely from earthquake records going back to 1953 (e.g., Rothé, 1969; Seismological Notes, *Bulletin of the American Seismological Society*). In both cases, the tables have been augmented by landslide descriptions furnished in the geologic literature. Many tabulated landslide "events" may actually have included hundreds or even thousands of individual slides, of widely varying morphologies. Although these tables certainly fail to record every landslide which has occurred in the Andes over even the past 25 years, they provide a sense of landslide frequency distribution, and a first-order approximation of the distribution landslide hazards in the Andes.

2.2. EARTHQUAKES AS LANDSLIDE TRIGGER MECHANISMS

In order to explore possible regional relationships between seismicity and landslide events, we obtained data on major earthquakes (magnitude 6.5 and above, or that resulted in loss of life and/or significant damage) that have occurred in the Andes since 1953. Excluded were those earthquakes lacking data regarding focal depth, magnitude, and/or epicentral coordinates. The data sets consulted also include information on the nature and extent of reported landslides related to an earthquake. Table 1 provides a subsampling of the 246 total events, restricting the data set to those earthquakes reportedly associated with landslides.

As a means of providing a qualitative assessment of regional trends in seismicity and landslide frequency, we have plotted major Andean earthquakes on Figures 2a, 2b, 3a, and 3b. Earthquakes have triggered many of the most catastrophic slides, which have killed as many as 20,000 people in a single event. Earthquake-triggered landslides are concentrated along the coastal regions of western South America. This reflects the high rate of occurrence of large shallow earthquakes within the inter-plate boundary and at relatively shallow depths within flat-slab portions of the subducted plate near the inter-plate boundary. Additional earthquake triggering occurs inland in Peru and central Argentina in the region of high crustal seismicity located above the shallow dipping segments of the subducted plate. No earthquake-triggered landsliding is reported along the steep slopes of the eastern cordilleras of northern Argentina, Bolivia, and southern Peru, in accordance with the low seismicity found for those regions. Earthquake-triggered landslides occur inland also in Ecuador and Colombia, but none have been reported in Venezuela.

Intermediate-depth and deep earthquakes located beneath the Andes and regions east of the Andes, located with the subducted Nazca plate, trigger very few landslides compared to shallow earthquakes. With the exceptions of two deep, high magnitude (M_b=7.7 and M_w=8.2) events in Bolivia, all of the landslides were triggered by earthquakes with focal depths of about 100 km or less. However, above 100 km there is no clear relationship between focal depth and landslide likelihood. The efficacy of shallow earthquakes as landslide triggers is simply related to the fact that the amplitude of short-period (0.1-10 seconds) ground shaking falls off rapidly with respect to distance traveled by the seismic waves. An additional factor may be the efficient propagation of short period surface waves which are more strongly excited by shallow earthquakes than deep earthquakes.

One of the large deep earthquakes located beneath the Peru-Bolivia border, the June 9, 1994 deep earthquake with a magnitude, M_w, of 8.2, triggered landslides near the Chilean border at distances greater than 100 km from the epicenter (Table 1). As Keefer (1984) concluded in an assessment of 40 historical world-wide earthquakes, seismic shaking associated with deeper earthquakes tends to propagate over a larger surface area, triggering landslides at greater epicentral distances than shallower events.

The likelihood of landslides tends to increase with magnitude; in particular, those earthquakes of magnitude 7.0 or greater often generate mass movements. Because other factors (lithology, hydrology, etc.) can act to decrease slope stability beforehand, a minimum magnitude threshold may not exist (Berrocal et al., 1978; Belloni and Morris, 1991), although Keefer (1984) proposes a global threshold of 4.0. As shown in Table 1, the preponderance of earthquakes triggering landslides are of magnitude 6.5 or greater. The lowest magnitude quake associated with landsliding is 4.7, and only 4 of the 39 earthquake-triggered mass movements in our data set were attributed to earthquakes with magnitudes lower than 6.1.

From available news reports in the NEXIS computer database (May 1986 - May 1996) and accounts in the geologic literature, we compiled a list of 29 earthquake-triggered landslides, including dates of occurrence, epicenter, magnitude, and depth of the earthquake's epicenter, and the nature and extent of the associated landslides (Table 1). Figures 2a to 3b show the locations of all of the known events. Three of the most catastrophic earthquake-triggered Andean landslides are summarized below.

2.2.1. Central Chile, May 1960

The great earthquake of May 22, 1960 is the largest magnitude ($M_w = 9.5$) earthquake recorded in this century. Suprisingly, no damage occurred in the vicinity above the location of the epicenter just off the coastline at about 38° south. The closest damage reported was in the Valdavia basin about 200 km to the south of the epicenter. Wischet (1963) found that earthquake damage coincided with tectonically depressed basins and areas underlain by weakly consolidated sediments. The following description of landsliding triggered by this earthquake is taken from Davis and Karzulcvic (1963) and Wright and Mella (1963). Seismic activity in this region began with a severe (M_S=7.8) earthquake on 21 May, 1960 and culminated with the magnitude (Mw) 9.5 event on the following day. The earthquake of May 22 triggered landslides which blocked Lago Rinihue (Central Valdivia) in three places. Altogether, approximately 38 million m^3 of material moved in these events. The largest slide comprised about 30 million m^3 of lake clays, which moved primarily by block gliding and lateral slumping.

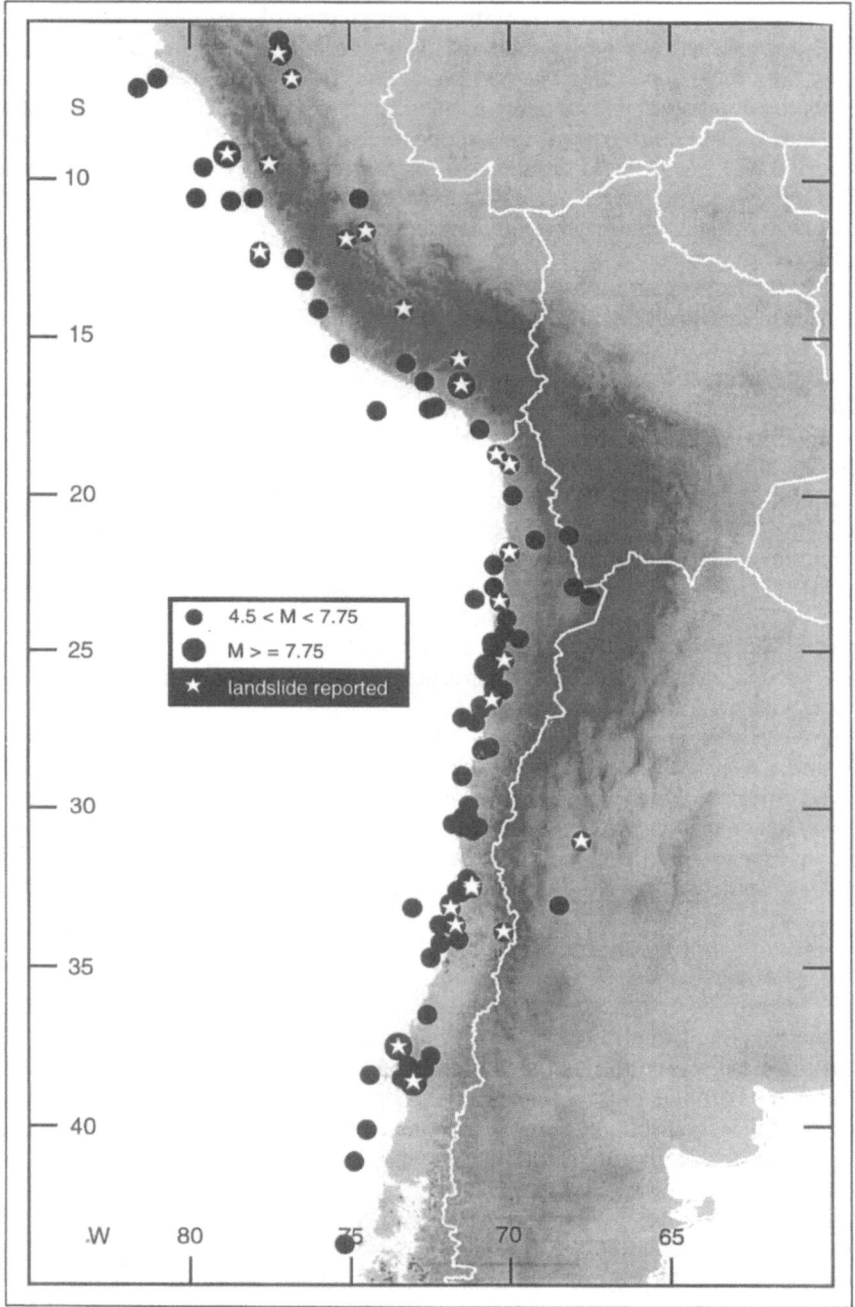

Figure 2a. Distribution of shallow to intermediate (depth < 80 km) earthquake epicenters in the central Andes with Richter magnitudes reported to be greater than 4.5. A star indicates that landslides were reported in association with the earthquake.

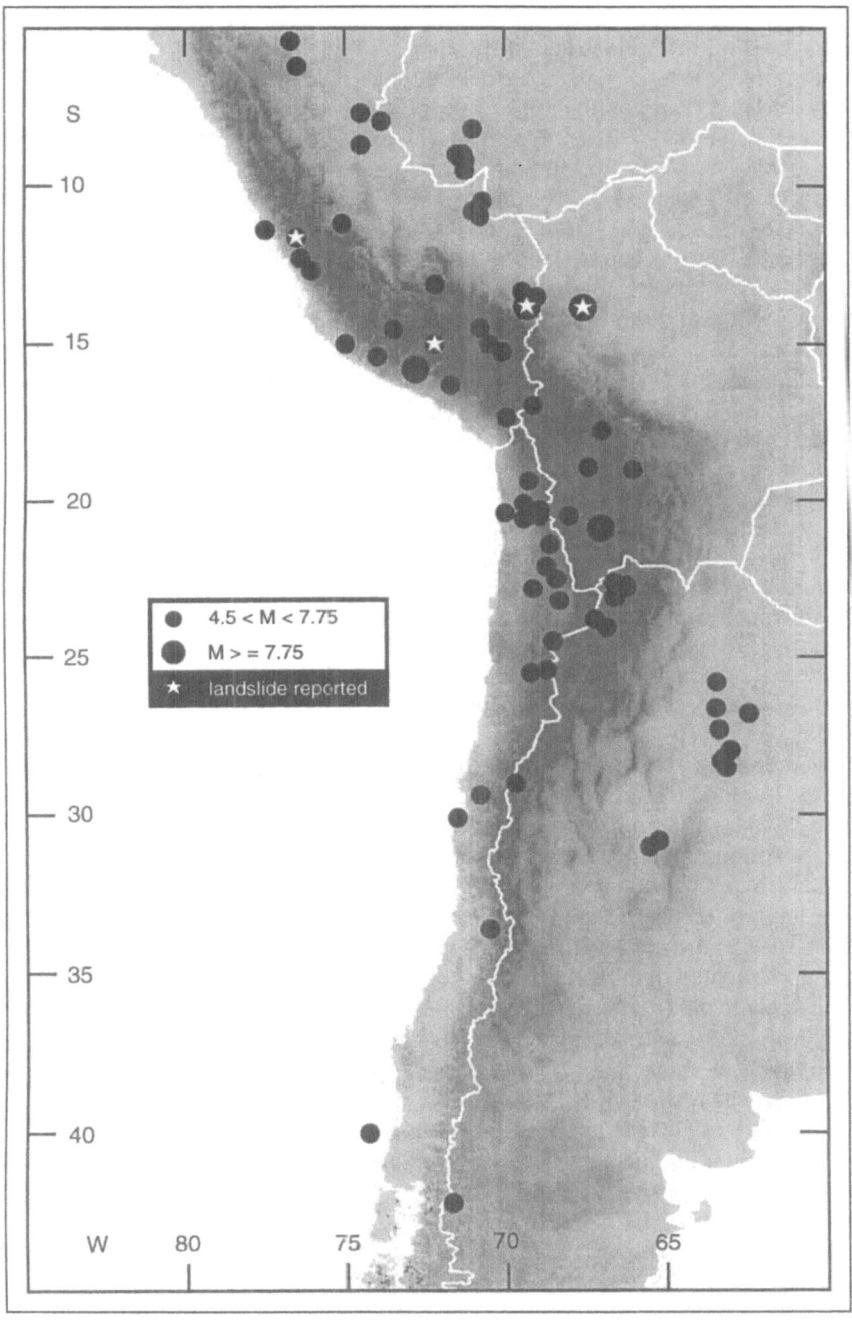

Figure 2b. Distribution of deep (depth >80 km) earthquake epicenters in the central Andes with Richter magnitudes reported to be greater than 4.5. A star indicates that landslides were reported in association with the earthquake.

Springs flowing from the base of the main scarp provide evidence that the basal sediments were saturated at the time of the slide (although the event predated the start of the winter rainy season).

Also on May 22, approximately 125 people were killed by landslides at the eastern end of Lago Rupanco. The settlement of Gaviota was practically obliterated. These slides followed heavy rain on 20 May. Initiated in areas cleared for agriculture, the mass movements advanced an average of 1,000 m down slopes averaging 40 degrees. On burned slopes, slides began near the crests; on more "natural" slopes, slides began at the highest point reached by land clearing operations. Mudslides were produced by converging slides, presumably as a result of downward pressure squeezing water out of the porous pumice and scoria fragments that comprised more than 70% of the mobilized material.

In Pellaifa, extensive landslides occurred on the steep slopes which border the head of Lago Calafquen, and also on the steep land around Lago Pellaifa. About 80 people, mostly farmers, were killed. Like the slides at Lago Rupanco, these slips mostly took place on slopes mantled with coarse, porous volcanic ash. Most also occurred where the forest had been cleared and were occupied by grass and fern. Burned-over forest slopes faired slightly better. Often entire patches of cleared land slipped away. On the upper slopes, debris avalanches developed along the drainage pattern. Mudflows were absent, possibly because the soils had lower percentages of porous materials and were drier than Rupanco at the time of the quake (there had been no rain at all during the previous week)

2.2.2. Huascarán debris avalanche, May 1970

On May 31, 1970, an earthquake with magnitude M_S=7.7 and a focal depth of 56 km triggered several thousand slope failures in the mountainous parts of central Peru (Cluff, 1971; Plafker et al., 1971). The mass movements included falls, slides, and flows involving various proportions of bedrock, unconsolidated deposits, and snow. The overwhelming majority of these slides took place on the steeper slopes of the Cordillera Blanca and Cordillera Negra, and the area of deeply incised drainage along the western slopes of the Cordillera Negra.

The largest, most devastating event was the Huascarán debris avalanche, which moved approximately half of the 100 to 200 million m^3 of material involved in all the mass movements put together. In terms of destructiveness, height of fall, velocity, and probably volume, it exceeds any other historic avalanche. The original slide mass probably involved about 50 to 100 million m^3 of rock with some ice. The avalanche gained velocity as it slid over Glacier 511, descending 1 km over 2.4 km of slope, at an average velocity of 360 km/hr. The debris split into two tongues. The majority of the material became a highly fluid debris flow which advanced 16 km from Huascarán to the Río Santa, damming the river. A smaller tongue of debris buried the city of Yungay, killing 15,000 of its 17,000 inhabitants. Altogether, approximately 18,000 lives were taken by the avalanche. The Huascarán avalanche deposits included many boulders, the largest of which weighed over 7,000 metric tons.

Only nine landslides involving at least 1 million m^3 of material were classified as slumps and rotational slumps. All occurred in unconsolidated fluvioglacial deposits or poorly consolidated pyroclastic materials. The largest rotational landslide, situated on the east bank of the Río Santa at Recuay, involved between 8 and 20 m³ of fluvioglacial sediments, with a slump depth of perhaps 15 m. The upthrust toe of the slump blocked Río Santa, forming a lake upstream. A smaller rotational slump (also in fluvioglacial deposits) disrupted a 150 m segment of a main road, destroying several homes.

Many ground failure events attributed to the earthquake occurred at epicentral distances of 135 to 180 km, with rockfalls extending beyond 180 km from the quake's epicenter. Nearly all of the mass movements were rockfalls, rockslides, and debris slides along steep valley walls, streambanks, and roadcuts. Approximately two thirds of the observed landslides of this class were in the Cordillera Blanca. This condition is attributed by Plafker et al. to the region's high rainfall and seasonal freeze-thaw. In particular, a reported record rainfall in the Río Santa valley during the 1969-70 wet season (with rains as late as May 17) probably enhanced landsliding considerably.

2.2.3. Northern Ecuador, March 1987

Two earthquakes (M_s=6.1, focal depth of 3 km and M_s=6.9, focal depth of 12 km) triggered landslides in northern Ecuador on March 5, 1987 (Belloni and Morris, 1991; Nieto et al., 1991). The greatest landsliding intensity occurred in the region of Reventador Volcano, and involved residual soils and highly weathered rock. More than 90% of these slides began as shallow slips or slides on either the main valley topslopes or slopes of the lower-order tributaries. Failures typically began on slopes steeper than 30-35 degrees. Average landslide thicknesses were from 1.5 to 2.0 m, with a range of a few decimeters to 5 m. A number of these slides extended to unweathered bedrock. These shallow landslides transformed into debris avalanches and finally debris flows as they moved downvalley and eventually downchannel. It was observed that the debris flows which started higher up on valley walls were more likely to reach the floodplain. Overall, approximately 110-120 m³ of material is calculated to have been transported into the main rivers from landslide activity. Belloni and Morris (1991) concluded that these landslides were the result of a synergy of heavy rains in the preceding months and the earthquakes, neither of which could have caused the slides by themselves.

2.3. VOLCANIC ERUPTIONS AS LANDSLIDE TRIGGER MECHANISMS

Volcanoes are located mainly in the volcanic arc that comprises the Western Cordillera (Figure 1). In regions with active volcanoes, risk from landslides caused by eruptions or tremors is high. Landslides in these areas are likely compounded by tremors, eruptions, and volcanic soils which are prone to sliding. Several shallow earthquakes, attributed to the movement of magma associated with active volcanoes, have been known to trigger landslides (Mileti et al., 1991). Loose volcanic soil is especially susceptible and the risk is compounded by the presence of snow and ice caps which complete the recipe for catastrophic mudflows, or lahars. Although these lahars are relatively infrequent, they can be highly devastating. The only recent lahar (according to our data review) is summarized below, abstracted from an account provided by Mileti et al, (1991). Nevado del Ruiz volcano is located approximately 120 km west of Bogata, Colombia. (location indicated on Figure 3a). The peak, situated at about 5,400 m above sea level, is covered by a perennial ice cap that is between 10 and 30 m thick and has a volume of about 337 million m³. On the evening of November 13, 1985, a small eruption triggered 5 lahars.

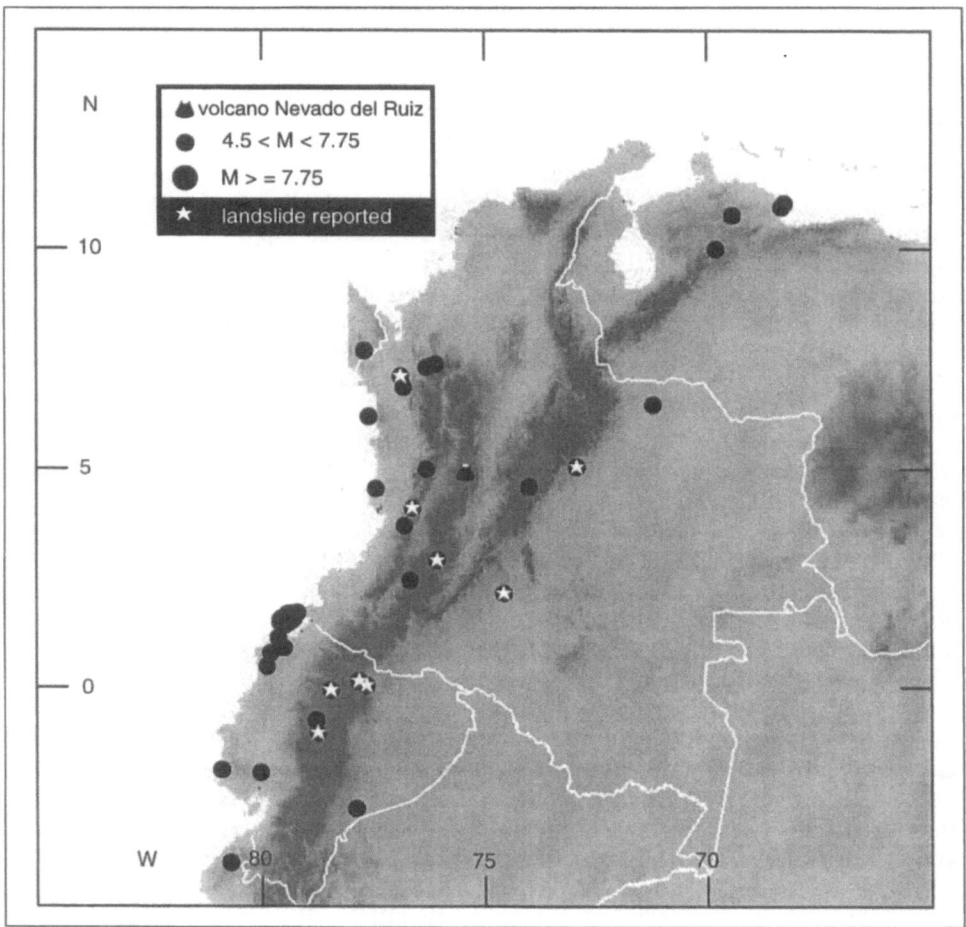

Figure 3a. Distribution of shallow to intermediate (depth <80 km) earthquake epicenters in the northern Andes with Richter magnitudes reported to be greater than 4.5. A star indicates that landslides were reported in association with the earthquake.

The catastrophic mudflows began on the 10-12 degree upper slopes of Nevado del Ruiz, entering the headwaters of several streams which drain the volcano. The most destructive of the lahars was an approximately 6-m thick series of warm to hot waves of mud, moving at a velocity of about 40 km/hr. This lahar obliterated the town of Armero and killed most of its 24,000 inhabitants.

Mileti et al. (1991) propose the following sequence of events as triggering the lahars: 1) ice cap melting was initiated due to a rapid accumulation of hot pyroclastic material; 2) falling ash and pyroclastic debris added mass to pre-existing sediments; 3) continued ice-melt and heavy rainfall added more water to the material; and finally, 4) ground vibration associated with the eruption increased pore-water pressure and led to failure.

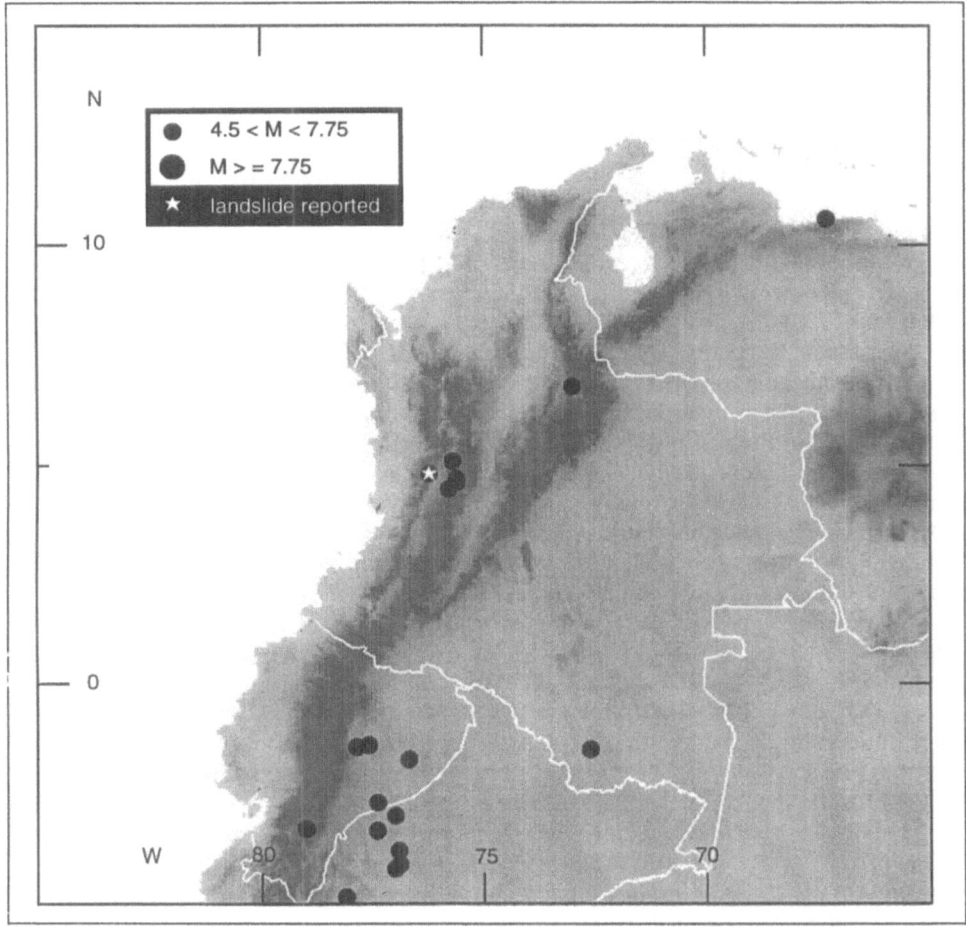

Figure 3b. Distribution of deep (depth > 80 km) earthquake epicenters in the northern Andes with magnitudes reported to be greater than 4.5. A star indicates that landslides were reported in association with the earthquake.

Mileti et al. (1991) have concluded that ice-melt alone was insufficient to trigger the mudflows; earthquake-generated ground motions, resulting in soil liquefaction, were also necessary to initiate them.

2.4. RAINFALL, SNOWMELT, AND FLOODING AS LANDSLIDE TRIGGER MECHANISMS

According to previous studies, landslides occur more frequently during snowmelt and in the rainy season than in the dry season (Michelena, 1989; Stalin, 1989; Blodgett et al., 1996). However, because many of the regions within various Andean countries receive high precipitation during different parts of the year, evaluating how often landslides occur during the wet season for each country is problematic. From available news reports found in the NEXIS computer database (May 1986 - May 1996) and accounts in the geologic literature, we compiled a list of 61 hydrologically triggered landslides, including dates of occurrence, descriptions of the event, and resulting casualties and/or damages. The most catastrophic landslides (here defined as those which incurred 15 or more casualties) are detailed in Table 2, while all of the known events are mapped in Figures 4a and 4b. No clear trend in the timing of landslide events by month for each country is apparent. Furthermore, hydrologically triggered slides are not restricted to the areas receiving the most precipitation, but also occur in more arid regions that have experienced a period of unusually heavy precipitation. Unusual deluges in arid regions of the Central Andes are often associated with El Niño events. Landslides triggered by heavy rains on the coast of Peru are examples of these (1983, 1987, and 1988).

In assessing the potential of landslide hazards for an area, other factors besides precipitation rate and distribution should be taken into consideration. For example, shallow landslides cannot occur unless the overburden reaches a critical depth. When pore pressure in the soil overcomes the frictional and cohesive forces holding the soil in place, slope failure results. The nature and density of vegetation determines to a great extent the cohesive strength of the soil. Human activity which involves the destruction of vegetation therefore reduces the threshold needed for a slide to occur. Although a deeper-seated slide may fail below root level, vegetation also helps transpire soil moisture, thereby reducing pore pressure and the weight of the soil.

Rotational slumps tend to be restricted to unconsolidated sediments, while shallow slips and slides occur with greater frequency on steep slopes underlain by resistant bedrock. In steep forested regions, landslides are not confined to swales, but occur over a range of hillslope morphologies and slopes. Landslides may be triggered not only by loading of the overburden, but also by rivers undercutting their banks and destabilizing the valley slopes. Human activity, including road-building, can similarly undercut a hillslope and contribute to failure.

Rockfalls or slides whose fault scarp occurs within the bedrock are not likely to be affected by vegetation, but groundwater may be a contributing factor. Examples of rock failure appear to be more common above the treeline where weathering rates are low and glaciated terrain has oversteepened the valley walls.

Demographics must also be considered when assessing landslide hazards in the Andes. In certain urban districts (such as Caracas, Medellin, and La Paz), shantytowns perched precariously on devegetated steep slopes have been subject to numerous hydrologically triggered landslides. Hydrologically triggered slides tend to be more common but individually less destructive than lahars or earthquake-triggered mass movements. They also particularly affect rural villages, farms, and roads, where land use practices have devegetated and/or undercut already steep, unstable slopes. However, such slides are also a natural feature of regions of high relief in the Andes. In a relatively uninhabited watershed in the Bolivian Yungas valleys, for example, 5% of the watershed as observed in air photos and field surveys has been visibly scarred by landslides.

A few of the most catastrophic hydrologically triggered landslides are summarized below, based upon descriptions drawn from geologic literature and news reports.

2.4.1. Mantaro landslide, Peru, April 1974

On the evening of April 25, 1974 approximately 1.0 to 1.3 x 10^9 m^3 of soil and rock slid into the Mantaro river, in a sequence of slides (Berrocal et al., 1978). The landslides buried the village of Mayunmarca and several farms, killing about 400 people. The trigger is unknown. It is known that the unconsolidated materials that slid had been saturated, as a result of a rise in the water table and an increase in underground flow. The landslide area had had very low seismicity, with only two intermediate (approximately 100 km) events in the previous 70 years. It is possible, though, that a very small earthquake (with a magnitude not recorded by local seismic stations) may have caused the slide.

2.4.2. Central Andes, Chile, November 1987

High snowfall followed by rapid snowmelt triggered a rockslide and debris flow of about 5.5 x 10^6 m^3 of material on November 29, 1987, on the Río Colorado of the Chilean central Andes (Cassassa and Marangunic, 1993). Unstable sedimentary rocks perched on a 60 degree slope slid about 1 km on a 40 degree slope, destroying a large part of a hydroelectric project and killing 29 people. The slide was about 40 meters in thickness.

2.4.3. Guanay, Bolivia, December 1992

On December 8, 1992, torrential rains triggered a landslide that carried hundreds of thousands of tons of mud and rock onto the mining camp of Llipi, approximately 55 km north of Guanay (from news reports). The mining camp is situated along the Tipuani River, in a valley scarred with large tracts of barren soil from large and small mining operations. In many areas, entire mountainsides have been deforested through strip mining.

2.4.4. Nambija, Ecuador, May 1993

On May 8, 1993, tons of rocks, boulders, and dirt slid down El Tierro Mountain, following heavy winter rains (from news reports). Part of a small mining town in Guayaquil was buried by the slide, and more than 300 persons were killed. According to a representative of an Ecuadorian mining agency, the village was particularly susceptible to landslides as a result of the primitive subterranean gold mines lacing the region.

3. Conclusions

Landslides are locally common in the high relief regions of the Andes. Based upon our review of reports on Andean landslides, by far the greatest number of mass movements tend to have been hydrologically triggered. On the other hand, the landslides most destructive to life and property were triggered by earthquakes; such landslides are particularly abundant in the coastal regions of Chile and inland in Peru. Landslides can be triggered by earthquakes with a wide range of magnitudes and depths, but larger, shallow quakes are more likely to cause landslides. Volcanoes rarely erupt, but when they do, they can serve as a third trigger mechanism for landslides in the Andes.

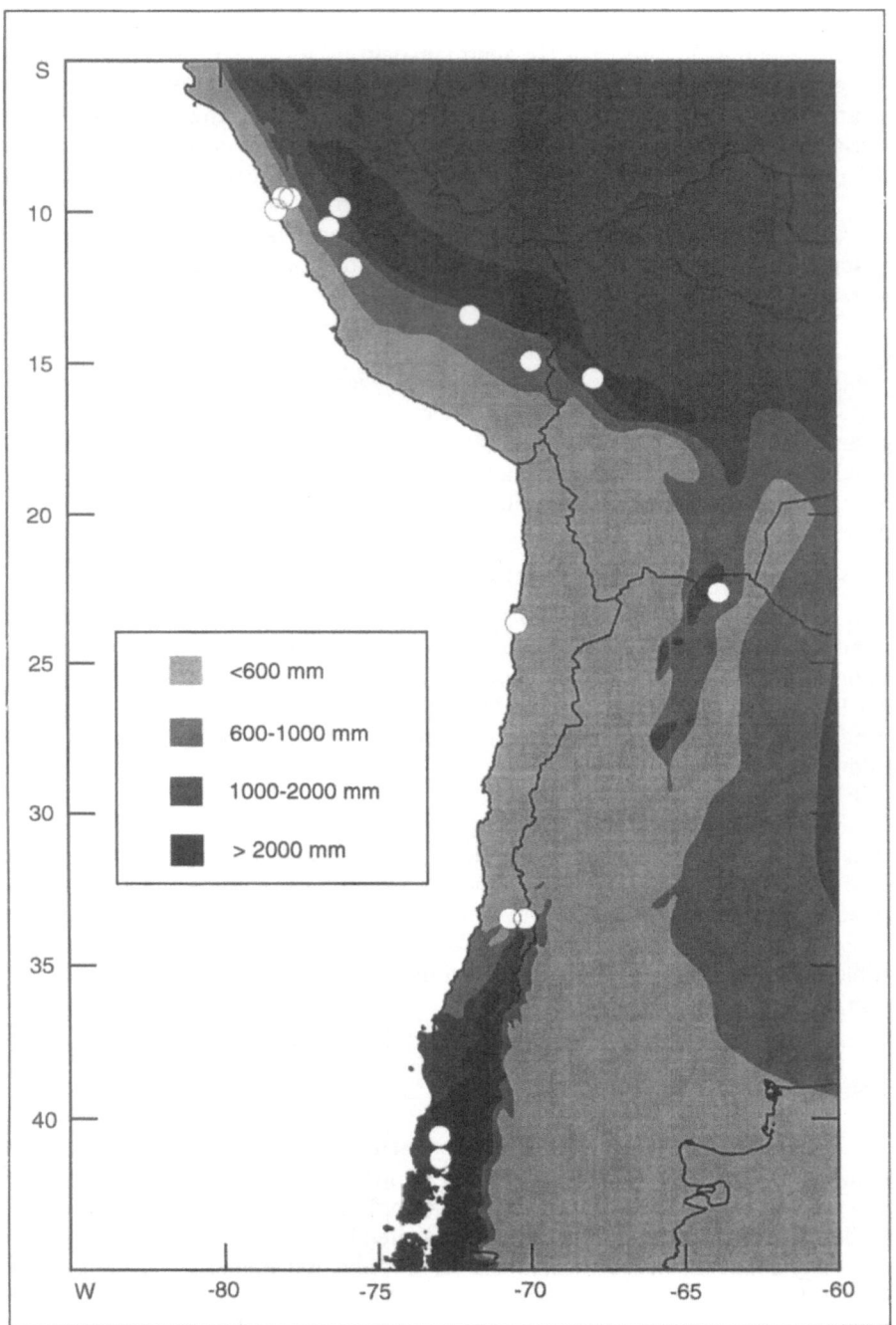

Figure 4a. Distribution of hydrologically triggered landslides and annual precipitation in the northern Andes.
Note landslides are most prevalent in regions which experience intermediate amounts of precipitation annually.

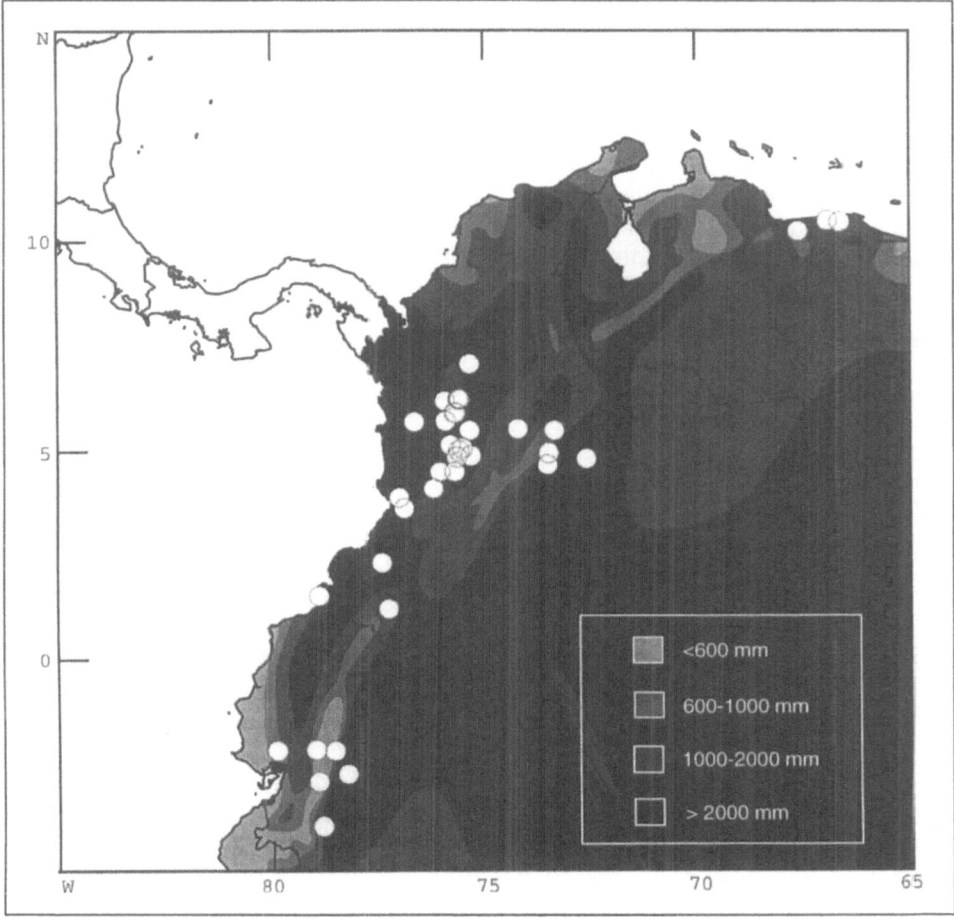

Figure 4b. Distribution of hydrologically triggered landslides and annual precipitation in the central Andes.
Note landslides are most prevalent in regions which experience intermediate amounts of precipitation annually.

Extremely hazardous conditions particularly exist in cases where the volcano supports a cover of snow and ice. Earthquakes associated with magma movements may also contribute to slope failure. In addition, volcanic soils are particularly prone to sliding.

Because a variety of factors contribute to landslide development, in our opinion no simple formula would be useful in predicting the likelihood that landslides will occur in a region. Instead, data regarding many factors, natural and anthropogenic, are necessary to assess the hazards that landslides pose to a particular area. Since landslide hazards by definition must include how the phenomena affect people, demographics are an important consideration. Reports from extremely wet areas and extremely dry areas are fewer because population centers tend to develop in regions with moderate amounts of moisture. Such reports probably do not reflect the gross magnitude of erosion by landsliding that is taking place.

While tectonics and climate are largely outside of the realm of human influence, in many cases anthropogenic factors increase landslide risk or magnify the destructive impact of such a slide. It is possible that many landslides could be prevented, or the damage markedly reduced, by modification of human settlement patterns, transportation networks, or land use practices.   Caracas, for example, suffers repeatedly from landslides because shanty towns are perched precariously on deforested, unstable hillslopes above the city.   Although physical modifications, such as the relocation of population centers, are easy to propose for cases like the shantytowns of Caracas, political and socioeconomic factors make the institution of such changes problematic.

References

1. Alexander, D. (1992) On the causes of landslides: Human activities, perception, and natural processes, *Environmental Geology and Water Science* 20 (3), 165-179.

2. Belloni, L. and Morris, D. (1991) Earthquake-induced shallow slides in volcanic debris soils, *Geotechnique* 41 (4), 539-551.

3. Berrocal, J., Espinosa, A. F., and Galdos, J. (1978) Seismological and geological aspects of the Mantaro landslide in Peru, *Nature* 275, 533-536.

4. Blodgett, T. A. (1996) Erosion attributed to landslides in the Cordillera Real, Bolivia, *Eos, American Geophysical Union 1996 Spring Meeting* 77 (17), suppl., Abstract T22B, S261.

5. Cahill, T. and Isacks, B. L. (1992) Seismicity and shape of the subducted Nazca plate, *J. of Geophys. Res.* 97 (B12), 17,503-17,529.

6. Casassa, G. and Marangunic, C. (1993) The 1987 Río Colorado rockslide and debris flow, Central Andes, Chile, *Bull. of the Assoc. of Eng. Geol.* 30 (3), 321-330.

7. Chinn, D. S., Isacks B. L. and M. Barazangi (1980) High-frequency seismic wave propagation in western South America along the continental margin, in the Nazca plate, and across the Altiplano, *Geophys. J. R. astr .Soc.*, 60, 209-244.

8. Cluff, L. S. (1971) Peru earthquake of May 31, 1970, Engineering geology observations, *Bull. of the Seismological Soc. of Am.* 61 (3), 511-533.

9. Davis, S. N. and Karzulovic, K. (1963) Landslides at Lago Rinihue, Chile, *Bull. of the Seismological Soc. of Am.* 53 (6), 1403-1414.

10. Ericksen, G. E., Ramirez, C. E., Concha, J. F., Tisnado, G., and Urquidi, F. (1989) Landslide hazards in the central and southern Andes. *In* Brabb, E. E. and Harrod, B. L. (eds.), *Landslides: Extent and Economic Significance.*, A. A. Balkema, Rotterdam, 111-117.

11. Hewitt, K. (1983) Seismic risk and mountain environments: The role of surface conditions in earthquake disaster, *Mountain Res. and Dev.* 3 (1), 27-44.

12. Isacks, B.L., 1988, Uplift of the central Andean plateau and the bending of the Bolivian orocline, *J. Geophys. Res.*, 93, 3211-3231.

13. Jordan, T. E., Isacks B. L., Allmendinger R. W., Brewer J. A., Ramos V. A., and Ando C. J. (1983) Andean tectonics related to geometry of subducted Nazca plate, *Bull. Geol. Soc. Amer.*, 94, 341-361.

14. Keefer, D. K. (1984) Landslides caused by earthquakes, *Geol. Soc. of Am. Bull.*, 95, 406-421.

15. Michelena, R. E. (1989) Landslides in Peru. *In* Brabb, E. E. and Harrod, B. L. (eds.), *Landslides: Extent and Economic Significance*, A. A. Balkema, Rotterdam, 119-121.

16. Mileti, D. S., Bolton, P. A., Fernandez, G., and Updike, R. G. (1991) *The Eruption of Nevado del Ruiz Volcano, Colombia, South America, November 13 1985*, Natural Disaster Series, Vol. 4, National Academy Press, Washington, D.C.

17. Mojica, J. and Eslava R., J. A. (1987) Aspectos geologicos y meteorologicos de los deslizamientos en Rondon (Departmento de Boyacá, Colombia) y en especial de los ocurridos en Junio-Julio de 1986, *Geología Colombiana* 16, 65-79.

18. Nieto, A. S., Schuster, R. L., and Plaza-Nieto, G. (1991) Mass wasting and flooding. *In* Schuster et al., *The March 5, 1987 Ecuador Earthquakes: Mass Wasting and Socioeconomic Effects*,. Natural Disaster Studies, Vol. 5, National Academy Press, Washington, D. C., 51-82.

19. Plafker, G., Ericksen, G. E., and Concha, J. F. (1971) Geological aspects of the May 31, 1970, Perú earthquake, *Bull. of the Seismological Soc. of Am.* 61 (3), 543-578.

20. Preciado, A. P. and Llinas, R. D. (1990) El deslizamiento de Coloradales - El Salitre en la zona de Paz de Rio, Boyacá, *Geología Colombiana* 17, 169-182.

21. Richter, C. F. (1958) *Elementary Seismology*, Freeman, San Francisco.

22. Rothé, J. P. (1969) The Seismicity of the Earth; 1953-1965, UNESCO, Paris.

23. Stalin, B. A. (1989) Landslides: Extent and economic significance in Ecuador. *In* Brabb, E. E. and Harrod, B. L. (eds.), *Landslides: Extent and Economic Significance*, A. A. Balkema, Rotterdam, 123-126.

24. Varnes, D. J. (1984) Landslide Hazard Zonation: A Review of Principles and Practice, UNESCO, Paris.

25. Weischet, W. (1963) Further observations of geologic and geomorphic changes resulting from the catastrophic earthquake of May 1960, in Chile, *Bull. of the Seismological Soc. of Am.* 53 (6), 1237-1257.

26. Wright, C. and Mella, A. (1963) Modifications to the soil pattern of south-central Chile resulting from seismic and associated phenomena during the period May to August 1960, *Bull. of the Seismological Soc. of Am.* 53 (6), 1367-1402.

Authors

Blodgett T.A., Blizard C., Isacks B.L.
Cornell University
Department of Geological Sciences
Snee Hall, Ithaca, NY 14853
U.S.A.

FLUVIAL HAZARDS IN A STEEPLAND MOUNTAIN ENVIRONMENT, SOUTHERN BOLIVIA

JEFF WARBURTON, MARK MACKLIN, DAVID PRESTON

1. Introduction

Mountainous environments are highly active geomorphologically and increasing human use of mountains has lead to an increase in natural hazards (Messerli and Ives, 1984). Heuberger and Ives (1994) argue that mountain environments are undergoing ever intensified development and because of the high geomorphic activity of mountain environments this development may create increased stress. In the context of fluvial hazards in mountain environments „Mountain rivers become a hazard only when they threaten human life or property, by inundation, erosion, sediment deposition or destruction" (Davies, 1991). There are a large range of fluvial hazards that occur in mountains (e.g. glacial outburst floods (Driedger and Fountain, 1989), Alpine debris flows (Lewin and Warburton, 1994), etc.) however they all generally result from extreme temporal and spatial variability in fluvial and hydrological processes. Unfortunately, catastrophic events and large-scale slope movements, triggered by human activities in mountain environments, are not fully understood or adequately incorporated into development planning. Possibly the most important contribution of fluvial geomorphology to flood hazard assessment in mountain environments is in recognising the relationships between control of flood waters and the control of sediment and river channel change. An appreciation of the geomorphology of fluvial landforms and deposits can yield valuable information about flood magnitudes, sedimentation and channel stability over long time periods and between river basins (Dunne, 1988). Because of these factors, mountain environments pose unique problems for fluvial hazard assessment, prediction and mitigation (Heuberger and Ives, 1994).

Erosion in mountain areas depends on climate and runoff; landscape character; degree of human interference; relief; recent tectonic activity and the underlying geology (Dedkov and Moszherin, 1992). In Latin America farm systems in the mountainous areas suffer environmental stress from a number of natural hazards such as floods, droughts, storms and landslides which all contribute to the breakdown of traditional agricultural practices. Bolivia, in particular, is characterised by low-levels of agricultural production per unit area with large parts of the land, sometimes up to 41% by area, being affected by soil erosion (Brockman, 1986; Woodward, 1994). An understanding of hazards is needed if these areas are to expand and develop (e.g. building of bridges, irrigation channels, etc.) or avoid irreversible degradation of georesources. In the mountain regions of Bolivia, detailed hazard information does not exist, however it is hypothesised that mountain river catchments are particularly sensitive to environmental change arising from both natural and anthropogenic causes. The aim of this paper is to identify the key fluvial hazards affecting the steeplands of southern Bolivia and

J. Kalvoda and C.L. Rosenfeld (eds.), Geomorphological Hazards in High Mountain Areas, 229-243.
© 1998 *Kluwer Academic Publishers.*

develop a framework for evaluating the frequency, magnitude and spatial impact of flood events.

Figure 1. Fluvial hazards in mountain environments.

LINK PROCESS (POTENTIAL HAZARDS)	SEDIMENT STORES (GEOMORPHIC ZONES)			
	Slopes	Gully systems	Alluvial fans and cones	Valley floor
Slopewash	●	○	●	
Debris flows		●	●	
Tributary floods		●	○	
Mass movements	●	○	○	○
Main valley floods			○	●

● Area of greatest impact   ○ Secondary impact

2. Approaches to fluvial hazard evaluation in mountain environments

Schumm (1988) defines geomorphic hazards as any landform change, natural or otherwise that adversely affect the geomorphic stability of a place. This can be the result of catastrophic events; a progressive change leading to an abrupt change; or a progressive change with slow and progressive results. Hazard identification and prediction therefore relies on determining the sensitivity of landforms to change, understanding the complex nature of geomorphic change and recognising the great variability (or singularity) in landforms and their susceptibility to change. This last condition of 'landform singularity' is the main obstacle to the accurate definition of geomorphic hazards (Schumm, 1988). This is especially true in mountain environments where the distinctive feature of fluvial processes is extreme variability in time and space (Davies, 1991).

The overall aim of a hazard assessment is to provide information on the magnitude, frequency and aerial extent of geomorphic events (Davies, 1991; Slaymaker, 1996). Hazard assessment in mountain regions has traditionally relied on two basic approaches (Aulitzky, 1994):

1) Historical approach - Accumulation of past histories of historical events e.g. using documented sources, old photographs, or oral histories.

2) Geomorphological mapping and interpretation - Gathering of geomorphological evidence using indicators of past disasters e.g. mapping and dating of flood deposits.

In the absence of detailed historical records geomorphology is the logical starting point for investigating past events and providing a basis for hazard planning. Baker (1988) has outlined a systematic approach to geomorphic event reconstruction. This involves the discovery in the field of key indicators of the occurrence and magnitude of

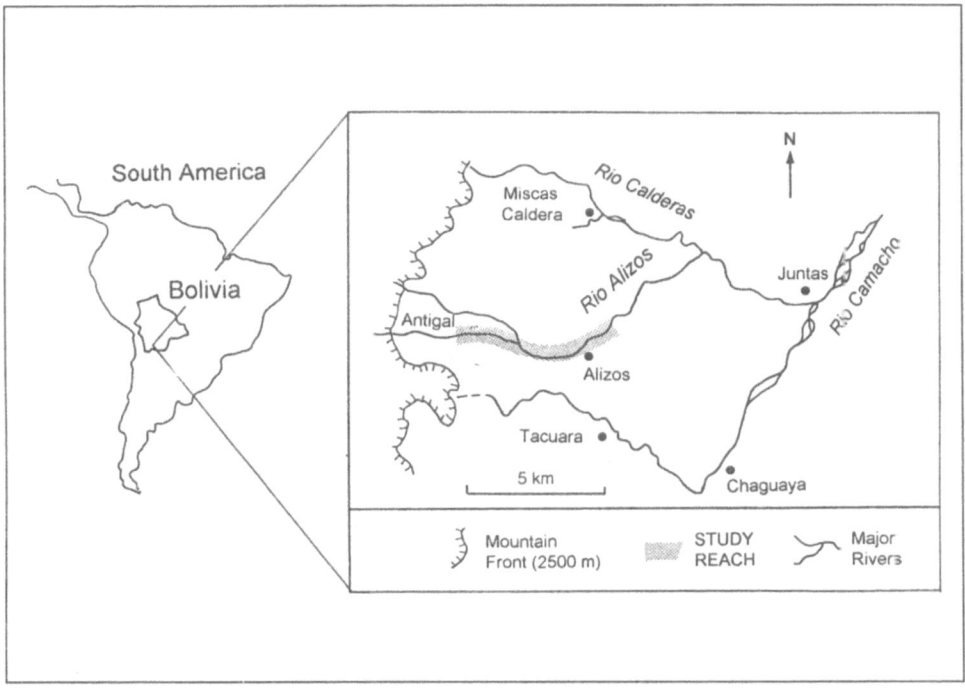

Figure 2. Location of the River Alizos catchment in southern Bolivia.

the geomorphic event; palaeomagnitude analysis of the key processes (e.g. flow recon-struction); stratigraphic studies and dating of events; and magnitude-frequency analysis. The obvious limitation of this approach is a full record of geomorphic events may not be preserved in the geomorphological and sedimentological record. This pro-blem can be minimised by investigations at multiple sites whereby a full record of geomorphic history can be reconstructed. Once event histories are established these can be used to interpret deposits in catchments with only fragmentary records.

Even if long historical records of geomorphic events exist sample data sets are ge-nerally too small to justify statistical analysis and, in situations where similarity in the generating mechanisms for moderate and large events cannot be guaranteed, the appli-cation of extreme value theory cannot be applied (Baker, 1988). For example, changing climate and land-use make predictions of catchment response based on relatively short records of past history very unreliable. Therefore it must be assumed that the probabi-lity of large events occurring is low but unknown. In this situation the aim of hazard assessment should be to determine the maximum likely event magnitude and the extent of the area threatened (Davies, 1991).

Davies (1991) recommends that the magnitude of past events should be estimated from sediment deposits, erosion scars and vegetation soil-age distributions, and palaeo-flood analysis and methods for the rapid monitoring of catchment and sediment source conditions should also be developed.

Figure 3. View upstream of Reach 2 towards the mountain front. The river flood gravels and terraces can clearly be seen in the floodplain.

A useful starting point for analysing hazards in mountain areas is to define the interrelationships between geomorphic processes· and potential sediment sources (Figure 1). Analysis of sediment sources is fundamental to understanding fluvial hazards in mountain areas because events of sufficient magnitude to represent a major hazard require large amounts of sediment. The likelihood of such an event, given the appropriate antecedent and prevailing meteorological conditions, should be assessed using a survey of potential sediment sources (Davies, 1991). Schumm (1988) in his assessment of geomorphic hazards distinguishes four main zones:

1) Drainage networks including lower order streams - sediment source areas.
2) Hillslope areas between channels and drainage networks.
3) The main river channel.
4) Piedmont and plain areas including fans and areas of accumulation.

These four zones approximate the zones shown as sediment sources in Figure 1. The symbols in Figure 1 show the processes that link the geomorphic zones and indicate the areas of greatest geomorphic activity. The diagram is useful in defining linkages between sediment sources and geomorphic processes and providing an initial framework for evaluating sediment transfers and potential hazards. For example, mass movements transfer sediment from the slopes to the gully systems and all processes deliver sediment to the main river valley. The exact nature of the links in Figure 1 will vary depending on the particular mountain environment.

The remainder of this paper will be devoted to developing methods appropriate to fluvial hazard evaluation in steepland mountain environments.

3. Study area and methods

The main study area is the west-east trending Alizos River Valley, in the Eastern Cordillera of Southwest Bolivia. This valley is within the Sub-Andean Zone on the eastern flank of the Andes (Figure 2). This area is characterised by high relative relief (>2000 m), spectacular badland gully systems and very active steep (0.033 m m^{-1}) boulder and gravel-bed braided rivers (Figure 3). Soil erosion and flooding are major problems in this region but at present it is unclear whether these are controlled primarily by land-use change or variations in climate. The Rio Alizos is a tributary of the Rio Camacho (Alizos catchment area 51.3 km^2). The valley contains a deep Holocene valley fill and suite of terraces made up of unconsolidated and semiconsolidated pebbles, gravels, sands, silts and clay. In the upper catchment Cambrian quartzites, sandstones, siltstones and conglomerate outcrop. Mid-valley Pre-Cambrian phyllites and schists occur, whilst in the lower valley Ordovician and Silurian siltstones and shales are exposed. Within the valley clear geological divisions correspond to distinct structural zones bounded by Quaternary reverse faults which trend in a north/south direction, parallel to the mountain front. The area is characterised by a semi-arid climate with a mean annual precipitation of 800 to 900 mm. Most of the rain is associated with intense storms in October through to February. Soils are generally poorly developed, mainly sandy clay loams (Woodward, 1994). Hillslope vegetation is relatively sparse, mainly bushes and trees, with intensive multi-cropping agriculture concentrated along the valley floor and on the river terraces.

Table 1. Characteristics of the four study reaches

	Reach 1	Reach 2	Reach 3	Reach 4
Distance from mountain front (km)	2.2	4.0	6.3	7.9
Length of mapped reach (m)	750	900	700	600
Active channel slope (m m^{-1})	0.043	0.033	0.031	0.025
Local bedrock geology	Shales & siltstones	Shales, siltstones & sandstones		
Number of units mapped in each reach	47	42	22	23

Fluvial hazards were assessed at two scales: the valley scale (Figure 4) and the reach scale (Figure 5). Geomorphological mapping at the valley scale (Figure 4) provided an overview of the fluvial system. A simple fluvial geomorphological hazard map was used to examine the degree of sediment transfer coupling between the slopes and tributary streams; between tributary streams and the main river valley; and determine the distribution of georesources and settlements at risk. This was completed from the most recent 1983 air photographs with field checks in 1995 to verify features on the ground. In many mountain environments this is the principal mode of hazard assessment but should never be undertaken without field verification. For example abundant large scale rills and shallow gullies on the south side of Alizos appear to be highly active however, inspection on the ground reveals that the slopes are relatively stable, they are poorly connected to the main stream network and sediment is accumulating in the gully beds. This map (Figure 4) was used as a basis of erosion assessment of the tributary catchments.

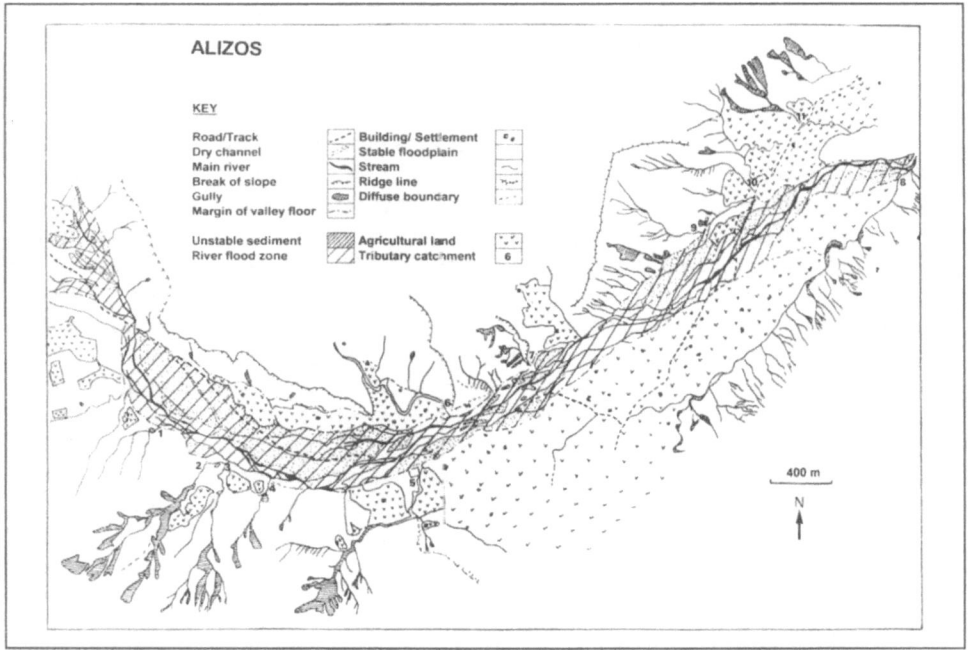

Figure 4. Fluvial geomorphological hazard map of the study area (see Figure 1).

Detailed assessment of flood magnitude and frequency was also made at the reach scale. Lichenometric dating (using *Rhizocarpon geographicum agg.*) and sediment size measurements at four sites were undertaken on a series of flood units characterised by boulder berms, splays and sheets (Table 1).

Four reaches were selected, spaced at approximately 2 km intervals away from the mountain front (Table 1). The four reaches were mapped in the field and a series of flood units delimited. Lichenometry was used for estimating the age of coarse flood deposits. This is a well established dating technique in glacial geomorphology and has more recently been applied very successfully in fluvial geomorphology (Macklin et al., 1992). The technique is based on the assumption that the size of lichens present on a deposit is directly proportional to the time elapsed since deposition. This assumes that that lichen colonisation occurs relatively quickly (estimated at 2 to 5 years in this study) and reworked boulders containing older lichens can be readily distinguished from fresh deposits.

The crustose lichen *Rhizocarpon geographicum agg.* was used in this study. An age-versus-size curve was constructed by measuring the diameters of lichens found on gravestones from five graveyards in the local area. This gave an average growth rate of 0.26 mm yr^{-1}. The age of flood deposits was estimated from the average of between 10 and 20 largest lichens measured on the surfaces of large cobbles and boulders. The intermediate axes of 20 largest boulders were measured on each of the flood units as a means of estimating relative flow magnitude. Local channel slopes were measured with an Abney level.

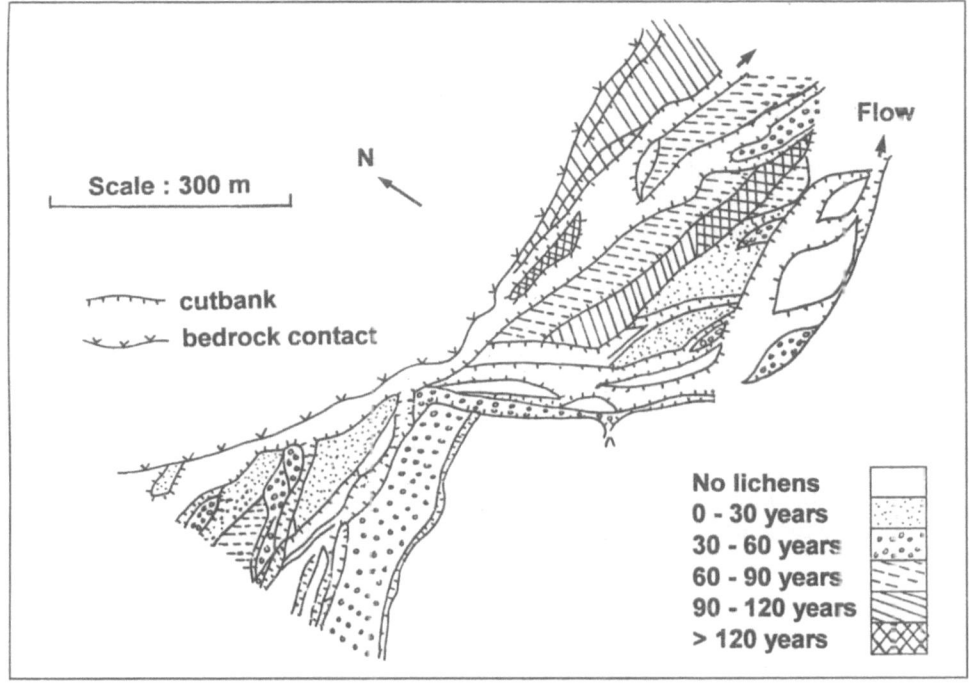

Figure 5. Morphological map and lichenometric-age estimates of flood units in Reach 2.

4. Fluvial hazard assessment: Alizos valley, southern Bolivia

Assessment of fluvial hazards in the Alizos Valley was undertaken at the valley and reach scale. At the valley scale a hazard map was produced, comparisons between 1967 and 1983 air photographs were made and an analysis of the erosion potential of tributary valleys was undertaken. Reach scale investigations provided detailed evidence on the timing, magnitude and style of valley flooding.

4.1. FLUVIAL GEOMORPHOLOGY HAZARD MAP

The objectives of mapping were to determine the form of the fluvial system; define sources of sediment and areas of erosion (bare or unconsolidated sediment which is potentially unstable); determine the maximum extent of floodplain erosion and sedimentation (recent cut banks); record the distribution of georesources (agricultural land, improved pasture); and map roads and settlements (human infrastructure).

A key to these features is supplied along with Figure 4. The hazard map in Figure 4 shows two main things; the zone of potential valley flooding and the tributary catchments which contain large areas of apparently unstable sediments. Catchments to the south-west of the valley seem to be the most erosion susceptible and most closely coupled with the main river valley. Given the braided nature of the river and evidence of gravel splays distributed across the full width of the floodplain most of the valley floor is at risk from flooding. Most settlement is at little risk from flooding, however, large areas of agricultural land, particularly in the south east, could be subjected to lateral erosion and the main valley road would be severely disrupted following a flood. Figure 4 illustrates the general point that in mountain environments the greatest risk from geomorphic hazards occurs along stream courses and on steeper slopes (Kienholz et al., 1984) however, because of the difficulty in predicting geomorphic events in mountain environments there is still considerable uncertainty in any hazard assessment. With a map of this type uncertainty arises for a number of reasons. In a reconnaissance survey, it is not possible to produce a detailed zonation of hazards because information on the individual degrees of hazard cannot be assessed without a detailed understanding of both the natural processes and the socio-economic values attached to the land. Furthermore the hazard map produced in this way represents a static impression of a dynamic system and should be regarded as a working document (Verstappen, 1983). Assessment of erosion from air photographs only provides an estimate of apparent erosion. Deeply incised gullies on the air photographs may be relatively stable features on the ground, as shown by stable vegetation and lichen covers, despite these provisos this general map provides a useful basis for more detailed investigations.

Table 2. Catchment characteristics of the principal tributaries of the upper Alizos Valley. Twelve catchment were analysed and these are numbered on the geomorphological hazard map (Figure 4).

Catchment Number	Area km²	Area eroded km²	Erosion / unit area	High point m	Low point m	Relative relief m	Catchment length m	Slope	Erosion stability *	Runoff peak flow **	Erosion potential ***	Erosion rank
1	0.038	0.0004	0.01	2300	2100	200	327	0.612	0.006	0.211	0.001	9
2	0.159	0.0280	0.18	2340	2090	250	655	0.382	0.067	0.884	0.059	3
3	0.074	0.0110	0.15	2320	2090	230	597	0.385	0.057	0.411	0.024	6
4	0.039	0.0044	0.11	2200	2090	110	327	0.336	0.038	0.217	0.008	8
5	0.150	0.0260	0.17	2390	2070	320	770	0.416	0.072	0.834	0.060	4
6	4.100	0.8200	0.20	2310	2060	250	3100	0.081	0.016	22.796	0.368	1
7	0.022	0.0037	0.17	2250	2040	210	231	0.909	0.153	0.122	0.019	7
8	0.800	0.0400	0.05	2300	2000	300	1200	0.250	0.013	4.448	0.056	2
9	0.047	0.0000	0.00	2250	2040	210	404	0.520	0.000	0.261	0.000	10
10	0.049	0.0000	0.00	2230	2010	220	558	0.394	0.000	0.272	0.000	10
11	0.064	0.0100	0.16	2240	2010	230	366	0.628	0.098	0.356	0.035	5

Notes:
* Erosion stability is defined as the erosion /unit area multiplied by the slope index of the tributary catchment.
** Asssumes a runoff coefficient of 0.4 and a rainfall intensity of 50 mm hr⁻¹ (Woodward, 1994).
*** Erosion potential is defined as erosion stability multiplied by the runoff calculated from the Rational Method (Shaw, 1994).

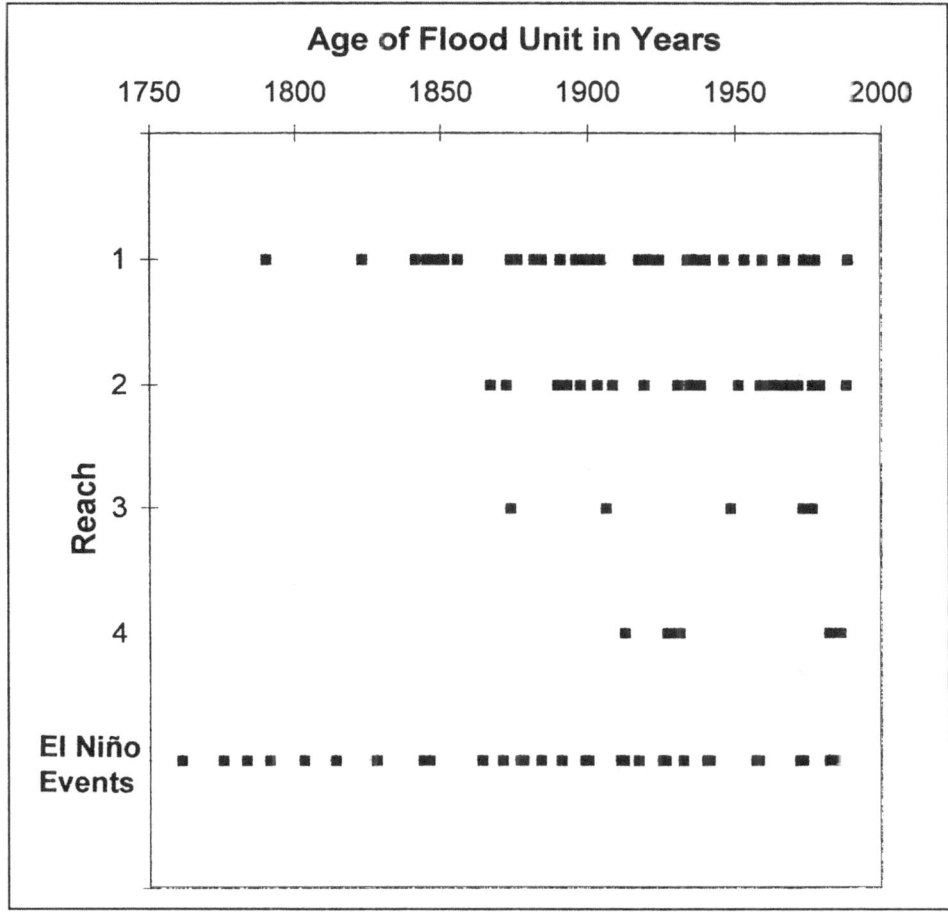

Figure 6. Sequence of Alizos Valley flood events. Reach one is closest to the mountain front Age estimates are based on a lichen growth rate of 0.26 mm yr^{-1}. Record of El Niño events from Quinn and Nea (1992).

4.2. COMPARISON OF 1967 AND 1983 AIR PHOTOGRAPHS

Historical documents of landscape change in Latin America are relatively few. This applies to southern Bolivia where only two sets of air photographs were obtained for the Alizos Valley. The first set taken in 1967 tend to be of relatively poor quality and lack definition. The most recent available photos, taken in 1983 are much better and were used for hazard mapping. Despite differences in quality it is possible to detect certain changes in the four study reaches. Reach 1 shows no apparent change in the extent of the gravel floodplain with many of the 1967 cutbanks still evident. Some fresh gravels were noted. Reach 2 is broadly similar between 1967 and 1983 however there is more fresh gravel by 1983 and there appears to be approximately 15-20 metres of erosion on a low terrace on the right bank below the main stream confluence (Figure

5). Reach 3 shows little change there is no evidence of floodplain widening although some fresh gravel has been deposited. Reach 4 shows signs of fresh gravel deposition and minor bank erosion (1-2 metres) on the left bank and some stripping of floodplain vegetation. Therefore over this 16 year period although there is evidence of extensive flooding from fresh gravel deposits apart from local bank erosion recent flooding was confined to the floodplain area.

4.3. EROSION IN TRIBUTARY CATCHMENTS

River channel confluences are often the foci for geomorphic hazards. Small, steep tributary catchments are characterised by high erosion rates and can potentially supply large volumes of sediment to the main river system. For example, there is an association between heavy rainfall and geomorphic activity; landslides tend to occur on slopes of 30 to 40 degrees (Corominas and Alonso, 1990) and starting zones for debris flows occur on slopes of 25-38 degrees (Zimmerman, 1990). Figure 4 identifies the catchments in Alizos which act as tributary sediment sources. These catchments were mapped (Figure 4, catchments numbered 1-11) and their erosion potential estimated (Table 2). Of the 11 catchments eight showed evidence of erosion with up to 20% of the catchment area affected in some instances. The erosion stability of the catchments was defined as the erosion per unit area multiplied by the slope of the tributary. Estimating the steepness of the catchment and the proportion of the area eroded provides an estimate of stability because it is assumed that steep catchments with abundant sediment supplies would be more unstable. Combining this estimate of stability with an estimate of runoff would provide an estimate of the erosion potential. In ungauged catchments peak flow (Q_p) can be estimated using the Rational Method (Shaw, 1994, 316-317). Q_p is calculated from the formula:

$$Q_p = b \ C \ i \ A \qquad (1)$$

where b is a conversion factor for metric units, C is a runoff coefficient (related to catchment characteristics, varies from 0 to 1), i is rainfall intensity (mm hr^{-1}) A is catchment area (km^2). This method works reasonably well for small uniform catchments however if little is known about the catchment characteristics this can only give a crude approximation of the peak flow. Erosion potential is therefore a measure of sediment availability and the ability to transport that sediment. In other words, steeper, larger catchments with abundant available sediment have a greater erosion potential. Results (Table 2) show that the larger catchments (6 and 8) dominate the tributary sediment transfer system, however some of the smaller catchments (2 and 5) have high erosion potential. These catchments tend to occur on the right bank of the Alizos valley.

4.4. TIMING, MAGNITUDE AND STYLE OF VALLEY FLOODING

Crucial in interpreting the significance of floods in mountain environments are attempts to evaluate the magnitude and frequency characteristics of the events. Estimates of flood magnitude can be determined from morphological and hydraulic reconstructions related to erosion features, trash lines and/or sediment characteristics indicative of the competence of the flood e.g. grain-size of flood deposits. Frequency or timing of flood events is often much more difficult to estimate. In this study boulder size and lichenometry are the principal methods used as a basis for reconstructing the recent flood history of this small mountain catchment.

The age distribution pattern of flood units is useful in assessing the relative stability of floodplain zones. Figure 5 shows the pattern for Reach 2. The oldest deposits occur to the centre of the reach and on the left bank. Units are fairly fragmentary and elongated downstream. Young deposits form gravel tracts typical of braided cobble-gravel rivers. River bank erosion is evident along the right bank where fresh gravels impinge on the floodplain margin. Relative relief across the floodplain is low and all units could be subject to flooding.

The number and age of the preserved flood units decreases systematically downstream away from the mountain front (Table 1). The largest number of flood events (the oldest of which dates from the late eighteenth century) are evident in an entrenching fan system (Reach 1) at the head of the valley while only three (all post 1900) are found less than 6 km downstream (Reach 4) (Figure 6). There are marked downstream changes in fluvial morphology and sedimentation styles along the study reach. In particular, a transition from a steep entrenched boulder fan at Reach 1 to a boulder gravel-bed braided system at Reach 4 (Table 1) accompanied by a downstream reduction in maximum boulder size (Figure 7) and a decrease in the number of identifiable flood units (Figure 6). There is also a reduction in valley-floor relief and channel slope. Boulder-size measurements indicate that the magnitude of floods has decreased over the last 200 years or so though the frequency of major floods has remained relatively constant Figure 6 and Figure 7). In terms of historical flood frequency and magnitude the valley floor has been significantly modified by flood events every 10 to 15 years since the early nineteenth century. It is noteworthy that sampling only Reach 1 or Reach 4 would have produced two very different flood histories. This highlights the importance of an appreciation of geomorphological setting and history when establishing historic flood series using the sedimentary record in mountain environments.

5. Discussion

Geomorphic hazards in mountain areas can be reduced through a better understanding of the main controlling factors. However, better awareness of the causes of natural hazards is only one means of reducing impact; recognition of the uncertainties of hazard prediction and a understanding the socio-economic system is equally important (Cannon, 1994).

It is hypothesised that flood hazards in the steeplands of southern Bolivia are linked to El Niño; the regional manifestation of the large-scale Southern Oscillation event (Quinn et al., 1987; Diaz and Markgraf, 1994). Periodic fluctuations in atmospheric pressure over the Indo-Pacific and the associated changes in wind fields, sea surface temperatures and oceanic circulation influence interannual climate variability on a global scale. These weather and climate anomalies with quasi-periodic occurrences of 1 to 5 years are referred to as El Niño/Southern Oscillation (ENSO) and have been identified as important influences on the continental hydrological cycle (Wells 1987).

Although linked to anomalous periods of floods and droughts (Baker, 1994), deciphering the links between the Southern Oscillation and the occurrence of large floods is not without difficulty. Three main factors hinder such a relationship being establis-hed. Firstly there is imprecision in determining the frequency of El Niño events. Quinn and Neal (1992) suggest the recurrence of moderate, strong and very strong events is about four years whilst for strong and very strong events it is nine years, whilst Bur-roughs (1992) suggests peaks in rainfall time series with 3.75, 7 and 20 year periodicities. Secondly, there are only weak correlations between variations in large scale ocean-atmosphere circulation and river flows (Diaz and Markgraf, 1992). This is not surprising given the complexity of the ocean-atmosphere-lithosphere interactions in producing runoff and the many filters which act to modify the climate forcing signal.

Thirdly, where documentary evidence does not exist, the fragmentary nature of the sedimentary record provides only a partial history of past events. For example, Wells (1987) using sedimentological and stratigraphic evidence from the Northern Coastal zone of Peru identified only 15 El Niño events during the Holocene which is equivalent to only one flood every 500 years which leaves behind significant deposits. In terms of floods constrained in a valley setting an incomplete sedimentary record could be the result of the action of later events removing the evidence of older episodes of flooding; sediment not being deposited on the surface (some events may be dominantly erosive whilst others may be depositional thus obscuring sediment histories); and all floods may not be leave deposits down the full length of the valley system.

Hazards in mountain areas have an important human dimension and understanding the socio-economic system is important (Cannon, 1994). Recognition of the problems of flood prediction and the futility of expensive engineering schemes has lead to a shift in emphasis away from traditional sectorial planning of complete hazard protection towards sustainable strategies involving indigenous adjustments (Haque and Zaman, 1994). Simple strategies can be highly effective in reducing loss e.g. in Switzerland bridges crossing debris flow channels are designed so that the middle sections will be lost during floods but can be easily replaced in the aftermath (Lewin and Warburton, 1994). In the Alizos valley, so long as current farming practices are not disturbed, the risk of erosion in this area seems small because present farming practices have evolved to cope with the climate, soils and slopes (Woodward, 1994). However, should the climate or soils change, by either climate fluctuations or slight changes in land-use practice, the landscape could very quickly suffer accelerated degradation. A scenario involving a slight increase in active gully erosion or the magnitude of valley flooding could greatly reduce the available land for agriculture. This could lead to intensified production and further environmental degradation.

6. Conclusions

In the mountains of southern Bolivia the greatest risk from geomorphic hazards occurs along stream courses and on steeper slopes. On the whole community land use practices are generally well adjusted to the local fluvial hazards. The risk to life is very low and loss of agricultural land is accepted within the current economic structure. If the valley is to develop further communications and irrigation would need to be exten-ded; this would increase the hazards particularly along main river valleys. Increased development may also trigger instability in sensitive tributary catchments which will have a knock-on effect through the fluvial system.

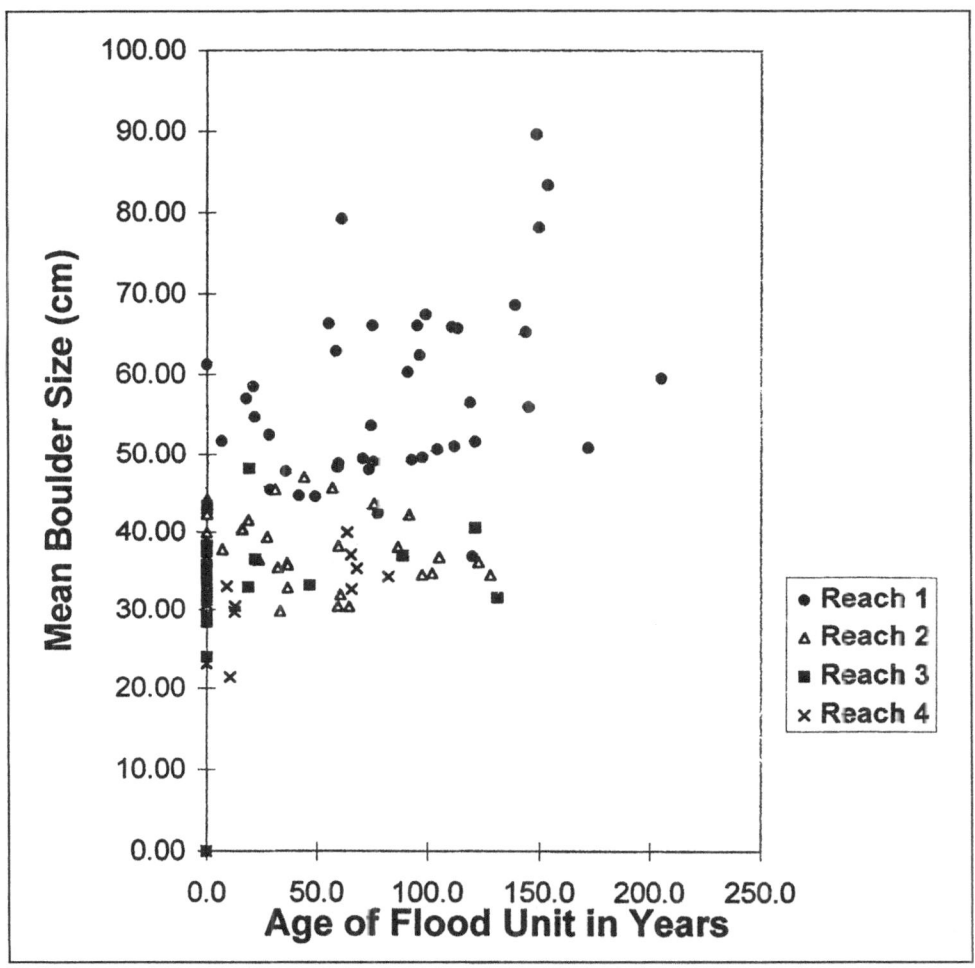

Figure 7. Age of flood unit plotted against boulder size for the four study reaches. The data show a downstream fining trend (Reaches 1 to 4) and towards the mountain front larger boulders are found in the older deposits.

The flood sequences identified in the Alizos valley show a similar periodicities to the ENSO climatic signal, however, further analysis is required to demonstrate how factors, not directly related to climate, complicate a simple correlation. These include land-use change, inaccuracies in dating control resulting primarily from a poorly constrained lichen growth curve, and a fragmentary record of the flood sediments. Further fieldwork and detailed analysis of regional rainfall data, together with an independent proxy climate record (e.g. lake deposits), is needed to resolve the various controls of flooding in more detail.

The hazard assessment methodology suggested here works well in mountain environments where little of the past history is known. Air photograph mapping and the calculation of erosion potential provide useful reconnaissance tools in determining areas at risk from fluvial hazards. The relative magnitude of flood events can be successfully estimated from morphological mapping and measurements of boulder size. Lichenometry provides a valuable dating technique in areas where no other forms of age control exist. Techniques can be refined with further fieldwork and new techniques e.g. survey of channel morphology will aid palaeoflow reconstruction and use of differential GPS will considerably speed up mapping of valley floor flood units.

References

1. Aulitzky, H. (1994) Hazard mapping and zoning in Austria: methods and legal implications. *Mountain Research and Development*, 14, 4, 307-313.

2. Baker,V.R. (1988) Cataclysmic processes in geomorphological systems. *Zeitschrift für Geomorphologie*, Suppl. Bd. 67, 25-32.

3. Baker, V.R. (1994) Glacial to modern changes in global river fluxes. In *Material Fluxes on the Surface of the Earth*. National Academy Press, Washington D.C., 86-98.

4. Burroughs, W.J. (1992) *Weather Cycles: Real or Imaginary?* Cambridge University Press, Cambridge.

5. Brockman, C.E. (1986) Resumen y Recomendación: Perfil Ambiental de Bolivia. La Paz, USAID.

6. Cannon, T. (1994) Vulnerability analysis and the explanation of 'natural' disasters. In Varley, A. (Ed.) *Disasters; development and environment*. Wiley, Chichester, 13-30.

7. Corominas, J. and Alonso, E.E. (1990) Geomorphological effects of extreme floods (November 1982) in the southern Pyrenees. In *Hydrology of Mountainous Regions. II-Artificial Reservoirs, Water and Slopes*. IAHS Publication 194, 295-302.

8. Davies, T.R.H. (1991) Research of fluvial processes in mountains - a change of emphasis. *Fluvial Hydraulics of Mountain Regions*. Armanini, A. and Di Silvio, G. (Eds.). Springer-Verlag, Berlin, 251-266.

9. Dedkov, A.P. and Moszherin, V.I. (1992) Erosion and sediment yield in mountain regions of the world. In *Erosion, Debris flows and Environment in Mountain Regions*. IAHS Publication 209, 29-36.

10. Diaz, H.F. and Markgraf, V. (Eds.) (1992) *El Niño: historical and paleoclimatic aspects of the southern oscillation*. Cambridge University Press, Cambridge, 476.

11. Driedger, C.L. and Fountain, A.G. (1989) Glacier outburst floods at Mount Rainier, Washington State, U.S.A. *Annals of Glaciology*, 13, 51-55.

12. Dunne, T. (1988) Geomorphologic contributions to flood control planning. In Baker, V.R., Kochel, R.C. and Patton, P.C. (1988) *Flood Geomorphology*. Wiley, Chichester, 421-438.

13. Haque, E. and Zaman, M. (1994) Vulnerability and responses to riverine hazards in Bangladesh: a critique of flood control and mitigation approaches. . In Varley, A. (Ed.) *Disasters; development and environment*. Wiley, Chichester, 65-79.

14. Heuberger, H. and Ives, J.D. (1994) Preface - Mountain Hazard Geomorphology. *Mountain Research and Development*, 14, 4, 271-272.

15. Kienholz, H., Schneider, G., Bichsel, M., Grunder, M. and Mool, P. (1984) Mapping of mountain hazards and slope instability. *Mountain Research and Development*, 4, 3, 247-266.

16. Lewin, J. and Warburton, J. (1994) Debris flows in an Alpine environment. *Geography*, 343,79, 2, 98-107.

17. Macklin, M.G., Rumsby, B.T. and Heap, T. (1992) Flood alluviation and entrenchment: Holocene valley-floor development and transformation in the British uplands. *Geological Society of America Bulletin*, 104, 631-643.

18. Messerli, B. and Ives, J.D. (1984) *Mountain Ecosystems: Stability and Instability*. International Mountain Society, Boulder, Colorado.

19. Quinn, W.H., Neal, V.T. and Antunez de Mayolo, S.E. (1987) El Niño occurrences over the past four and a half centuries. *Journal of Geophysical Research*, 92, C13, 14449-14461.

20. Quinn, W.H. and Neal, V.T. (1992) The historical record of El Niño events. In Bradley, R.S. and Jones, P.D. (Eds.) *Climate since A.D. 1500*. Routledge, London, 623-648.

21. Schumm, S.A. (1988) Geomorphic hazards - problems of prediction. *Zeitschrift für Geomorphologie*, Suppl. Bd. 67, 17-24.

22. Shaw, E.M. (1994) *Hydrology in Practice* (Third edition). Chapman and Hall, London.

23. Slaymaker, O. (1996) Introduction. In Slaymaker, O. (Ed.) *Geomorphic Hazards*. Wiley, Chichester, 1-7.

24. Verstappen, H. (1983) Applied geomorphology: geomorphological surveys for environmental development. Elsevier, Amsterdam.

25. Wells, L.E. (1987) An alluvial record of El Niño events from northern coastal Peru. *Journal of Geophysical Research*, 92, 14463-14470.

26. Woodward, J. (1994) *Some soil characteristics of a subAndean agricultural system: The Alizos valley Bolivia*. Farmer Strategies and Production systems in Fragile Environments in Mountainous Areas of Latin America, Project Working Paper 94/02

27. Zimmermann, M. (1990) Debris flows 1987 in Switzerland: geomorphological ands meteorological aspects. In *Hydrology of Mountainous Regions. II-Artificial Reservoirs, Water and Slopes*. IAHS Publication 194, 387-393.

Authors

Jeff Warburton
University of Durham
Department of Geography
South Road, Durham, DH1 3LE,
United Kingdom

Mark Macklin, David Preston
University of Leeds
School of Geography
Leeds, LS2 9JT,
United Kingdom

GEOMORPHOLOGICAL RESPONSE OF NEOTECTONIC ACTIVITY ALONG THE CORDILLERA BLANCA FAULT ZONE, PERU

VÍT VILÍMEK, MARCO ZAPATA LUYO

1. Introduction

The Cordillera Blanca Mts. and the neighbouring Cordillera Negra Mts. belong to the Cordillera Occidental Mts., located in the northern part of Peru. The Santa River separates the Cordillera Negra Mts., situated more to the west, from Cordillera Blanca Mts., situated more to the east (Fig. 1).

The area of the Cordillera Blanca Mts. belongs to a very active region, as to neotectonic uplift and seismicity (Bonnot 1984 and Silgado 1978). The aim of the research was to characterize geomorphological manifestations of neotectonic uplifts of the mountains, mainly in the zone along the main fault zone.

Regarding such an orientation of research we directed our attention mainly to a section of SW slopes of the Cordillera Blanca Mts., notably to the zone of fault slopes. Geomorphological mapping and selection of key notes reflects the character of such a research, oriented to geodynamics rather than to general geomorphological mapping.

Geomorphological research in the Cordillera Blanca Mts. was performed within the Prague Charles University grant project "Geomorphological Hazards in Cordillera Blanca Mts." in a close Cupertino with the Glaciology and Hydrology Resources Unit - Electroperu s.a., Department of Geology and Glaciology, Huaraz, Peru. Special thanks belong to Ing. César Portocarrero Rodrigues, head of the Unit, for his universal support.

Our interpretation of black and white aerial photographs used a complex screening made during the year 1948 and the period 1962 to 1967. A limited area was covered by infrared aerial photographing during the year 1970, when NASA monitored the top area of the Cordillera Blanca Mts. Besides, the historical genetical method of research and the field geomorphological survey were used in connection with the in-situ geomorphological mapping in the scale 1:25,000.

The research, supposed to cover a long time period, was based on the use of the fissure gauge TM-71. Field selection of a suitable locality for its installation had to be made. The gauge is able to monitor displacements on a fault in three dimensions (Košťák et al. 1988).

2. Geology and tectonics

The western margin of South America is an example of an active continent margin with Cordillera tectonics (Frutos 1981). This margin has developed along a deep-sea Peruvian-Chilean Trench. This is a zone of near-shore terraces and elevation zone of Andes with intensive magmatism and volcanism. The Wadati-Benioff zones, connec-

J. Kalvoda and C.L. Rosenfeld (eds.), Geomorphological Hazards in High Mountain Areas, 245-262.
© 1998 *Kluwer Academic Publishers.*

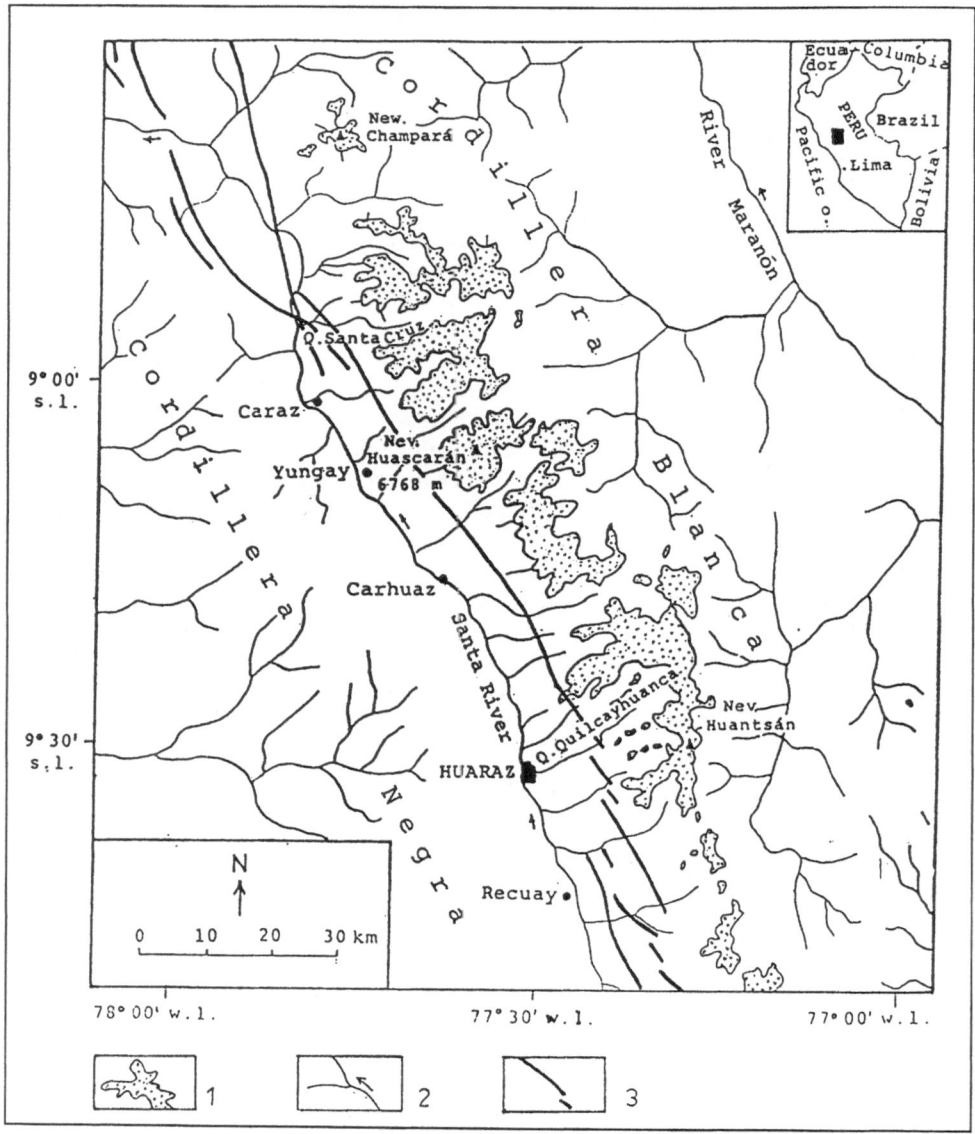

Fig. 1. Schematic plan of the investigated region.

Key: 1 - present glaciation, 2 - permanent water course, 3 - main fault.

ted with the earthquake foci, are sinking beneath the continental plate which is understood as a shifting unit. The volcanism and the seismic activity are directly connected
with the tectonic development of the oceanic plates (Noble et al. 1974). Quoting Mísař

(1987): "For instance the impulse of the Upper Eocene volcanism can be correlated in time with the change of the spreading velocity, as well as with the orientation of the Pacific plate rotation".

Because of the belt character of the Andes formation, the main fault systems are oriented parallelly with the western margin of the continent. This basic structural articulation is evidently manifested also in the origin of the mountain ranges. Elevations are followed by deeply incised valleys, grabens or plateaux. The contact of cordilleras and depressions is typically accompanied by shift-type disturbances or by vertically running faults (for instance the Titicaca Lake Fault), as indicated by Mísař (1987).

Perpendicularly to the so-called lengthwise orientation of fault structures a system of transversal faults originating from deep-seated faults has developed. At the same time, it reflects oceanic structures and limits within the continental platform. The system of transversally running deflection zones divides the South America into several partial segments (Jaroš 1975). Equally Loczy (1969 in Deza, Carbonel 1979) supposes that the deflection zones are huge faults crossing from the oceanic to the continental core. The limits defining the region of the Peruvian part of the Andes, including the Cordillera Blanca Mts., are the Huancabamba and the Arica-Santa Cruz deflection zones (Mísař 1987). The one situated more to the north is perpendicular to the Andes system roughly in the region of the border between Peru and Ecuador. The southern deflection zone of Arica-Santa Cruz limits this segment of the Andes in the region situated south of the Titicaca Lake (Fig. 1). This geological classification of the Andes is only partially corresponding to the orographical one. Jaroš (1979) gives other examples of transversal classification.

The Cordillera Blanca Mts. have a core of Middle to Late Tertiary granodiorite and related granitic rocks. Mesozoic marine sedimentary rocks lie on their flanks and locally cap the granodiorite in the center of the range (Wilson et al. 1967b). The age of intrusive rocks is given by Bonnot (1984), and reads 3-16 million of years.

The same author concludes that the batholite of the Cordillera Blanca Mts. intruded the Tertiary sedimentary formation of "Calipuy" and the sediments of the Upper Jurassic - Upper Cretaceous. These sediment were deformed by the Incan tectonic phase in the period of the Upper Eocene. The Western limit of the Cordillera Blanca Mts. is formed by a 200 km long system of normal faults, well visible from a satellite. The fault system resulted in formation of fault slopes which are dissected to facets by glacial valleys carved in the batholite.

According to Bonnot (1984) the formation of "Calipuy" is covered discordantly by another sedimentary series of "Lloclla", which outcrops in the erosional trenches. These are fluvioglacial superficial cones, lacustrine sediments and debris flows sediments.

The formation "Lloclla" in its lower part shows sedimentation under conditions of tectonical instability (Bonnot 1984). It came to sinking of the Santa River basin, as well as to uplifting of the basin margins. Continuous sedimentation represents lacustrine sediments in alteration with coarse gravel sedimentation. Due to active tectonics a longitudinal dewatering did not take place. It can be said that geodynamic development is dominant in the Pliocene under simultaneous sinking of the basin and uplifting of the Cordillera Blanca Mts.

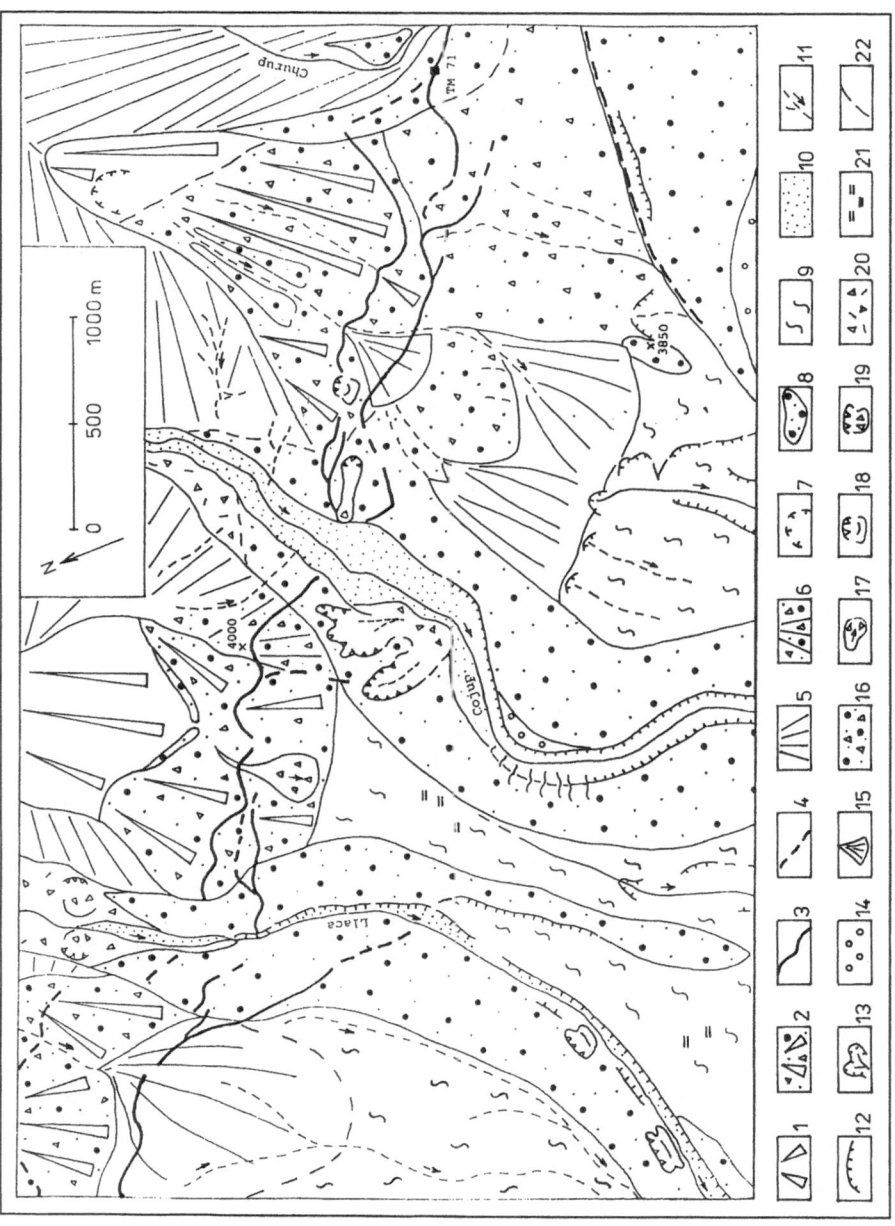

Fig. 2. Geomorphological map of the region where the fissure gauge TM-71 was installed

Fig. 2. (←) Key: 1 - fault slope, 2 - fault slope covered by Quaternary sediments, 3 - fault, 4 - uncertain fault, 5 - erosional slope, 6 - erosional slope covered by Quaternary sediments, 7 - glacial cirque, 8 - moraine, 9 - fluvioglacial sediments, 10 - sediments of present river bed, 11 - permanent water course, 12 - erosional edge, 13 - area of intensive fluvial erosion, 14 - river accumulation terrace, 15 - dejection cone, 16 - polygenetic sediments (e.g. glacial, deluvium), 17 - debris flow, 18 - landslide, 19 - rockfall, 20 - deluvium, 21 - swamp, 22 - uncertain border.

3. Geomorphological research of the fault zone

3.1. GENERAL CHARACTERISTICS OF GEOMORPHOLOGICAL CONDITIONS

The development of the relief of the SW Cordillera Blanca Mts. slopes can be characterized as a result of two principal processes, which are neotectonical uplift of the mountains and the Quaternary glaciation (Photo 1). The glaciation is covering an extensive area. Now, it is in a period of recession, however (C.R. Portocarrero, personal communication). Erosional and slope processes can be found only as secondary effects. The relief of the SW slope of the Cordillera Blanca Mts. is characterized by a fault-slope which evolved on granodiorites as a rule, and lower sections show glacial, glaciofluvial and fluvial sediments, as well as debris flow or mudflow accumulations. Individual sediments are found in interposition.

The main fault-slope evolved at contact of the granodiorite batholite of the Cordillera Blanca Mts. and its Quaternary cover. In the northern range, near the town of Carazo, the fault zone appears to be a boundary zone between the granodiorites and the Tertiary sedimentary volcanic group of "Yungay". Farther to N, granodiorites are bordered with a Jurassic formation of "Chicama" shales (Wilson et al. 1967a).

Bonnot (1984) estimated the total uplift by 4500m, 1000m of which apportioned to Quaternary. Then, Pliocene uplift would be 3500m. He gave an actual example from the valley of Querococha, where it was 600m at minimum. A relative height of fault-slopes was up to 1000m in our investigated localities.

The fault-slope is divided by transversal valleys (in respect to the Santa River flow). The valleys are very narrow at its lower margin. An example is the valley of Santa Cruz in the northern section of the mountains (Photo 2). The majority of the valleys have been remodelled by Quaternary glacial tongues. Thus the right-sided tributaries of the Santa River dissect the SW fault-slope of the Cordillera Blanca Mts. to a series of triangular facets, the contours of which are well identifiable (Photo 3). In relation to the youngest deglaciation in the ridge section, the upper parts of several fault-slopes have not a thicker Quaternary cover, but remnants of moraine ramparts, which give evidence of the last stage of glaciation. An example can be found in the fault-slope between the Llaca and Cojup valleys E of Huaraz (Fig.2). The lower parts of fault-slopes have usually a Quaternary cover in a form of deluvium or a blend of moraine and deluvial sediments.

In the side walls of transversal valleys a course of fault planes parallel with fault-slopes can be observed. In some of the valleys even antithetical faults come out from the relief, as for example in the lower part of the Santa Cruz and Cojup valleys.

Fault-slopes are modelled partially by erosional processes. However, erosional gullies or ravines are neither frequent nor deep. The gullies and ravines are extended to accumulations of dejection cones (Fig. 2). In spite of relatively steep slopes, slope movements are not most typical for fault-slopes, $(30^0 - 38^0$, Photo 4), although they develop locally, related mostly to the youngest faulting. This will be discussed later. Debris flow accumulations appear at places along slope foots. In the section between the foots of fault-slopes and the bottom of the Santa River valley (Callejon de Huaylas) slope deformations appear more frequently due to an intensive side and deep erosion of water courses in glaciofluvial and fluvial sediments.

From aerial photographs and the following field survey a zone of the youngest faulting in the Quaternary cover of the SW fault-slope of the Cordillera Blanca Mts. can be well identified (Photo 5). The description of the individual localities will be given.

Traces of the Cordillera Blanca Mts. glaciation are very fresh, namely in mouths of individual valleys on the SW slope of the mountains. They can be found along the lower edges of fault-slopes or in lower altitudes. The age of these moraines was specified by Clapperton (1972) to the level of Würm glaciation after the Alpine classification (Photo 6). The internal sections of individual valleys display some moraines, however, they are not frequent because the valleys are cleared out by erosional activity of flows. A characteristic U-profile of glacial valleys is obscured by debris accumulations, mainly in places of a more intensive interference with transversal tectonic zones. It is also a local occurrence of rockfalls which contributes to polygenetic accumulations. Valley bottoms are filled with fluvial, glaciofluvial and lacustrine sediments. The youngest moraines which frequently retain lakes, have been preserved at the valley heads. Problems of hazard processes linked with pounded lakes were described in detail e.g. by Zapata (1977), Reynolds et al. (1988), and Lliboutry et al. (1977a, 1977b).

Oozing of mineral springs which are practically aligned to one line parallel with the belt of fault-slopes found at a lower altitude closer to the bottom of the Santa River marks the foots of the Cordillera Blanca Mts. Several springs directly appear in the Santa River bed, as near Monterrey. The water temperature ranges from 48^0 C to 74^0 C (Peňabera C., 1989), which testifies to their relatively deep origin. The warmest one in Chancos ($70^0 - 74^0$ C), is located about 10km SE of Carhuaz. The spring mineralization provides usually carbon-sulphur compounds with iron and chloride additives.

3.2. THE COURSE OF THE YOUNGEST FAULT BELT

Regarding the fact that the course of the main fault belt, along which the Cordillera Blanca Mts. have been uplifted, is not continuous in its full length, but consists of several sectional segments, our geomorphological survey was oriented to several separate localities.

Our field survey dealt mostly with the lowest parts of the transversal glacial valleys of the Cordillera Blanca Mts. and the fault facets sited among the valleys, because the youngest fault zone runs along the foots of fault-slopes or in their lower parts. The whole valleys have been surveyed in some cases only.

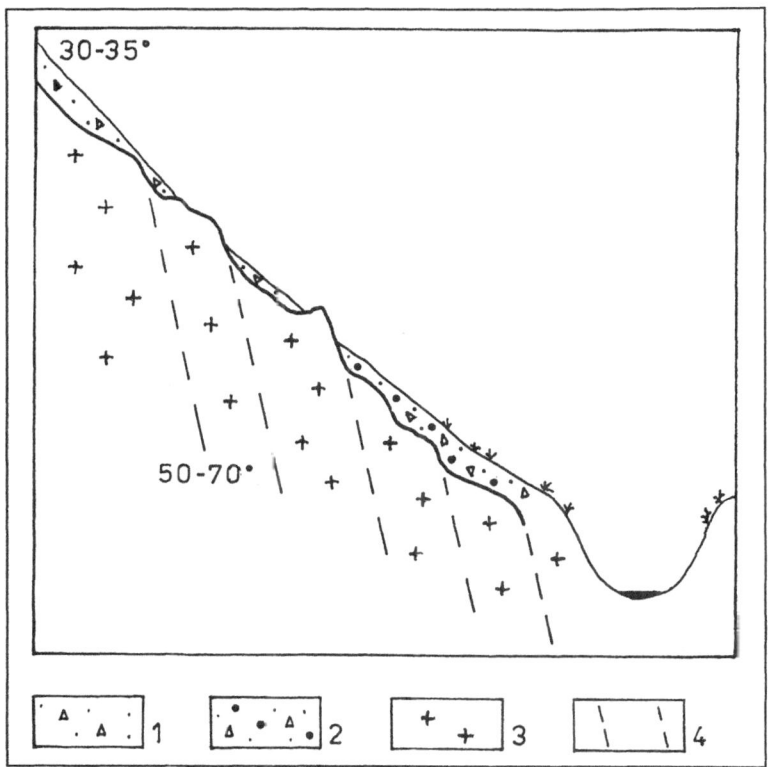

Fig. 3. Schematic structure of a fault slope.

Key: 1 - deluvium, 2 - polygenetic sediments (e.g. glacial, deluvium), 3 - granitic rocks, 4 - partial faults

3.2.1. Paron Valley

At the mouth of the Paron Valley on the left bank there is a fault developed clearly in a relative continuous course. Its line is locally doubled. It is most obvious at places directly bordering the Paron Valley. Here it is tripled at places and distorts a side moraine. Due to consequent slope movements the moraine is disturbed at a length of about 20m.

The course of the fault on the right bank directly in contact with the river bed could not be well evidenced. It probably continues at an angle higher on the fault slope in two parallel lines to N. One of its branches has been opened by erosion to cut a ravine 5m deep in average. Farther down it penetrates an artificial barrier, which may have its origin even in the a pre-Incan period (Nelson, personal communication). On the right bank of the ravine the barrier shows a shift of about 0.5 to 1m towards the valley, and what is more important, it is rotated so that it does not extend the structure of the left bank. This may be due to a displacement along the fault, although a slope movement effect cannot be excluded. In the upper part the ravine is branching out and the orientation of the observed slickensides on the fault plane was diverse. The fault disruption then cuts through a relatively large block type slope movement accumulation. Regarding the deformational undulation of the slope surface the course of the fault cannot be well traced on. The landslide scar does not coincide with the youngest tectonic faulting, although it is likely that the slope movements are due to a steep slope formation. The youngest tectonic faulting as indicated in two localities, one of which is the ravine, has following characteristics: azimuths between 122^0 and 171^0, dipping 36^0 to 50^0 to SW. They appear therefore to be very diverse, however, the presence of landslides may explain a higher differentiation of data.

A total inclination of the slope in this part of the Cordillera Blanca Mts. reads 29^0 to 35^0. The Quaternary cover is free of moraine admixture and is composed by a deluvium approx. 0.5m thick, at places 10 to 20cm only, however. Rock fragments are sharply angular, without explicit roughing. The vegetation cover is sparse, therefore any human interference within the slope results in an immediate movement of colluvial deposits. This was just the case of a path construction for the geodetic survey on the slope. Man-made caves are there as a sign of older investigations to find expected deposits of raw materials. One of the caves gave a chance to investigate the character of fault planes. None of them bore signs of fresh polish. On the contrary, there was a polish with apparent striations on the surface under the Quaternary cover. A new platform made for geodetic work in the central part of the slope has exposed apparent signs of mylonitization at a length of 5m at least. We assume that it is a result of a block-type slope movement. It is a fissure several meters deep, to which a part of the road has caved in, and other places of observable signs of gradual caving, that reveal the block-type slope movements.

There may be a discussion whether the polished faces of the granodiorite fault-slope originated due to glacial erosion or whether they are tectonic fault polish. Quaternary cover movements were undoubtedly unable to produce fault polish. It is supposed that during the uplift of the Cordillera Blanca Mts. the closest block of granodiorites remained several hundred meters below at the foot. The fault-slope representing an uplift of up to 1000m had to be produced gradually and the fault plane produced by such a process could not keep smooth polish with striations for a long time. This contradiction induced discussions about the effects of glacial erosion on fault-slopes.

3.2.2. The Cojup Valley

The youngest fault zone adjoining the Cojup Valley NE of Huaraz, is found divided frequently in two parallel branches in a total width of up to 250 m, and it runs partially at the foot of the fault-slope and partially in its lower half. The height of the fault-slope reads 700 to 1000 m.

Between the Quilcayhuanca and Cojup Valleys the fault zone runs in two or three lines in the Quaternary sediments, most likely copied through from the crystalline rocks. The azimuths of the youngest fault plane were found in expositions and read 144^0 to 162^0, dipping of 30^0 to 54^0 to SW.

Fault tectonics activated smaller slope movements at this place. A slope deformation scar makes the course of the fault step more evident. The step was originally only about 2 to 3m high. This exaggeration due to secondary slope processes makes the evaluation of the length of the original tectonic uplift more complicated. The youngest fault cuts the fault-slope through lateral moraines, as well as through a relatively young dejection cone. Places where the fault zone cuts through moraines show sliding

Photo1. The highest peak of the Cordillera Blanca Mts. - Huascaran (6,768m) formed in granodiorites. A less known view of the glaciated double peak of Huascaran from N.

Photo 2. Mouth of the canyon like valley of Santa Cruz. The fault-slopes on both sides are only moderately modelled by young erosional processes. The youngest fault belt runs most probably at the foot of the main fault-slope.

Photo 3. The general view of the Cordillera Blanca Mts. Valleys modelled by the glacier and dissecting the SW fault slope to facets. The snow line reaches 5,000m a.s.l., approximately.

Photo 4. Quaternary accumulations are deformed by landslides more frequently at cut-banks of water courses (see right side of the photo) or at places of human activity (the road on the left) than on fault slopes directly. The snap was taken from the lower part of the Ishinca Valley.

Photo 5. The youngest tectonic fault disturbs the Quaternary cover at the contact with the batholite of the Cordillera Blanca Mts. Its course is well identifiable at places of cutting through moraine ramparts.

Photo 6. The lateral moraine at the mouth of the Cojup Valley. On the left side of the photo one can see the youngest faulting split to two parallel lines disturbing partially the moraine sediments (left side of the photo), and partially a dejection cone (central part of the photo).

processes in consolidated glacial accumulations, although at other places they are able to be stable at very steep angles. According to the Clapperton's (1972) classification, such moraines can be classified as first category moraines from last glacial stage.

The fault zone on fault-slope between the Cojup and Llaca Valleys, and farther to NW, is developed in one or two branches. Some sections cannot be unambiguously identified. There is a road to the valley of Llaca rising on this fault-slope, the cutting of which exposed two parallel fault planes, about 3m apart (Photo 7). One of them was found on the contact with a mylonitized zone. The fault planes with striations close to the edges of outcrops are covered with Quaternary sediments, i.e. with a mixture of deluvium and glacial sediments. The soil cover of this fault-slope is either very thin or completely missing.

3.2.3. The Santa Cruz Valley

Aerial photographs gave an impression that the fault-slopes bordering the Santa Cruz Valley are very fresh. It is namely a slope farther to the south which displays a straight foot, practically no erosional trenches, and an even inclination of the fault plane. The Santa Cruz valley itself in its lowest section is very straitened. A majority of the Cordillera Blanca Mts. are like that, and in the case of the Santa Cruz Valley it is even more apparent. Here the Santa Cruz Valley becomes like a canyon. The mouth of the valley is cut through transversally by two faults, which induced two morphological steps gradually disappearing higher in the slope. The youngest fault zone runs most probably at the foot of the main fault-slope.

Wilson et al. (1967a) in their map of Santa Cruz Valley present a fault transversal to the main fault belt of the Cordillera Blanca Mts. ·This view is supported by an abnormal intensity of erosion. The deepest section of the valley was not probably affected by glacial erosion like the above described valleys because the accumulations of the valley mouth are either fluvial or sediments of debris flow. Also, the canyon-like shape of the valley does not support the view that a glacial activity has remodelled the valley.

The fault-slope in its lower part is extended by giant accumulations cut through by present river erosion. Due to that the accumulation surface appeared in relatively high levels, out of the reach of present floods or debris flow sedimentation. It is likely that it occurred in connection with a young uplift of the range and the consequent intensive erosion. Alternatively it might be due to sinking represented by the very low erosional level of the main flow draining Callejon de Huaylas, i.e. of the Santa River.

An earthquake of a magnitude M = 3 (Nelson, personal communication) occurred in the Cashapampa Village region in January 1996. The earthquake intensity is unknown to us. Nevertheless, it produced landslides, cracks in soil cover, and damaged houses in the above mentioned village. The town of Caraz, not very far away, was not damaged. Considering the magnitude and the superficial effects one can conclude that it was a very shallow earthquake. Regarding the geometry of the subduction zone in this part of Peru, it could not have been a so-called subduction earthquake but a resulting effect of the internal stress release of the Cordillera Blanca Mts.

3.2.4. The Ishinka Valley

The course of the recent fault zone near the mouth of the Ishinka Valley is partially different from the previously described valleys. Studies of aerial photographs indicated that the recent fault zone on the left bank of the Ishinka Valley cuts also into granodiorites of the Cordillera Blanca Mts. The field survey did not support such an assumption, however. On the contrary, it became clear that none of the fault zones in granodiorites shows signs of young movements which would be observable as fresh polish with striations. There were only some similar signs at one rock block, which separated from the main massif. In this case the polish was due to slope movements, however. The youngest fault zone has developed at the foot of the fault-slope, again, and formed a continuous wall about 10m high in the Quaternary sediments, at one place. The wall is interrupted with stony accumulation of a flow which runs down the fault-slope at present.

3.3. PRELIMINARY   CONCLUSIONS   OF   THE   GEOMORPHOLOGICAL INVESTIGATIONS

The problem to which extent the documented slickensides in granodiorites are due to tectonic movements or due to glacial erosion can be discussed as follows. On the basis of studying fault-slopes and polish inclinations in expositions we conclude that the fault-slope is not represented by one single plane but by a system of granodiorite plate blocks of an inclination up to 24^0 which is a little higher than the present general dip of the slope (Fig.3).   Exogenetic processes and Quaternary accumulation have equalized the slope to the present relatively uniform slope shape.   The polishes were produced at the contact of such plates and there is no necessity to look for the effect of the nearest granodiorite block at the foot. The fault-slope of this composition was covered by Quaternary material and the tectonic polish in expositions can be observed only locally, like in the case of road cuttings or in the walls of erosional trenches.   The Quaternary cover which is locally extremely thin (up to 0.5m only), could not produce the tectonic polishes. Theoretically, glacial erosion could strengthen the process. However, the occurrence of mylonitization in connection with the polishes and the obvious relation between the polishes with striations and the system of the youngest tectonic faulting supports the idea of tectonic origin. Moreover, planes produced during the youngest tectonic faulting are always steeper than the general fault slope inclination. The idea of tectonic slope composed of secondary plates is supported even by rock blocks on fault slopes and by inclination of their fault systems, although they are relatively rare.

Another question can be raised about the conclusion that the step in the fault slope represents the youngest fault zone and how the amount of displacement on it was derived. Regarding the fact that no exposition of shifted blocks in granodiorites, that could be clearly assigned to the youngest fault zone, was uncovered, we are not able to evaluate the displacement value definitely. On the other hand, the apparent step in the fault slope can be trustworthily considered as tectonic, regarding its clearly continuous linear course. This is in spite of some modellation by slope movements, which can be seen as secondary effects only. Moreover, similar signs are found in other areas where no slope movements have been observed. The step appears at places even across older slope deformations.

Photo 7. The fault plane next to the road rising to the Valley of Llaca. Fault plane with striations is in the upper part covered by Quaternary sediments (a mixture of deluvium and glacial sediments).

4. Installation of the TM-71 device

As reported before in the introduction, the geomorphological survey of the fault belt comprised selection of a suitable place to install the fissure gauge TM-71. The monitoring should satisfy the needs to quantify the present tectonic activity, and to detect if it is continual or what is the extent of discontinued shifts due to earthquakes.

The vertical shift in the youngest faulting has been found seriously variable, between 1 and 10m, and the exact detection of the uplifting step in the Quaternary sediments becomes complicated by slope deformation processes, while their scarp plane coincides sometime with the fault plane. Generally, we do not expect the youngest tectonic step to be produced by a single earthquake. This is in respect of very high earthquake magnitudes that should be responsible for such high vertical shifts (see Lamar et al., 1973). The question can be solved not solely by precise measurements of the present fault shifts but also by more detailed field investigations into the fault zone. Notably, it is a question of the extent in which the scar of the slope deformation contributed to the fault step.

The fissure gauge was installed in July 1996 at a locality close to the mouth of the Quilcayhuanca Valley, 10km E of Huaraz (Photo 8). A final decision about the representative locality preceded a relatively complex process of site selection after a study of aerial photographs and successive field surveys, in which individual sites were rejected. After not having succeeded in finding place, where a recent fault would directly penetrate granodiorites of the Cordillera Blanca Mts., we decided to accept a consolidated side moraine cut through transversally by a fault forming a step 4 to 5m high. The fissure gauge was laid across the fault plane under the terrain. Free ends of steel holders connected to the gauge were fixed into Quaternary sediments by concreting.

5. Conclusions

There is a reason to consider the youngest fault zone as of recent origin, mainly due to the observable fresh forms. There is a step well identifiable on aerial photographs, and in the relief directly. The fault zone cuts through not only slope and moraine accumulations, but also dejection cones. There is only one case, at the mouth of the Ishinca Valley, where we have found an accumulation of a present flow going across a fault step.

The activity of slope deformations on fault-slopes was connected just with the youngest fault zone, but with some exceptions. However, this fault belt cannot be confused with a belt of slope deformation scar planes because it is effective even in places without slope deformations, and has a clearly continuous character on almost all the neighbouring fault-slopes. Polishes with striations, and mylonitized zones at places, support also the existence of faulting.

The youngest faulting in plan view has an undulated course and bifurcates into several parallel lines, mostly two or three.

The youngest fault belt in the investigated localities can be found developed always at the contact between the granodiorite batholite and the Quaternary cover represented by deluvial or moraine accumulations or their mixture, at places by fluvial or glacio-fluvial accumulations, and rarely by accumulations of slope movements.

Dips of the fault plane outcrops were found higher than a general fault-slope inclination by 20^0 in average. The effect of exogenetic processes gradually reduced the original fault-slope inclination.

Exceptionally, there were discovered fault planes which, regarding striations, can be seen as relatively young, and which were parallel to the main fault-slope, yet a direct relation with the youngest fault zone, that could be documented in the relief, was missing. Therefore, it was decided to exclude such localities from the selection of the fissure gauge TM-71 possible sites. Such is the case of the left bank of the Cojup Valley.

A modification of the course of faulting, as well as a local increase of the actual height of faulting step by successive slope processes, as suggested by Read (1979), can be confirmed e.g. in the area adjacent to the Cojup Valley.

According to Bonnot (1984) the uplift of the Cordillera Blanca Mts. can be dated as back as to the period before 5 million years. Contrary to that Clapperton (1972) looking to the absence of all the moraines, assumes that the major part of the uplift occurred in the younger Quaternary. In any case, the question about the main period of the Cordillera Blanca Mts. uplift, including the intensity of present movements, remains open, and needs detailed studies not only into the main fault zone as a whole but also into individual sections, including the problem of correlation with the Quaternary glaciation. It can be confirmed, therefore, that the identification of Quaternary and recent movements calls for a research into the interference between tectonic and climatic processes that used to be a reason of fast changes in the relief (Kalvoda 1995).

Photo 8. A trench had to be made 5m long and 1m wide to a depth of 1 to 6m, to install the fissure gauge TM-71. The gauge was installed exactly above the fault plane and its tube holders fixed to concrete blocks at the edges of the trench.

Methods of archaeological dating of fault plane movements have not been used, because ancient monuments that can be found directly on the fault cannot be reliably dated. These are at the mouth of Paron Valley.

Read (1979), looking to a general distortion of the main fault zone and to its irregular course, assumes that this is the case of a overfault of the Cordillera Blanca Mts. He does not provide any concrete profile to prove such affirmation, however. Contrary to that the inclination of the fault planes verified in several valleys, e.g. in the Quilcayhuanca Valley, leads to an opposite conclusion.

Bonnot (1984) assumes the average uplift rate of the Cordillera Blanca Mts. to be 0.7mm per year. Schwartz (1983) quotes Yonekura´s et al. (1979) estimate of 2 - 3mm per year (possibly 0.86 - 1.1mm per year) with a note that no systematic study exists to document that. We can assume that further detailed studies into tectonics of dated faulted Quaternary sediments, together with results of contemporary movement monitoring using the fissure gauge TM-71, will result in more concrete findings.

The uplifts along individual sections of fault planes rarely exceed 10m according to Bonnot (1984). The youngest uplift in the Queroccocha Valley after Schwartz (1983) reads 2.5m. Faulting of a moraine in a region between the Llaca and Cojup Valleys characterized by Lliboutry et al. (1977) by the value of 10m, was corrected by Read (1979) to 6m. Our own recent highest observation of fault uplift at the mouth of the Ishinca Valley did exceed 10m.

Regarding the values of maximum possible shifts of the Earth surface given by Lamar et al. (1973), if related to earthquake magnitude, it does not seem likely to find a recent fault zone due to a single earthquake. Nevertheless, this question calls for a more detailed investigation of the recent fault zone in question of vertical shift, and consequent slope movements. An estimate of maximum possible intensity of prehistoric earthquakes, and final dating of the youngest fault plane should be taken into consideration, too. It is also necessary to consider the contribution of the transversal fault in the development of the relief of the Cordillera Blanca Mts. in individual sections, because we assume that the uplift of such a high intensity could not occur without transversal faulting.

References

1. Bonnot, D. (1984) Néotectonique et tectonique active de la Cordillère Blanche et du Callejon de Huaylas (Andes nord-péruviennes), thèses Université de Paris-sud, Centre d'Orsay, Paris.

2. Clapperton, M. (1972) The Pleistocene moraine stages of west-central Peru, *Journal of Glaciology*, 11, 62, 255-263.

3. Dez E. and Carbonel, C. (1979) Regionalizacion seismotecnica preliminar del Peru, *Bol. Soc. geol. Peru*, 61, 215-227.

4. Frutos, J. (1981) Andean tectonics as a consequence of sea-floor spreading, *Tectonophysics*, 72, T21-T32.

5. Jaroš, J. (1975) *Structural Record of the Neoidic Tectogenesis in the South American Cordilleras*, Academia, Praha.

6. Kalvoda, J. (1995) Prologue: An introduction to the dynamic geomorphology of tectonic active zones, *Acta Univ. Carol. Supplem.*, 30, 9-20.

7. Košťák, B. and Avramova-Tacheva, E. (1988) A method for contemporary displacement measurement on a tectonic fault, *J. Geodyn.*, 115-125.

8. Lamar, D. L., Merifield, P. M., Proctor R. J. (1973): Earthquake recurrence intervals on major faults in southern California in: Geology, Seismicity and Environmental Impact (Moran D. E. et al. Eds) University publishers, 265-276, Los Angeles.

9. Mísař, Z. (1987) *Regionální geologie světa*, Academia, Praha.

10. Noble, D.C., Mc Kee, E.H., Farrar, E. and Petersen, U. (1974) Episodic Cenozoic volcanism and tectonism in the Andes of Peru, *Earth Planet. Sci. Let.*, 21, 213-220.

11. Pañabera, C. (1989) *Atlas del Perú*, Ministerio de Defensa, Instituto Geografico National, Lima.

12. Read, S.A.L. (1979) Geological report on visit to the Huaraz area, northern Peru, unpublished report, Electroperu s.a. Unidad de Glaciología y Recursos Hídricos, Huaraz.

13. Schwartz, D.P. (1983) Evaluation of seismic geology along the Cordillera Blanca fault zone, Peru, unpublished report for Hidroservice, Engenharia de Projetos Ltda, Lima.

14. Silgado, E.F. (1978) Historia de los sismos mas notables ocurridos en el Peru (1513-1974), *Boletin No.3 Instituto de Geologia y Mineria, Serie C, Geodinámica e Ingeniería Geológica*, Lima.

15. Wilson, J., Reyes, L. and Garayar, J. (1967a) Mapa geologico de los Cuadrangulos de Mollebamba, Tayabamba, Huaylas, Pomabamba, Carhuaz y Huari, *Servicio de Geologia y Mineria, 1: 200 000*, Lima.

16. Wilson, J., Reyes, L. and Garayar, J. (1967b) Geología de los Cuadrangulos de Mollebamba, Tayabamba, Huaylas, Pomabamba, Carhuaz y Huari, *Servicio de Geologia y Mineria, Bol. 16*, Lima.

Authors

Vít Vilímek
Faculty of Science, Charles University, Prague
Department of Physical Geography and Geoecology
Albertov 6, 128 43 Prague,
Czech Republic

Marco Zapata Luyo
Glaciology and Hydrology Resources Unit - Electroperu s.a.
Department of Geology and Glaciology
Avenida Confraternidad Internacional Oeste No 167, Huaraz,
Peru

GEOMORPHOLOGICAL HAZARDS AND RISKS IN THE HIGH TATRA MOUNTAINS

JAN KALVODA

1. Introduction

For its diversified landscape and the quantity of research problems it offers, the quaint nature of the High Tatra Mountains merits the permanent attention of natural scientists. The territory of the Tatras National Park in the Slovak Republic (Fig. 1) displays a unique natural environment in Central Europe both by its geomorphological evolution history and unique biosphere complex. The topic of this paper is to emphasize the natural hazards of the alpine-type relief manifested mainly by different types of slope movements and anthropogenous share of geomorphological hazards and risks in the Tatras National Park.

The Carpathians mountain system's evolution is connected with the Alpine orogene from a chronological and morphostructural viewpoint. Important palaeoclimatic changes that went on in the High Tatra region in the Pliocene and the Pleistocene were, together with the geological features of the mountain vault surface and with the neotectonic activity, the main factors enabling the evolution of high mountain landforms. The uniqueness of the High Tatra landforms evolution is expressed mainly by their alpine-type relief (Photo 1), which originated prevailingly from the glacigenous, nival and cryogenous process in the Quaternary.

The granodiorite core of the High Tatra Mts. arch is an expressive landscape which dominates the northern part of the West-Carpathian mountains. The main part consists of quartz diorites to granodiorites (Gorek 1959, Gorek, Kahan 1973) formed in the Tatras during the main stage of the Carpathian system orogeny. An intensive development of the high-mountain landforms was conditioned by the Palaeogenic vaulting and by the Miocene and younger tectonic dissection of the mountain system. By the end of the Pliocene, the High Tatras were a tectonically differentiated and asymmetrically inclined elevation with no signs of glacial modellation (Lukniš 1973a, b). The sedimentation process was interrupted by stage uplifts with the modellation impact of fissure disintegration of High Tatra granites in the central part of the mountains. The High Tatra uplift continuing in the Quaternary has been reanimating destruction processes and stressing denudation of the crystalline core of the mountains, initially covered by upper sedimentary complexes. Large mylonite zones, located within them and generated mostly by rock crushing during slow tectonic movements, are the most weathered ones.

The original Pleistocene glaciation brought a qualitative change into the High Tatra geomorphological development. The glaciers' activity developed the system of glacigenic destruction and accumulation landforms founded on the pre-glacial relief. In valleys enlarged and deepened by glaciers and in the foothills, moraines covered older

263

J. Kalvoda and C.L. Rosenfeld (eds.), Geomorphological Hazards in High Mountain Areas, 263-284.
© 1998 *Kluwer Academic Publishers.*

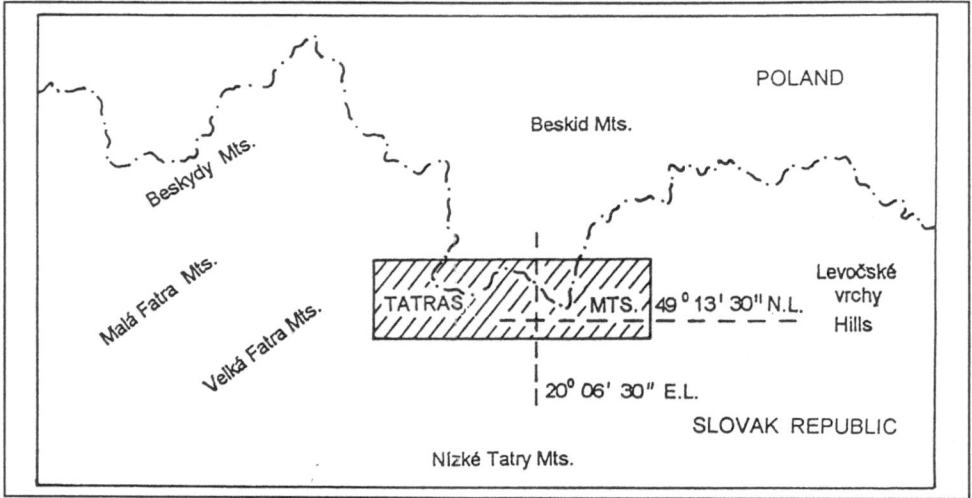

Figure 1. Geographical position of the High Tatra Mts (the Slovak Republic) in the context of the main Western Carpathian orographical units.

river alluvial accumulations and the other remnants of the weathered material (Mazúr, Lukniš 1956, Mičian 1959, Zaťko 1961). Glacial erosion and gelivation in the extraglacial territory affect the older ridge part of the mountains. After the relatively warm interstadials and interglacials the glaciers found during their progression an underground freshly affected by periglacial processes. In the Riss, the glaciers reached far into the foothills.

The rock slopes of the Tatras mountains represent a specific group of steep slopes with a high intensity of relief forming processes (Kalvoda 1970, 1974, 1994, Kotarba 1976). It is necessary to carry out a systematic research of their stability conditions. A detailed monitoring of configuration, present modellation processes and the stability of the high rock slopes (comp. Photo 2) in the territory of the Tatras National Park will assist the prognosis and determining of the risk of catastrophic slope movements.

The paper by Kalvoda (1994) briefly describes the development of the High Tatra alpine-type relief with their high rock slopes. Geomorphological analysis of the crest part of the mountains showed that the development of the high rock slopes on crystalline rocks was determined mainly by their morphostructural position, in detail by the local arrangement of tectonic discontinuities and by the extent of glacigenic and cryogenic modellation in the Pleistocene (Photo 3). Postglacial and recent changes of the rock slopes of the High Tatra crest part are continuing on mainly in periglacial conditions (Photo 4), there are frequent rockfalls, slidings and slope sediments movements. The theoretical snow line today is above the highest mountain tops at an altitude of about 2,800m (Lukniš 1973a).

The determining morphostructural elements of the rock relief configuration on the Tatra's granodiorites are orientation, inclination and the frequency of fissures and mylonite zones. Their arrangement is evidence of a complicated breaking of the granodiorite mountain core. The papers by Kalvoda (1974, 1994) illustrate selected configuration types of geomorphologically significant fissures in the High Tatra rocky relief characterized by a diversity of landforms and stability conditions of the steep mountain slopes and crests.

Photo 1. Large cirques of the Pleistocene mountain glaciation accumulation area in the Velká Studená dolina Valley. Granodiorite massifs of the High Tatra alpine-type relief are separated by fault zones of crushed granites and mylonites. The orientation of the main fissure systems is also followed by structural plateaux directly in the ridge part of the mountains. They also include the S inclined structural denudational surfaces of rock slopes of the Hranatá veža Tower (left) and the Javorový štít Peak (right, 2,417m). Photo: Jaromír Kalmus

In recent times, the crest part of the mountains has produced favourable climatic conditions for the development of landforms conditioned by periglacial processes activities, and mainly for the granular disintegration of rocks, formation of debris and for its movement on slopes, for small soil periglacial landforms (Sekyra 1954a, Šmarda 1956) and for soil and vegetation disturbances due to regelation. Postglacial modellation processes have been slowly modifying the whole glacial accumulation landforms in valleys, flattening the valley deepening and, mainly in lower positions, remodelling glacial and glacifluvial sediments. Further on, these processes have been to a different degree eroding the rock landforms which were situated in the Pleistocene mainly in the extraglacial zone above the glacier masses (Sekyra 1954b, Lukniš 1973a). The differences exist not only in the progress of gelivation and related processes, but also in the occurrence and intensity of different modellation processes.

2. Alpine-type relief hazards

A conspicuous sign of geomorphological hazards in the area of high rock slopes are rockfalls. On rock walls, that is in places of detachment of rockfalls, the new wall surface is affected by weathering. Smooth, flat to slightly curved surfaces of granite plates bear significant traces of frost fissuring, of modellation by snow, melting or rain water, wind and lichens. Their margins are limited by tiny fault zones (Photos 2 and 3). In the crests' top parts they form overhangs above the mylonite col and fissure zones. Relics of tectonic mirrors are mostly irregular and never cover the whole surface of the rock plates.

Localities of rockfalls breakings, on which no rock plates are formed by sharp crossing and frequency of fissures, are strictly limited and without a shallow surface weathering crust (the older ones have a 1 - 2mm thick crust) with freshly rooted lichens or smoothed. They are mostly found in a steep or overhanging position above the entry into smaller gullies transporting debris down to the main transport way in the wall. Rockfalls due to lightning appear in the walls next to occasional waterfalls, deep couloirs and narrow steep gullies, and on crests then above overhangs; they are whitish with grey to black convergent smudges, sharply bent and without any evident relation to the frost breaking. The largest ones have been determined in the N wall of the Hlinská veža Tower with a surface area of 20 x 30m, and smaller ones can be seen on the northeastern Kriváň Mt crest, under the Terianská veža Tower, southwards from the Baštové sedlo Col, on the crest between the Malá Končistá and the Prostredná Končistá Mts, on a rock bar of the Smrčinové sedlo Col, in the N wall of the Jahňací štít Peak and the Malá ľadová veža Tower and elsewhere. Opening couloirs as frost fissures in primary tectonic discontinuities show the progress of rockfalls breaking away in strongly weathered walls.

According to the degree of weathering, to the fissures frequency and their location in the wall massifs, three types of rock plates can be distinguished:

1. The oldest rock plates have a rough irregular surface, locally with 10 to 15cm large garland projections. Tiny granite grits, on which tiny vegetation grows, are maintained in hollows. The entire surface, when not in proximity of regular avalanches of snow sliding trajectories, is covered by lichens. At the margins there are open frost fissures. In the strongly crushed zones, the wall stability is partly disturbed. Running water creates tiny rainwash furrows between garlands which are usually firm, with quartz veins and phenocrysts. The most accessible example of this generation of plates is a 25 x 40m wide trapezoid one in the middle part of the NW wall of the Volová veža Tower. Larger typical localities are also above the Veľká Zahrádka site in the Triumetal wall, in the W wall of the Východní štít Peak above the Železná brána Gate, in the W wall of the Dračí hlava Mt., in the N wall of the Vysoká Mt., in the lower part of the Gánok Gallery, in the E wall of the Gerlachovský štít Peak, on the Granátová veža Tower, in the N wall of the Javorový štít Peak, on the Ľadový and the Jahňací štít Peaks, etc. The plates inclination is between 60 to 90°.

Photo 2. The upper part of the Menguśovská dolina Valley trough (left) below the rock steps of fossil glacier cirques in the Žabie plesa Lakes area with large steep debris cones below the E walls the Satan Peak crests (2,432m) and the Baštové veže Towers. In the background on the right the Kriváň Mt (2,493m) can be seen on the horizon. Photo: Jaromír Kalmus

2. The most frequent rock plates are of younger generation situated in the rocky crest part of the mountains. They are slightly weathered rock surfaces with lichens and frost fissures without any rough tiny relief. Their inclination oscillates between 15 and 90°, locally with a several cm long bench after the splitting of the weathered layer. Similarly as with the oldest type of plates, they are grey to brown with the exception of a strip due to occasionally running water and the places whitened by snow. The weathering effects are smaller under overhangs where the quantity of running water on the rough surface of the rock plate is smaller as it is captured by lichens.

3. The youngest rock plates provide evidence of the recent rock mass breaking away and frequently their surface has been exposed for less then 10 years. They have the following characteristics: a smoothing along the surfaces of separation, sharp passages between the crossing fissures, no or minimal fresh lichen covering, pale colour and detailed granodiorite granularity without a weathered surface layer (significantly white and yellowish feldspars, dark biotite and milky white, respectively pink quartz phenocrysts, sharply fitting limited quartz veins 2 to 30mm thick).

Typical examples are the rock plates in the middle part of the E wall of the Veľký Kežmarský štít Peak or those in the E walls of the Javorový hřeben crest and the Javorový štít Peak. They can be easily found in the majority of the High Tatra walls.

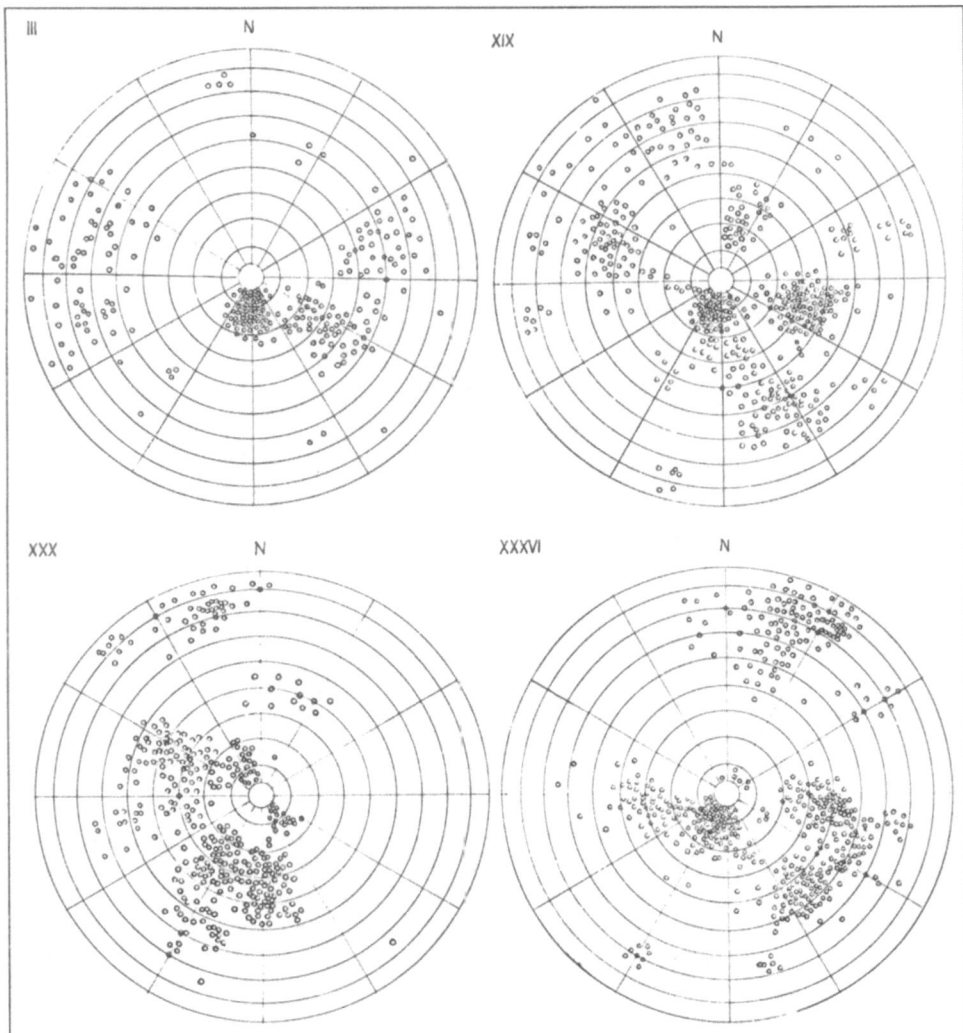

Figure 2. Distribution of geomorphologically significant fissures in the High Tatra Mts granodiorites. Examples of their configuration are demonstrated from crest areas: III - Karbunkulový hrebeň Crest and Jastrabia veža Tower, XIX - Kriváň Mt. massif from Špára Saddle to the Malý Kriváň Mt., XXX - crest between Východní Rumanovo sedlo Saddle and Gánekova štrbina Saddle including Kačací veže Towers, XXXVI - crest between Sedlo nad Czarným Saddle and Sedlo Váha Saddle including Mengusovský Volovec Peak and Rysy Mt. massif. Direction and inclination of the fissures are represented in the graph by projection of the inclination lines of the fissure planes on the Lambert areal projection of a lower hemisphere section in polar position. The main course of the crests (in degrees) and numbers of measurements (in parenthness) for segment diagrams: III. 110-290, (216); XIX. 175-355, 90-270, 140-220, (462); XXX. 70-250, 40-220, 120-300, 170-350, (275); XXXVI. 80-260, 10-190, 115-225, (526).

Photo 3. Rock relief of granodiorites on the crest part of the High Tatra Mts above the Mlynická dolina Valley with distinct features of slope debris movements. In the right background, the highest top is the Gerlachovský štít Peak (2,663m). Photo: Jan Kalvoda

The granite tectonic structure determines the extent, the degree of exposition in the wall area, the inclination and the degree of the rock plates' weathering (Fig. 2). The fitting of fissure surfaces, which are against the dip, into the wall causes the roof-like linking plate belts with lower overhanging margins; accordingly the inclination of the fissure surfaces with the principal wall inclination (maximal variations of 20 to 40°) accelerates rock surface disintegration and water leaking on the margin ledges into the subsurface fissures continuing on the lower belt of rock plates. Crossing and sharp fitting of fissure steep surfaces lead to the formation of broad cuttings limited by plates. One of the basic conditions for the rock plate formation is a low frequency of fissures, i.e. usually 3 -5 per m^2 and less, without traces of rock crushing. Tectonic mirrors and smoothed surfaces on tectonic disturbances slow the weathering of the plate surface.

Larger rock plate occurrences are minimal in the marginal rock mountain massifs. Rocks loosened by strong tectonic disturbances, and through ensuing weathering, tend to form inexpressive towers, short pillars and ribs. Most frequent are the large rock plates in the deepest denuded central part of the mountains in the segment of the main crest between the Český štít Peak and the Zmrzlá veža Tower and on the adjacent saddle-bows.

Larger plates and bulging rock blocks limit the pillar lines due to the regressive destruction of rock walls going on in cross or opposite directions. Transverse fissures and fault zones or mylonite belts form ramps or flat cross belts of plates covered by moving debris. If they are situated at different levels, continuous pillars get transformed into a system of triangle to trapezoid towers. Larger rock ribs and pillars are relics of short crests reaching down to valleys and destroyed by glacier erosion.

Gravitational deformations of mountain massifs along partial skidding zones, near-surface manifestations of creeping movements and fossil rock glaciers in the Tatras have been studied in detail by Nemčok (1972, 1982, Mahr 1977, Mahr, Nemčok 1977). In the High Tatra they have documented a series of gravitational deformations on slopes of the Gerlachovský štít Peak and the Javorinská Široká Peak massifs.

An exceptional region of recent activities of landforms hazard processes is the zone of the Tomanovská dolina Valley in the eastern part of the Western Tatras. It is a morphostructurally important zone in the frontal parts of the Mesozoic sediments thrust of the Červené vrchy Mts onto the crystalline rocks of the High Tatras (Gorek 1958). In the very varied set of landforms of the Tomanovská dolina Valley (Figs 3 and 4) active frost gravitational scars have been established hundreds of metre long on the Polská Tomanová crest, fissure chasms and grades karst on the steep slopes of the Červené vrchy Mts (Droppa 1957, 1965, Nemčok 1972, Kotarba 1979, Kos 1982). Landslides and rockfalls accumulations are frequent.

On the assymetrical crest of the Rozpadlý Grúň on the Mesozoic limestones and dolomites at altitudes of 1,700 to 2,000m there are large gulfs up to 3m wide, more than 10m long and dozens of metres deep (Fig. 3). They are gravitational depressions of the near-surface parts of the rock massif on steep slopes with a NE exposition which had been cut by glacial erosion in the Pleistocene. In the wide neighbourhood of these depressions and gulfs there are conspicuous rock outcrops of dissociable surfaces of fossil rockfalls. This mountain slope has been obviously affected by deep gravitational deformations with the potential hazard of other rapid slope movements. Periglacial processes in such tectonically and gravitationally disintegrated high-mountain karst have formed numerous other slope disturbances of the fissure chasm type (Droppa 1965), the highest of them being the Vyšná Kresanica one (2,089m) more than 70m deep.

The rock walls of the Tomanovský vodopád Waterfall (with the relative height of about 40m beginning at an altitude of 1,262m in quartzites of Triassic age) were formed as lateral steps of the glacially redeepened trough of the Tichá dolina Valley (Fig. 4). Nevertheless, the effects of deep erosion are apparent in the Tomanovský vodopád Waterfall defile. In the wider neighbourhood there are frequent rockfalls and landslides. The relief of of the Tomanovský vodopád Waterfall rocks steps is in a specific morphostructural position of the frontal part of the Neogene thrust of Mesozoic sediments of the Červené vrchy Mts onto crystalline rocks and seems to manifest certain evidence of the recent morphotectonic activity.

The way of destruction of the High Tatra mountain vault during the retreat of glaciation is shown by erosional denudational landforms and by slope modellation of the Upper Würm to Holocene debris accumulations. The character of weathering, which together with the structural geological elements of the granodiorites massif reflects the character of debris accumulations, is the decisive factor of the intensity and course of the changes in the development of the mountain crest part landforms. The retreat of the last glaciation glaciers have loosened huge rock masses on the sides of the upper part of troughs and in cirques under the form of rough boulder and block accumulations. Typical examples are the block accumulations in the small valley below the Špára in the Nefcerka Valley, the cirque of the Suchý potok Brook Valley, the uppermost part of the Zlomisková dolina and the Batizovská dolina Valleys under the Končistá and Zadný Gerlach Mts, as well as the cirques of the Kačací, Česká, Žabí, Bialovodská, Kolová and Čierná Javorová dolina Valleys. The formation of block accumulations has in general terminated the steepening of the crest walls and slopes by glacial activity and initiated the relatively rapid deposition of a thick cover of weathering materials in the High Tatra valleys with a less evident transport into the lower parts of the mountains, and eventually into the foothill.

The more than 10m thick accumulations are polygenetic debris, and/or periglacial debris cones on the periphery of the studied region, at the foothill of the fault conditioned S slopes of the mountains (below the Kriváň, Ostrva, Tupá, Gerlachovský štít, Slavkovský štít and Huncovský štít Peaks and the Lomnický hrebeň Ridge). They had already been formed after the formation of the Podtatranský zlom Fault. Below the rock steps of the trough lock in the Kôprová dolina Valley and below the rock step of the trough of the Za Handel and Suchý potok Brook Valleys (other ones are situated outside the studied territory), there are glacifluvial cones of the last glaciation retreat period. The glacifluvial cones, together with block accumulations, polygenetic debris and periglacial cones and moraines are the base for a progressive filling of the Tatra valleys by debris during the Holocene. The younger Holocene generation of debris cones and avalanches accumulations, rockfalls and slidings cover and penetrate in all cases these older debris. Polygenetic debris, glacifluvial and periglacial cones have been stabilized by forests, dwarf pines and a continuous grass cover and secondarily modelled by recent erosional denudational processes on their surface (comp. Pelíšek 1955, Plesník 1967, Linkeš 1981). Rainwash and erosional gullies, often curbed according to the granularity of the debris accumulations, are conspicuous.

Mountain debris has been formed by the free falling of weathered rocks from steep walls, by the crossing or connecting of debris cones' sides and by their secondary washing and sliding. Continuous debris zones in cirques and at trough sides are anything up to 15-32m thick, with their surface structured according to the character of the walls. At the feet of continuous walls without expressive active glens, or with inclinations inferior to 30°, the mountain wastes are slightly undulated with erosional and debris streams up to 1m deep (for instance in the Furkotská dolina Valley below the Velké Solisko Mt, below the Černý štít Peak in the Velká Zmrzlá dolina Valley, under the Svinky Mt in the upper part of the Čierná Javorová dolina Valley or below the Karbunkulový hrebeň Ridge in the Malá Zmrzlá dolina Valley). A major part of the mountain wastes are, however, based on older block accumulations with a regular passage onto the flat valley bed as a system of wreath-shaped debris cones, the layers of which are superimposed and penetrated on the sides.

The youngest debris cones, at present situated on the surface of the mountain wastes, are separated by irregular, mainly triangular surfaces with their bases at the feet of the walls and with their tops in the rainwash gully reforming the adjacent cone sides (for instance below the N wall of the Malý Kežmarský štít Peak, below the E wall of the Gerlachovský štít Peak, below the N wall of the Svišťový štít Peak, below the

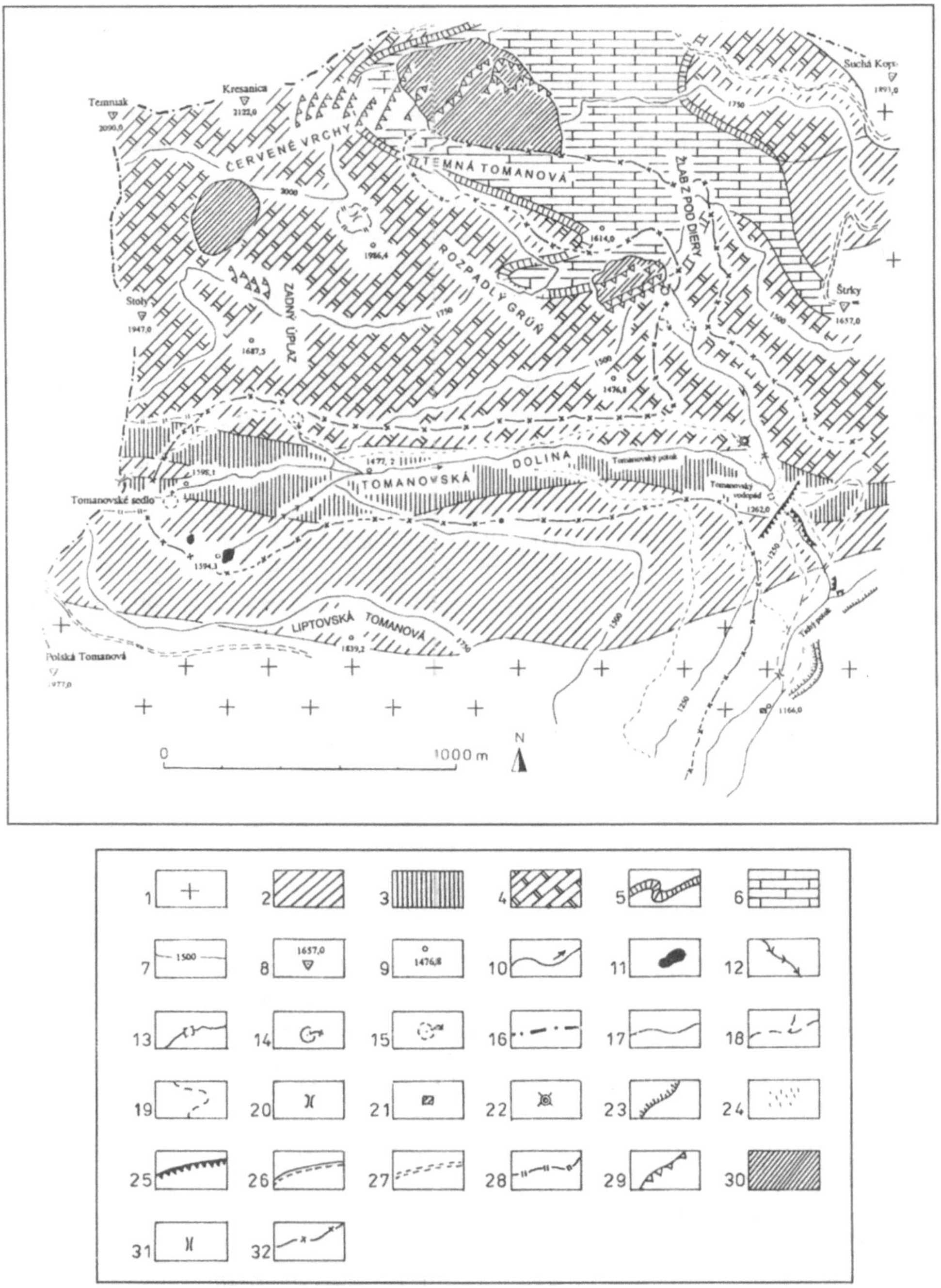

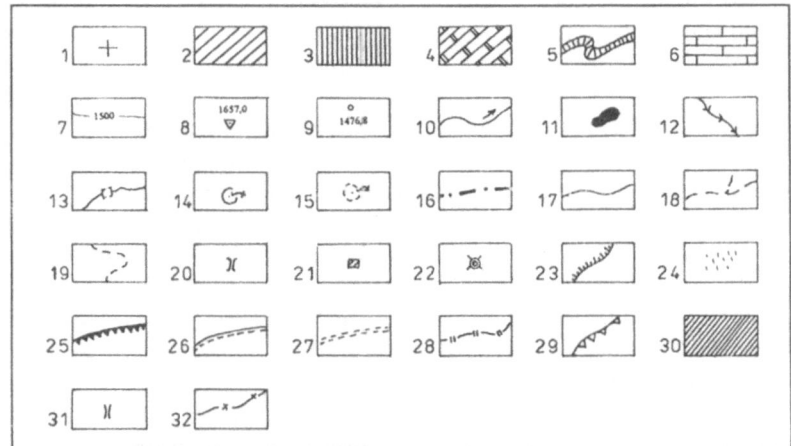

Figure 3. Geological and morphostructural situation in the Tomanovská dolina Valley, in the High Tatras Mountains (modified after Kos 1982).

Figure 3. (←) Key: 1-6 geological situation: 1 - granodiorites and gneisses of the High Tatra crystalline system, 2 - quartzites of the Lower Triassic, 3 - Werfenian slates, 4 - limestones and dolomites of the Middle Triassic, 5 - crinoid limestones, 6 - limestones of the Malm; 7 - contour lines, 8 - main tops, 9 - elevation points, 10 - brooks, 11 - lakes, 12 - cascades, 13 - waterfalls, 14 - rock springs, 15 - debris sources, 16 - state border of the Slovak Republic and Poland, 17 - roads, 18 - ways, 19 - paths, 20 - bridges, 21 - timber chalet, 22 - ombrometre, 23 - anthropogenous slope cuttings, 24 - dump, 25 - geomorphologically distinctive fault lines, 26 - geomorphologically distinctive contact of crystalline rocks and Mesozoic sediments, 27 - presupposed contact of crystalline rocks and mesozoic sediments, 28 - geomorphologically distinctive limits in rock lithology, 29 - structural rock steps, 30 - structurally inclined rock surfaces, 31 - structural cols, 32 - limit of accumulation and destructional landforms.

Gánok Gallery and elsewhere).At the feet of the walls and at the mouthing of glens. the inclination is generally inferior to 40°; an inclination of 40 to 50° has been measured below the Malá ľadová veža Mt, below the S wall of the Mlynár, below the N wall of the Východná Vysoká Mt, below the Karbunkulový hřeben Ridge in the Malá Zmrzlá dolina Valley, above the Žabí plesa Lakes at the main High Tatra crest in the Mengusovská dolina Valley, southwards from the Mengusovské and Hlinské sedlo Cols, below the Popradský Ľadový štít Peak in the Zlomiska Valley and below the N wall of the Kolový štít Peak. The average inclination of the mountain wastes in the upper part is estimated to be 36° to 38° according to 68 control measurings of typical localities, from 32° to 36° in the middle part and to 12° to 28° at their lower end. When the retreat of the cut walls is quick, the mountain wastes' form copies the system of glens and erosion gullies in the walls.

The debris cones in the High Tatra are a typical sign of weathered rock transported into valleys (Photos 2 and 3). Their formation is a direct consequence of tectonic structuring of the granodiorite into partial rock blocks, and of a complex hierarchy of fissure systems, because of their importance in the detailed structuring of the rock crest part. The shape, the size, the position and the frequency of debris cones are directly proportional to the number and configuration of fissures and smaller fault zones situated crosswise to the present surface of walls and steep slopes. The presence of mylonite zones and crushed granite ones more than 35 to 40m thick leads to the formation of continuous mountains wastes belts, or to stage transfers of the weathered mantle in the case when the presence of sandy and loamy components is greater than the internal friction between the sharp angular debris.

Less evident are the debris cones parting from several hundreds of metre long and 5 to 30m wide glens with an inclination of 30 to 55° on slopes which have not been significantly modelled by lateral glacier erosion. They are in contrast to the huge debris cones 150 to 200m long (from the internal mouthing in glens) formed at the feet of the 300 to 800m high walls closing the main system of the central glens of walls. These debris cones are always based on the oldest continuous mountain wastes and provide evidence of a rapid destruction of the walls at present. The opposite or parallel position of glens and couloirs in the walls leads to the formation of singular cones. On the contrary their convergence leads to the couloirs doubling and a superposition of layers. Examples of the first type are the debris cones of the Krčmárov žlab Glen below the E wall of the Gerlachovský štít Peak, below the Magurská lávka Bench in the Čierná Javorová dolina Valley, below the S walls of the Čubrina Mt in the Piagrová dolinka Valley, and below the Javorová špára Col in the Veľká Studená dolina Valley.

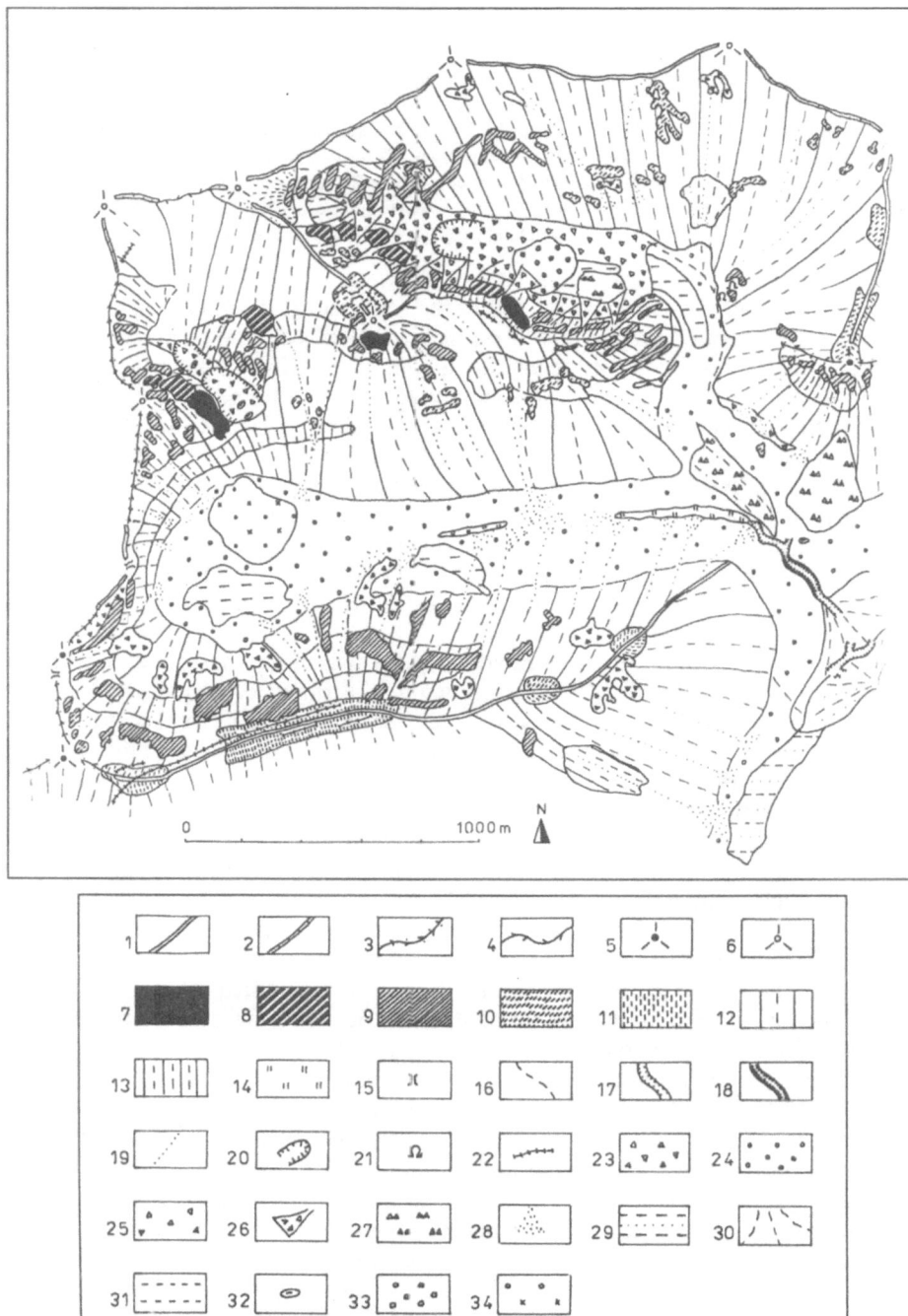

Figure 4. Geomorphological sketch map of the Tomanovská dolina Valley in the High Tatras Mountains (modified after Kos 1982).

Figure 4. Key: 1 - crests with rough sandy to stony deluvium, 2 - crests with rock outcrops and block deluvium, 3 - strongly eroded rock crests, 4 - rock crests, 5 - rock tops, 6 - round tops, 7 - continuous rock relief, 8 - dissected rock relief, 9 - strongly eroded rock relief, 10 - sporadic rock outcrops, 11 - mild denudational slopes with an inclination inferior to 15°, 12 - denudational slopes with an inclination of 15 to 45°, 13 - steep denudational slopes with an inclination superior to 45°, 14 - erosion gullies, 15 - cols, 16 - erosion gullies on the surface of Quaternary accumulations, 17 - erosion cuttings of a depth inferior to 3m, 18 - erosion cuttings of a depth superior to 3m, 19 - avalanche and debris avalanche gullies, 20 - cirques, 21 - karst caves, 22 - frost gravitational scars, 23 - block and stone slope sediments, 24 - denudational slopes on slope sediments, 25 - mountain wastes, 26 - debris cones. 27 - accumulations of rockfalls and slides, 28 - accumulations of snow-stone and debris avalanches, 29 - alluvial accumulations, 30 - fluvioglacial cones, 31 - firn moraines, 32 - sink in debris accumulations, 33 - huge block debris on slopes, 34 - accumulations of slope sediments, debris, avalanche and firn moraines.

The examples of the second type are the debris cones on the mountain wastes below the N wall of the Malý Kežmarský štít Peak, below the Veľká Javorová veža Tower in the Javorová dolina Valley, below the N wall of the Svišťový štít Peak, etc.

According to the character of the debris material, to the mobility of the debris and to the morphometric characteristics, the debris cones can be divided into four basic types:

a) Simple debris cones with a permanent supply of material have only one principal source of debris supply. The regular shape of the cone is mostly maintained only after the glen passage below the curbed line at the foot of the wall. Greater inclination induces on the form surface a more perfect gravitational differentiation of individual pieces of debris. A greater supply of debris is marked by irregular waves of contour lines at the front of the cones. After the emptying of the debris from the glens neither the surface nor the ordering of larger stones are stabilized in the upper and medium parts of cones. There are evident seasonal changes on the surface. The inclination in the uppermost parts of the cones is superior to 55°. A typical example is the debris cone below the wall of the Žltá veža Tower in the Dolinka pod Sedielkom Valley.

b) Composed debris cones with a permanent supply of material have at present only one main source of material supply, but their form and surface are evidence of a change of orientation of debris storage from emptied and today less active glens and couloirs near the debris cone, for instance below the wall of the Ušatá veža Tower in the Zelené pleso Lake Valley.

c) Stabilized debris cones of a) and b) types with a limited supply of material. A large amount of sand and tiny gravel debris, not held in the free spaces inside the debris cones between rough gravel, remains on the surface. The sufficient humidity and an inclination inferior to 36 to 38° in the central part (the movement of individual particles of debris of a diametre of 5 to 10cm having been stopped) enable the securing of a part or of the whole cone by grass. These cones are in irregular shapes locally covered by younger debris or by debris avalanche or avalanche streams (for instance debris cones below the Jastrabí zub and the Jastrabí veža Mts in the Zelené pleso Valley).

d) The oldest (fossil) debris cones of a) and b) type, the development of which has almost stopped and in which secondary modellation is prevailing. They are continuously covered by grasses or by dwarf pine in the corresponding vegetation belt. There are rainwash gullies on their surface and rough boulder and gravel material from rockfalls or debris streams suddenly poured out from glens and couloirs in the walls. The inclination at the foothill of these debris cones is 12 to 20° (e.g. the debris cone below the wall of the Zbojnické Towers in the Veľká Studená dolina Valley, below the Javorový roh Peak in the Rovienky or below the Svišťový roh Peak in the Svišťová dolina Valley).

An increased volume of weathered material transported from the walls and slopes during the spring snow melting (Kalvoda 1971) and in periods of more abundant rainfall is largely due to different kinds of debris avalanche streams, the volume of which could reach in rare cases as much as 2,000 to 3,000m^2. The water saturated to muddy tiny debris with an important part of sand and loam are loosened under the form of narrow mud streams from the rainwash eroded open free slopes, from rock glens, from the feet of the walls in the upper part of the mountain wastes and in funnel-shaped erosion gullies, and then deposited in a tongue form on older debris accumulations. The debris avalanche stream is loosened either by falling stones or by the breaking of the cohesion of small grained debris components and by the melting of the inner snow and ice. Stopping of the stream is conditioned by the loss of a great part of moving matter, by a smaller slope inclination, by a high friction against the under-layer or by a combination of these factors. In debris avalanche cones, this mechanism is completed by occasional seasonal transfer of weathered material by water stream. The debris avalanches mostly use the previously formed erosion trajectories. In the period of spring melting, sharp angular debris and larger rock blocks loosen in the collecting glens in the walls and only after the clearing of the glens of snow debris do avalanches appear. They often stop in the shadowy foothill parts of cuttings or immediately below the walls where snow remains long in summer, or flow on melting snow. After the melting of the snow the characteristic shape of the debris avalanche stream disappears and its material integrates into debris cones or mountains wastes.

The length of the debris avalanche streams that can be observed beneath the walls almost to the debris margin, is very variable. On the free E slopes of the Javorový hre-beň Crest, on the W slopes of the Predná Bašta Mt and on the S slopes below the Hrubô hreben Crest the debris avalanches reach a length of 250 to 300m. They are shorter (less than 100m) in enclosed basins and on the walls of overdeepened troughs. On the relatively quickly changing surface of expressively concave cones they convergently ramify according to the local gradient line course at the mouthing of steep glens. In the wall of the Malý Kežmarský štít Peak above the couloir of the Weber way, in the E wall of the Svišťovka and on the W wall of the Kozí kôpka Mt were observed debris avalanche streams falling several metres over walls like stone avalanches.

At the glen mouthings, in the lower parts of the slopes and at the feet of the walls there are unsorted accumulations brought from the regions of mostly regular avalanches trajectories. In good climatical conditions, each slope with an inclination superior to 12 to 15°, as well as the walls on which regular snow cover can be maintained, give birth to avalanches. Besides the slope inclination, the formation of avalanches is contributed to free debris and a lack of depth of vegetation on slopes and by grass in places with stabilized layers of debris or mountain soils (Andráši 1965, Kňazovický 1968). Unsorted debris with chaotic layers of sandy loams with grasses bedded on Holocene accumulations can be very well distinguished from younger firn moraines in territories with small-grained disintegration of granites. The debris material, initially bound in the snow mass of avalanches, is bedded according to the thickness and the original form of the displaced snow. The transport of weathered material by smaller powdery snow avalanches is minimal, the greatest ones caused by base avalanches reaching, in the place of loosening, the whole weathered rock under-layer of the snow cover. Older debris material displaced by avalanches, is being integrated by rainwash and by debris movement on the slopes into the continuous weathered mantle displaced from the walls and slopes into the valley.

The largest avalanche slopes with frequent avalanche trajectories are in the saddle bow of the Veľká Kopa and Krížna Mts., on the S slopes of the Valentková and Krížna Mts, on the S fault slopes of the mountains, on the W slopes of the Javorinská Široká Mt saddle bow, in the rocky crest part in the Nefcerka, on the S walls of the Hrubô Mt, in the Malá ľadová dolinka Valley, in the Mlynár Mt massif and in all the cirque headings of the main Tatra crest. They are less frequent in the steepest walls of the S saddle bows. The most dangerous ones, with a great amount of stone material, are the avalanches parting from cols and fissures in large glens on mylonite zones, for in-stance in the region of the Furkotská, Mlynická and Zlomisková dolina Valleys, in the E slopes of the saddle bow of the Satan Mt and Baštové veže Towers, in the Červená dolina Valley, etc. In the given region, the quantity of loosened material from the rock or debris underlayer is determined by the kinetic energy of avalanches given by the snow mass, the length of the avalanche trajectory and by the slope inclination.

Rockfalls of large masses (on areas maximally 50 x 20m) are typical for the lower parts of walls, cut by lateral glacial erosion during the Pleistocene glaciation, and for rock steps in cirques. If the directions and inclinations of faults in granodiorites roughly copy the general course of the wall surface, massive blocks get torn away and slide down on the layer surface of faults (Fig. 2). Above that zone continuous over-hangs get formed, their rock mass is progressively getting weathered and the wall destruction progressively rising (for instance the N wall of the Malý Kežmarský štít Peak, the S wall of the Malý Kolový štít Peak, the W wall of the Lomnický štít Peak, the E wall of the Gerlachovský štít Peak and of the Zadní Gerlach Mt, the segment of the main ridge between the Zlobivá and the Gánok Mts, and the N walls of the Javo-rový štít and the Mengusovský štít Peaks).

Lesser rock slides are nevertheless more frequent; they are formed at the crossing of fault surfaces anywhere in the rock massif of a wall with an inclination superior to 50-60°. These rock slides are combined with rockfalls, especially in the higher parts of the walls. Rockfalls are most frequent at the margins of fault zones and mylonites belts which exist in tectonically less-deteriorated granites. Characteristic elements of the deposited material of the rockfalls or slides are clearly distinct from the majority of older debris cones and mountain wastes with soft modellation of boulders or with typi-cal sandy to tiny gravel disintegration of material from the fault zones.

Accumulations of rockfalls are nearly always roughly fissured, non-wrought blocks (the largest one at the foothill of the Krátká Mt in the Nefcerka of the size of 10 x 12 x 6m) or sharp angular elements of flat or prismatoid form. The sharp-edged blocks are of different sizes and - according to the intensity of fissuring in the wall - they nor-mally do not exceed a diametre of 1m. The originally chaotic order during the fall is disturbed on steep walls by inertia according to the mass of different debris elements. The largest boulders reach the valley bottom and the feet of the mountain wastes soon after their loosening from the original bed, while the sand and gravel parts remain on ramps and benches in the wall, in glens or at the upper margin of mountain wastes and debris cones. The magnitude of transport orientation changes is proportional to the number and intensity of rockfalls. Frequent changes of transport trajectories have lead to the formation of continuous zones of debris at the feet of mountain, slightly undulated and curved according to the quantity of transported material in the given direction.

In the closed cirque headings and in the bends of valley sides, the different generations of debris cones are often superimposed, linked and cover the margins of older rockfalls. The initial transfer from the place of weathering is almost always of a gravitational character, the initial impulse being most frequently caused by snow melting in open fissures, loosening of snow cover, snow avalanches, melt water and rain, lightning fissuring, animals, man, by movements of debris from the upper parts of slopes and walls.

The impact of tectonic fissuring and crushing of granodiorites or the inclination and stratification of Mesozoic sediments are the most intense in the Tatras' walls. The frequency of fissures and mylonite zones of size 0.5 - 10m is proportional to the "capacity" of the corresponding glens, gullies, benches and couloirs and to the quantity of recently displaced debris material. For instance, vertical steep fissure systems (with view to the present shape of the wall) accelerate and advance the formation of zones of mountain wastes at the feet and tiny debris cones, and the larger fault zones lead to the formation of voluminous debris accumulations with long period of development. They are usually supplied by a secondary network of glens. On the contrary, the little inclined across the wall running structural steps (most frequently 2 to 8m large and several tens of metres long) slow down the formation of mountain wastes. In the highest High Tatra's walls (E wall of the Malý Kežmarský štít Peak, N wall of the Mengusovský štít Peak, of the Veľká Vysoká Mt, of the Rumanový štít Peak, of the Zlobivá Mt and of the Javorové štíty Peaks, the E wall of the Gerlachovský štít Peak, the NE Bradavice, etc.) there is an evident linking of several levels of the transfer direction, and that in close connection with changes of the fissuring system and the course of the mylonite belts.

It is necessary to stress the fact that the High Tatra alpine-type relief and geological structure form not only the basic elements of a high mountains landscape systems, but they also represent above all the essence of their appearance and are irreproducible. The High Tatra have an exceptional position in the Carpathian system, mainly for the extent and the effects of the climato-morphogenetical processes in the Pleistocene glaciations. The principal high mountain features are visible not only in crest parts (comp. Photos 1 and 4), but also in accumulation landforms in valleys and in the foreland. The most valuable relics of the former mountain glaciation are fossil moraines and their glacifluvial and fluvial equivalents. Also the region of valley accumulations in the High Tatra foothills (Mazúr 1955, Lukniš 1973a), composed of weathered and transported materials of the developing mountain vault, belongs by its origin to the High Tatra region.

3. Anthropogenous share in geomorphological hazards and risks

The basic idea of the Tatras National Park foundation consists of linking the original intact nature, locally influenced by human activity, to the needs of present-day society, including preservation of its aesthetic and scientific value. Social pressure to increase the use of the Tatras countryside for recreation, sporting and spa activities is nevertheless getting stronger and the danger of an irreversible damaging of typical landscape units in the Tatras region is very probable.

Photo 4. The High Tatra crystalline rock complexes prevailingly with alpine-type relief of granodiorites, are covered in the eastern part predominantly by Permian to Cretaceous limestones, dolomites and shales which constitute the Belanské Tatry Mts. On the horizon, the relief of the Palaeogenic flysch formations of the Spišska Magura Mts and the undulated Podhal pahorkatina Hills can be seen. Photo: Jan Kalvoda

In the ridge part of the High Tatra, the anthropogenous landforms are of small area (Kalvoda 1978) and of little importance for the rock landforms. Among the building localities, let us mention the small tourist chalet below the Rysy Mt, the Solisko chalet, the hotel Sport and the Bílikova chata chalet on the Hrebienok, the Zbojnícka chata chalet in the Veľká Studená dolina Valley and the Téryho chata chalet in the Malá Studená dolina Valley, the funicular with the top station building on the Lomnický štit Peak, the symbolic cemetery below the Ostrva Mt, ski lifts (Solisko, Hrebienck, Lomnické sedlo Col and other moveable ones), haylofts and hunters' timber chalets in the Bielovodská dolina and the Zadné Meďodoly Valleys. There is a granite emergency bivouac on the Široké sedlo Col next to the contacts of granodiorite folds of the Javorinská Široká with Werfen slates. Another bivouac is the Vlkov one in the Zlomisková dolina valley and a quantity of arranged natural shelter places (for instance near the Vyšné Temnosmrečinské pleso Lake, below the Žabí plesa Lakes in the Mengusovská dolina Valley, below the N wall of the Malá Snehová veža Tower in the

Čierna Javorová Dolina Valley, etc.). The high-mountain marked paths with displaced boulders, blasted larger stone blocks (ray and fan shaped granite splitting) with stabilization dams and chains in walls or neglected hunter paths are found in the majority of Tatras' valleys. On the paths leading to the most frequently visited summits and cols (Kriváň, Rysy, Východná Vysoká Mts. and Sedlo pod Ostrvou, Vyšné Kôprové sedlo, Sedielko, Prielom, Priečne sedlo, Poľský hrebeň Cols, etc.) the natural composition of the surface of debris slopes is disturbed. During snow melting and heavier rains, the paths quickly change into brooks.

Among the rare anthropogenous landforms, let us mention the mines on the W and SW hill-sides of the Kriváň Mt, the traces of ore extraction in the Medená dolinka Valley, wicker and other barriers against the sliding of easily moveable limestone debris on the SW slope of the Belanská kopa Mt, the arrangement of the S wall of the Lomnický štít Peak during the construction of the second funicular pillar, stone marks (on the col below the Ostrva Mt, on summits, out of which the biggest one being on the Jahňací štít Peak, in valleys, at the feet of walls, etc.) and deterioration of rocks and debris by the hooks of mountaineers and their drilling, the dislodging of debris and snow fields, by unvoluntary acceleration of slidings of rocks, snow avalanches and debris movements.

The reconstruction of touristic paths is necessary in order to preserve the relics of the natural character of slopes. It is also necessary to protect the paths from rockfalls, landslides, avalanches and falling stones. One of the dangerous localities between Hrebienok and Kamzík in the Studenovodská dolina Valley was described by Halouska (1992). The path runs at 1,300m a.s.l. and, for some hundreds of metres, is situated below the rock walls of the Slavkovský štít Peak (2,452m). The eastern spur of this peak, composed of biotite granodiorite and quartz diorite, is strongly and irregularly tectonically disturbed and weathered. The rocky slopes are inclined up to 60°, by periglacial processes with many rock towers, free blocks and stone deposits. Their lower parts are covered by dwarf-pine and coniferous forests. The instability of these steep rock slopes is evidenced by many open fissures, featuring slow debris movements, and occasionally by snow and stone avalanches, rockfalls and falling boulders. The foot of the described steep slopes and rock walls extends above huge morainic and stony-bouldary accumulations.

Little solidified or loose sedimentary formation, various by their granularity and lithological composition, relatively quickly respond to sudden changes in equilibrium between the natural modellation process and the present landform due to human intervention. This factor is reflected in worse foundation conditions by building, and maintenance of settlement units and ways. One of the unfavourable results of the anthropogenous elements accumulation in the landscape is also the aesthetic deterioration or deformation of its natural character (Podbanské settlement, Štrbské pleso Lake and its large neighbourhood, Popradské pleso Lake, Velické pleso Lake, Hrebienok site, etc.). From the viewpoints of both geomorphology and nature protection, these losses are irreparable.

An example of the direct anthropogenous process in the High Tatra is the modification of the foothill relief by the construction of settlements and communications connected with the artificial replacement of several metre thick soil layers and with the planned permanent deterioration of the original character of the locality. An undesirable example of an indirect anthropogenous process is the destructive effect of water erosion on deforested slopes, especially at the upper forest line (Ksandr 1953, Šmarda 1956, Zelina 1964). In this century, the line of vegetation belts has been seriously affected mainly in the Western and Belanské Tatras by wood extraction, excessive pasture, fires and natural calamities. Soil and vegetal communities are devastated by the deterioration of the balance during the formation and transport of weathered material (Pelíšek 1973).

Ravine water erosion is very often initiated by indirect human interventions. Erosion of the natural vegetation cover is followed by a rapid washing up of loamy and sandy soil particles, the connection with what lies beneath is lost and a greater occasional surface water outflow may easily tear away the whole weathered mantle. This disturbing process is facilitated not only by the establishment of building sites on slopes, but also by everyone who shortens his way by going up or down the slope perpendicularly on the paths following the contour lines. Also descent trajectories are changed during snow melting and more abundant rainfall into water beds. The resulting rainwash gullies then irreversibly remodel the natural form of denudational slopes and often deepen down to the rock underlayer.

The washing of weathered material down the slope and erosion are rapidly progressing mainly during building works on accumulation landforms; secondary modifications may slow the destruction process, but sometimes it is very difficult to stop it. In the given climatical conditions of the periglacial and the cold humid model-lation zone the most dangerous among the indirect anthropogenous processes are those connected with destructive activities of the underground and surface water. In winter, the anthropogenously disturbed slopes manifest an increased danger of avalanches, accompanied almost always by calamities. Building activities frequently deteriorate the seasonal regime of underground water by changing the outflow conditions and the degree of water saturation of the weathered mantle. In the Tatras lakes dammed by fossil moraine walls, this state may lead even to the disparition of lakes or to a progressive reduction of their volume.

Unpremediated interventions into the moraine bed near the Velické pleso and the Skalnaté pleso Lakes have damaged the sealing of fissures in block debris and under-layer fissures by soft sedimentation material (Zajíček 1957), which has increased the danger of outflow. The necessity to avoid larger interventions into the moraine dam of the Štrbské pleso Lake has already been stressed by Lukniš (1959). The whole Štrbské pleso Lake is situated in unsorted clay to block fractions of the disappeared glacier moraines in the Mlýnická dolina Valley above the more than 120m high steep margin slope of moraine accumulation. From the morphological point of view, the Štrbské pleso Lake is situated in a very unstable position with a small underground outflow which could be seriously increased by building activities and the water balance of the lake could be seriously endangered. Other negative aspects of human activities in the region have been suggested and described by Bugan (1972).

Human activities may accelerate gravitational modellation processes, as, for instance, anthropogenously initiated rock slides and rockfalls, slidings of little solidified or loose materials, and the gradual movement of soil, which may break the dynamic of the mountain slopes on denudational and accumulation landforms.

All anthropogenous processes, phenomena and resulting landforms which histori-
cally, genetically, biologically and aesthetically deteriorate the High Tatra relief, must
be considered as disturbing and noxious. Only that human creative activity which re-
spects the need to safeguard the natural landscape aspects and its structure, will enable
the fulfillment of the aim of the Tatra's nature protection (Kalvoda 1978, 1981). From
that point of view, it is possible to reliably distinguish the inevitable human interventions into the environment of the territory of the Tatras National Park from the
disturbing anthropogenous transformations due to a lack of understanding or the un-
derevaluation of the Tatras Mountains uniqueness and of the dynamism and rhythms
of the natural processes.

The possibilities of regeneration of the anorganic components of the natural environment in the present climatic, hydrological and modellation conditions are minimal
with the exception of the rock crest part of the mountains (Mazúr et al. 1985). In addition, it is probable that without an essential intervention against the steadily increasing
requirements on further enlargement of the zones affected by man, the intensity of
disturbing irreversible anthropogenous processes will entirely change in a number of
localities the original aspect of the landscape.

4. Conclusion

The destruction of the fossil alpine-type relief of the ridge part of the High Tatra
Mountains and the transfer of weathered material into valleys of mainly glacial origin
and into their foreland occur very intensively. The mountains as a whole have preserved marked features of glacial relief, the retreat of the ice and the transformation by
postglacial denudation of which are the latest stage of the High Tatra Mts polygenetic
development. In this unique state, when specific high mountain features of rock complex and relief have been affected by modellation of a periglacial and humid character,
anthropogenous activities in the High Tatra environment can have a very negative
influence on the dynamic balance between the present-day existing mountain landforms and the climato-morphogenetic processes.

A landscape which has been anthropogenously transformed in historical and
mainly in recent times has a specific aspect, regionally differing in the quality and the
intensity of landform changes. In a landscape affected by man, the intensity of anthropogenous modellation processes is incomparably higher than in areas influenced only
by natural geological and climatical relief-building processes. Given the relatively
small extent of the High Tatra not exceeding 900km^2, every greater local irreversible
anthropogenous change of their relief seriously affects the natural set of the mountain
landforms and at the same time devastates the natural catena of fauna and flora.

Exceptional attention must be paid to the stability of rock slope of the High Tatra
alpine-type relief. In a number of places in the ridge part of the mountains, it is possible to see the landforms of recent rockfalls, landslides and avalanches, and, in the same
time, manifestations of the preparatory phases of the formation of further rapid slope
and/or debris mass movements. The erosion of soil and parent material by water is also
very conspicuous, as well as avalanches partly from productive land including forest
and roads.

At present, morphotectonic observations in the region of the Tatras Mts show that some localities with manifestations of geomorphological hazards on mountain slopes can be connected with historical earthquakes, mainly as trigger agents, and with recent activity of fault zones.

References

1. Andráši J. (1965): Škody a nešťastia spôsobené lavínami vo Vysokých Tatrách v rokoch 1850 - 1960. - Zbor. Pr. TANAPu, 8, 285 - 302, Martin.

2. Bugan M. (1972): Vplyv majstrovstiev sveta v lyžování na tatranskú prírodu vo svetle potrieb a tendencií vzťahu moderného človeka k prírode. - Čsl. Ochr. Prír., 13, 1, 41 - 64, Bratislava.

3. Droppa A. (1957): Krasové zjavy na Kresanici. - Čsl. Kras, 10, 68 - 73, Praha..

4. Droppa A. (1965): Geomorfologický výskum priepastí v Červených vrchoch. - Slov. Kras, 5, 42 - 48, Martin.

5. Gorek A. (1958): Geologické pomery skupiny Červených vrchov, Tomanovej a Tichej doliny. - Geol. Zbor. SAV, IX, 203 - 240, Bratislava.

6. Gorek A. (1959): Prehľad geologických a petrografických pomerov kryštalinika Vysokých Tatier. - Geol. Zbor. SAV, X, 13 - 86, Bratislava.

7. Gorek A., Kahan Š. (1973): Prehľad geologického vývoja a stavby Vysokých Tatier. - Zbor. Pr. TANAPu, 15, 5 - 88, Martin.

8. Halouska R. (1992): Commentary notes to excursion routes. Locality Hrebienok - Kamzík route in the Studenovodská dolina valley. - Excursion guide-book, Intern. Symp. "Time, frequency and dating in geomorphology", Tatranská Lomnica - Stará Lesná, June 16 - 21, 1992. 18 - 20, Bratislava.

9. Kalvoda J. (1970): Geomorfologie hřebencvé části Vysokých Tater. - PhD. Thesis, Czech. Acad. Sci., 103 p. + 262 p., Praha.

10. Kalvoda J. (1971): Drobné tvary povrchu sněhové pokrývky hlavního hřebene Tater. - Sbor. Čs. geogr. Spol., 76, 2, 146 - 150, Praha.

11. Kalvoda J. (1974): Geomorfologický vývoj hřebenové části Vysokých Tater. - Rozpr. ČSAV, Ř. mat.- přír. Věd, 84, 6, 1 - 65, Praha.

12. Kalvoda J. (1978): Antropogénne narušenie reliéfu Vysokých Tatier. - Zbor. Pr. TANAPu, 20, 115 - 125, Martin.

13. Kalvoda J. (1981): Návrh na umístění Státních přírodních rezervací na území Tatranského národního parku z hlediska geomorfologie. - Expert. zpr. Úst. Geol. Geotechn. Čsl. Akad. Věd, 19 p. + 2 příl., Praha.

14. Kalvoda J. (1994): Rock Slopes of the High Tatras Mountains. - Acta Univ. Carol., XXIX, 2, 13-33, Praha.

15. Kos J. (1982): Geomorfologické poměry Tomanovské doliny v Západních Tatrách. - Dipl. pr., Katedra fyzické geografie a kartografie PřF UK, 97 p., Praha.

16. Kotarba A. (1976): Wspóczesne modelowanie weglanowych stoków wysokogórskich na przykladzie Czerwonych Wierchów w Tatrach Zachodnich. - Pr. geogr. IG PAN, nr. 120, Kraków.

17. Ksandr J. (1953): K otázce ochrany a boje proti vodní erozi v Tatranském národním parku. - Ochr. Přír., 8, 1, 30 - 31, Praha.

18. Kňazovický L. (1968): Lavíny. - Vydav. SAV, Bratislava.

19. Linkeš V. (1981): Geografie pôd Vysokých Tatier a ich predpolia. - Geogr. Čas., 33, 1, 32 - 49, Bratislava.

20. Lukniš M. (1959): Problémy Štrbského plesa a jeho ochrany. - Geogr. Čas., 11, 3, 241 - 257, Bratislava.

21. Lukniš M. (1973a): Reliéf Vysokých Tatier a ich predpolia. - Veda, Vydav. SAV, 375 p., Bratislava.

22. Lukniš M. (1973b): Reliéf Tatranského národného parku. - Zbor. Pr. TANAPu, 15, 89 - 143, Martin.

23. Mahr T. (1977): Deep-reaching gravitational deformations of high mountain slopes. - Bull. Int. Assoc. Enging. Geol., 16, 121 - 127, Krefeld.

24. Mahr T., Nemčok A. (1977): Deep-seated deformation in the crystalline cores of the Tatra Mt. - Bull. Int. Assoc. Enging. Geol., 16, 121 - 127, Krefeld.

25. Mazúr E. (1955): Príspevok k morfológii povodia Studeného potoka v Liptovských Tatrách. Geogr. Čas., 7, 1, 15 - 45, Bratislava.

26. Mazúr E. (1962): Príspevok k formám vysokohorského krasu v Červených vrchoch. - Geogr. Čas., 14, 2, 87 - 104, Bratislava.

27. Mazúr E., Lukniš M. (1956): Geomorfológia a kvartér vysokohorskej oblasti Slovenska. - Geogr. Čas., 8, 2-3, 95 - 100, Bratislava..

28. Mazúr E. et al. (1985): Krajinná syntéza oblasti Tatranskej Lomnice. - Veda, Vydav. SAV, 107 p., Bratislava.

29. Mičian L. (1959): Geomorfológia a kvartér Bielovodskej doliny vo Vysokých Tatrách. - Acta geol. geogr. Univ. Comen., Geogr., 1, 85 - 130, Bratislava.

30. Nemčok A. (1972): Gravitačné svahové deformácie vo vysokých pohoriach slovenských Karpát. - Sbor. geol. Věd, Ř. HIG, 10, 7 - 38, Praha.

31. Nemčok A. (1982): Zosuvy v Slovenských Karpatcch. - Veda (SAV), 319 p., Bratislava.

32. Pelíšek J. (1955): Výšková pásmovitost půd v oblasti Vysokých Tater. - Geogr. Čas., 7, 84 - 91, Bratislava.

33. Pelíšek J. (1973): Pôdne pomery Tatranského národného parku. - Zbor. Pr. TANAPu, 15, 145 - 180, Martin.

34. Plesník P. (1967): Vplyv geomorfologických pomerov na horní hranici lesa vo Vysokých Tatrách. - Geogr. Čas., 19, 2, 81 - 92, Bratislava.

35. Sekyra J. (1954a): K otázce recentnosti strukturních půd. (Z kryopedologického výzkumu v oblasti Veľká Kopa - Križné ve Vysokých Tatrách). - Věst. ÚÚG, 29, 21 - 32, Praha.

36. Sekyra J. (1954b): Velehorský kras Bělských Tater. - Nakl. Čas. Akad. Věd, 205 p., Praha.

37. Šmarda J. (1956): Vegetační kryt erozí obnažených půd v Tatrách. - Biolog. Pr., 2, 8, 5 - 44, Bratislava.

38. Zajíček V. (1957): Zanikání Skalnatého plesa ve Vysokých Tatrách a sanační práce k jeho záchraně. - Ochr. Přír., 12, 1, 19 - 25, Praha.

39. Zaťko M. (1961): Príspevok ku geomorfológii Furkotskej, Suchej a Važeckej doliny v západnej časti Vysokých Tatier. - Geogr. Čas. SAV, 13, 4, 271 - 295, Bratislava.

40. Zelina V. (1964): Erozívne javy v Belanských Tatrách. - Zbor. Pr. TANAPu, 7, 5 - 33, Bratislava.

Author

Jan Kalvoda
Department of Physical Geography and Geoecology,
Faculty of Science, Charles University, Prague
Czech Republic

GEOMORPHOLOGIC HAZARDS IN A GLACIATED GRANITIC MASSIF: SIERRA DE GREDOS, SPAIN

DAVID PALACIOS AND JAVIER DE MARCOS

1. Geomorphologic background

Sierra de Gredos has the greatest elevations in the Sistema Central (Fig. 1), and was built by the uplifting of Paleozoic blocks during Alpine tectonic activity. Although the highest peaks were considerably altered by glaciation during the late Pleistocene, most summit areas tend to be flat and very broad and expose remnants of the pre-Alpine erosion surface.

Sierra de Gredos is composed primarily of granites that form tectonic blocks, tilted northward and separated by a network of fault lines that run N-S and E-W. The tectonic history of the area is responsible for the most important geomorphologic characteristics of Gredos: a steep southern wall that is incised only by short ravines and a shallower northern slope that is cut by longer and deeper valleys. In the late Pleistocene, some glaciers reached up to 9km in length and were located at the heads of the northern valleys. In contrast, the glaciers on the southern slope were much shorter and rarely exceeded 2km.

Studies on glaciation in Sierra de Gredos began in the 1850's and centered on identifying and describing the major areas of glacial activity during the Pleistocene. Between 1862 and 1970, many studies were published which defined, delimited, described and interpreted the glacial relief of Sierra de Gredos. Early research (Prado 1862; Schmieder 1915; Hernández-Pacheco 1957; Huguet del Villar 1915 and 1917; Obermaier and Carandell 1916 and 1917; Vidal 1932, 1934, 1936 and 1948) identified a number of significant regional characteristics that can be summarized as follows: glacial features were usually restricted to gorges on the north side of the range; moraines marked multiple retreat phases; a regional late Pleistocene snowline appeared at 1,800-1,900m; and glaciers advanced to 1,415m. The sites of earlier, small niche glaciers were later found on the south side of Gredos (Hernández-Pacheco 1962 and Asensio 1966), along with moraines that appeared above 1,400m.

During the 1970's, a basic model of the glacial processes of Sierra de Gredos was established. Martínez de Pisón and Muñoz (1972) defined the important lines of glaciation in Gredos and provided an exhaustive description of the glacial forms on the northern slope. Preexisting fluvial valleys were responsible for channelling the glaciers. The intense periglaciation that occurred during the glacial period also takes place today. Arenillas and Martínez de Pisón (1977) and Sanz Donaire (1977 and 1979) studied the gorges on the south side of the range, and much ongoing geomorphologic research is still concerned with analyzing specific processes and developing detailed maps. (Pedraza and Lopez 1980; Alonso et al. 1981; Sanz Donaire 1981; Pedraza and Fernández 1981; Acaso 1983; Ruiz and Acaso 1984; Centeno 1989; Rubio 1990; Rubio

J. Kalvoda and C.L. Rosenfeld (eds.), Geomorphological Hazards in High Mountain Areas, 285-307.

et al. 1992; Carrasco and Pedraza 1992; Franco 1995; Martínez de Pisón y López 1986; Muñoz et al. 1995; Parrilla y Palacios 1995; and de Marcos and Palacios 1995). There is also recent synthesis (Martínez de Pisón 1990, Arenillas 1990, Pedraza 1993, Martínez de Pisón and Palacios 1996).

2. Climatic characteristics

The available climatic studies on Sierra de Gredos are incomplete, primarily because there are too few weather stations located at high elevations that can provide more information. Consequently, it is difficult to determine exactly how climate is influencing in the geomorphologic dynamics of the area. The existing data is collected at stations in towns that surround the Gredos area and are located at elevations between 400 and 800m on the south face and 1,100 to 1,500m on the north. Although scarce, the data reveals that Sierra de Gredos influences atmospheric circulation and causes significant climatic consequences.

The stations on the south face are at lower altitudes and record high amounts of precipitation. Candeleda (430m), for example, has a yearly average of 1,009 mm, while Arenas de San Pedro (510m) receives 1,414mm. The town with the highest elevation on the south side is Guisando (766m), and its yearly average is 2,271mm. No other stations on the south face are located at altitudes higher than Guisando, but by extrapolating the existing data, one can predict conditions at 1,600m, where the estimated yearly average precipitation would exceed 3,000mm.

The northern stations are higher, but they receive less precipitation. Bohoyo (1,142 m), for example, has a yearly average of 884mm; Hoyos del Espino (1,440m), 906 mm; and Navarredonda (1,525m), 986mm. The only high altitude station in Sierra de Gredos is located on the Gredos Platform at 2,200m but has recorded data only from 1973 to 1988. Readings from this station indicate very irregular, average annual precipitation since 1973: 1,431; 1,673; 945; 630; 1,823; 1,800; 1,206; 697; 990; 1,137; 1,578; 1,549; 1,676; 765; 1,687; 2,272mm.

The readings reflect the mountain range's effect on atmospheric dynamics. Because of its E-W orientation, Sierra de Gredos acts like a giant barrier against storms that approach the Iberian peninsula from the Atlantic from the southwest (García Fernández 1986).

In fact, precipitation is most intense on the south side where warm polar weather fronts cause wet tropical air masses to collide with the 2,000m wall that forms Sierra de Gredos' south face. When warm fronts pass frequently and rapidly, precipitation may be so heavy that it can provoke catastrophic erosion processes such as slides, washouts or severe flooding at the mouths of gorges. The slides and floods that occurred in December 1989, were caused while a persistent warm front released more than 700mm of rainfall on Candeleda in less than 40 days. In January 1996, similar repetitive weather patterns affected the southwest side of the range causing catastrophic mass movements on a major part of the southern face of Gredos. Rainfall rate was so intense that the town of Candeleda, for example, received more than 150mm in four days.

Readings from the weather stations on the north side indicate that the shadow effect is true only with regard to rain and not to snow. Since data is collected at rain stations and not snow stations, the readings may not be reliable for all types of precipitation. Seventy seven percent of the precipitation on Gredos Cirque from 1992-94 was snow, and most of it was the result of W-NW storm conditions. Snow accumulates mainly on the lee side of the eastern slopes of the north and south faces of Gredos. The topography of the south face is much more abrupt with only a few protruding rocky spurs. Snow accumulates only in the small niches that face east and are protected by the spurs. Also, since the southern orientation increases the effect of insolation and the slopes on this side are steep, there is little chance for snow to accumulate and when it does, it disappears quickly. On the north face, however, the situation is very different and snow lasts all spring. This side is dominated by long ridges and broad high slopes where snow accumulates easily.

3. Geomorphologic characteristics and current erosive processes

The contrasting north-south climatic conditions in Sierra de Gredos are also accompanied by contrasting geomorphology. The north face consists of a series of long glacial valleys that cut the top surface of a horst that tilts slightly northward (Photo 1). The south face is steep and sculpted by short gorges. The non-glaciated areas are covered with a continuous layer of chemical weathered mantel that is a few number of meters thick. Sierra de Gredos can be divided into three sectors that have common geomorphologic characteristics: the middle and upper reaches of the northern glacial valleys; the headwall of the valleys marked by steep glacier cirques; and the southern gorges, some with headwalls that have been shaped by glaciers. One area from each of the three geomorphologic units was chosen for the study.

3.1. MIDDLE AND UPPER REACHES OF THE NORTHERN GLACIAL VALLEYS: GREDOS GORGE AND PINAR GORGE

Gredos Gorge is a glacial valley that originates in Gredos Cirque and runs N-S to an elevation of about 1,400m, where the last of the morainic sediments appear (Fig. 2). The interfluves that create a boundary for the gorge form flat and very broad uplands that gently slope toward the north and have a relatively thick weathered mantle.

The upper parts of the valley walls are traversed by well preserved ridges of lateral moraines. The small closed depressions that form where the morainic ridges intersect the slope are the site of seasonal lagoons. These basins are found more often on the left lateral moraine.

The morainic deposits that form in the area of contact between the lateral moraine and the slope often become saturated and form numerous debris flows. Streams later occupy the channel carved by the flow and cut so deeply into the rock that they drain the upper basins. Wide alluvial fans form at the mouth of the channel.

The left lateral moraine invades a tributary valley and is subdivided into a number of ridges (Photo 2). The waters of the obstructed stream pool materials that precipitate to form thick deposits. Also, stream action in this tributary valley is currently responsible for transporting these deposits to the inner part of the main glacial valley, where a great alluvial fan is forming.

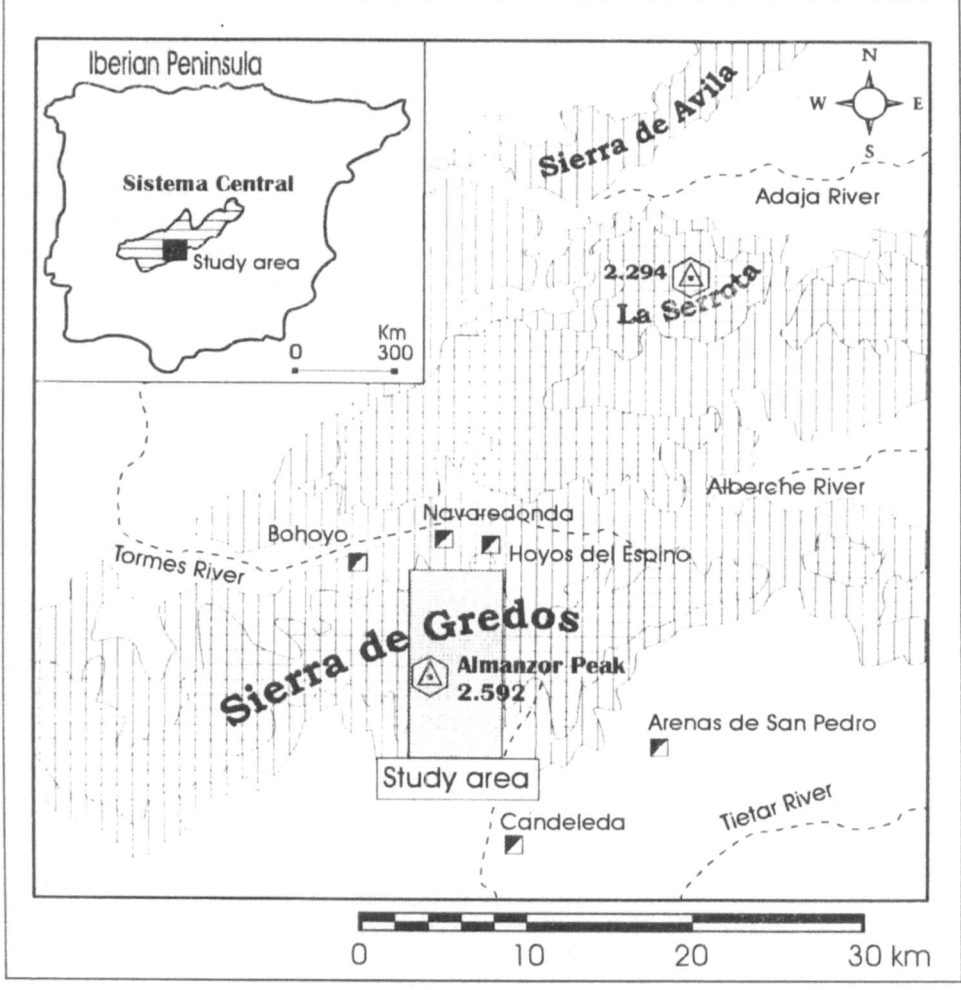

Figure 1. Sierra de Gredos: Location map of the study area.

Figure 2. (→) Key: 1. Weathered mantle area, 2. Non-glaciated rocky outcrops, 3. Fluvial valleyhead, 4. Scattered morainic blocks on slopes, 5. Morainic ridge, 6. Glaciated rocky outcrops, 7. Glacial threshold, 8. Glacial rocky spur, 9. Debris flows, 10. Rockfall cones, 11. Inactive alluvial fans, 12. Active alluvial fans, 13. Solifluction lobes, 14. Lacustrine area, 15. High alluvial terraces, 16. Low alluvial terraces, 17. Alluvial plain

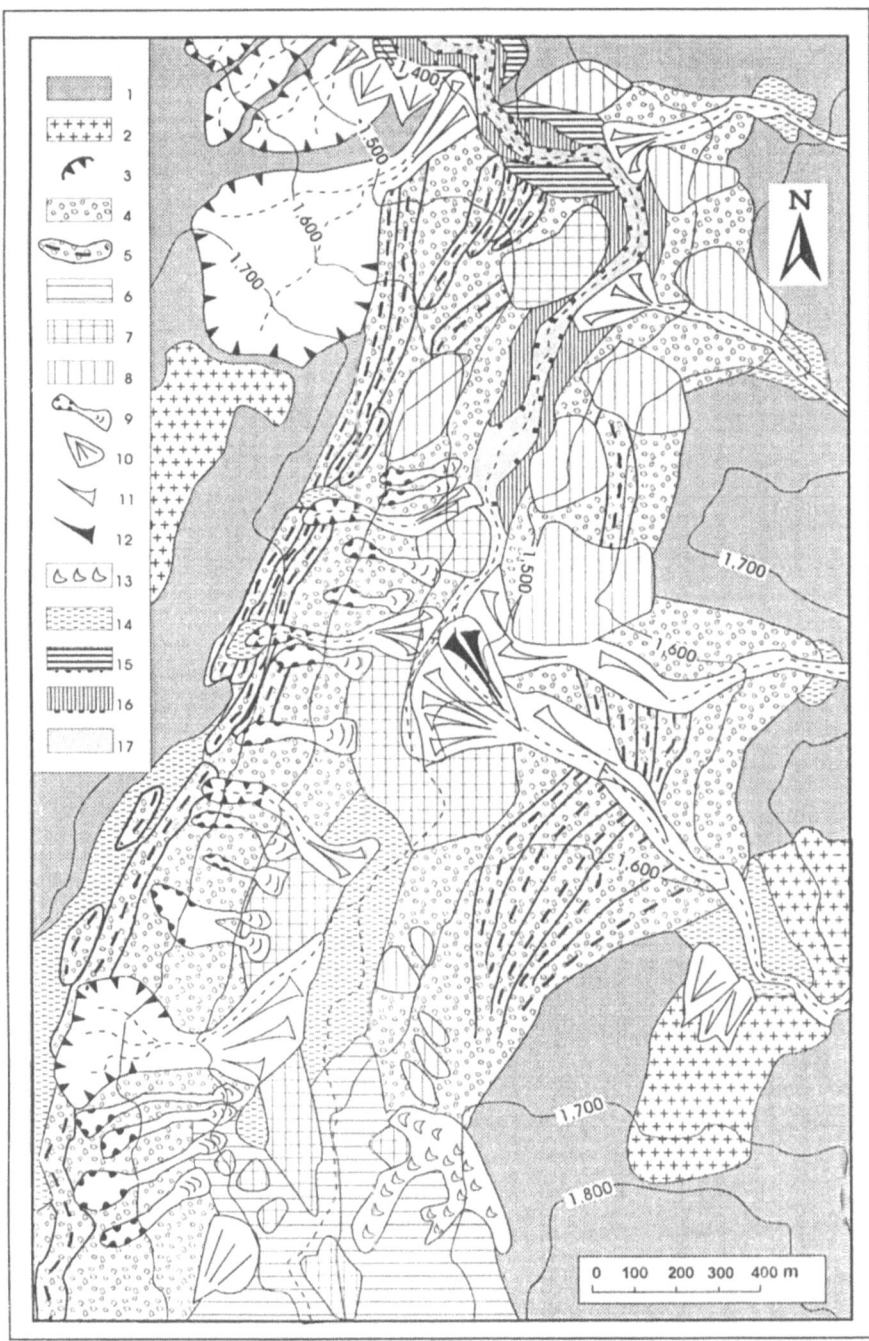

Figure 2. Geomorphologic map of Gredos Gorge (with the collaboration of Martínez de Pisón E., Muñoz J., and Parilla G.).

Photo 1. The glacial valley of Pinar Gorge with clear evidence of glacial quarrying and well defined lateral moraines.

The previous discussion suggests that Gredos Gorge can be divided into geomorphologic stable and unstable areas (Fig. 3). The stable areas include the flat uplands that are covered by a weathered mantle, and the non-glaciated slopes. The glaciated slopes dotted with rocky outcrops, the floor of the glacial valley and the inside of the well drained basins formed by obstruction (not the poorly drained ones that often flood) are also stable areas.

All of the slopes covered with morainic material are unstable, since they are more prone to rockfall and debris flows. The flat flood plain of the present river has the classic characteristics of a high risk area dominated by streams that swell during spring snowmelt. Severe erosion on the slope by rillwashing can create badlands topography. Alluvial fans are also unstable, because they change indiscriminately in response to water streams.

Solifluction creep sometimes occurs where the weathered mantle on the summits and the glacierized slope converge, and is associated with the destabilization of the weathered mantle. Active rockfall cones appear in only a few sectors, especially at the base of rocky walls that have been fractured by tectonics.

Figure 3 (→) Key: 1. No risk area on weathered mantle , 2. No risk area on glaciated rocky floor of valley, 3. No risk area on glaciated rocky slopes, 4. Risk of flooding in lacustrine area, 5. Risk of rockfall and debris flows on morainic materials, 6. Risk of flooding on alluvial plain, 7. Risk of gully erosion, 8. Risk of flow in solifluction lobes, 9. Risk associated with not very active alluvial fan, 10. Risk associated with very active alluvial fan, 11-Risk of rockfall on rockfall cones, 12. Risk on debris flows with active stream incision, 13. Risk on debris flows with non active stream incision.

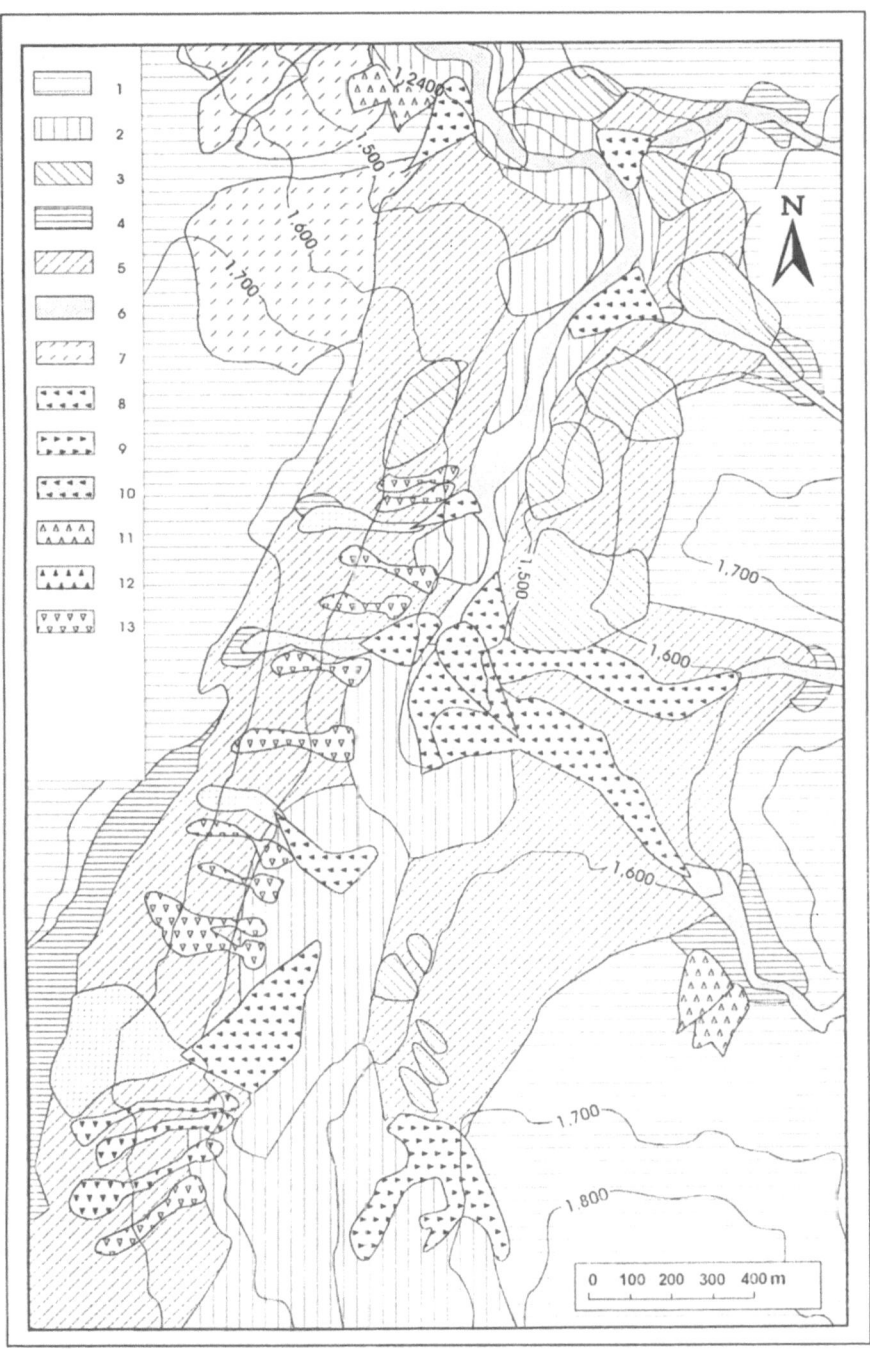

Figure 3. Risk map of Gredos Gorge.

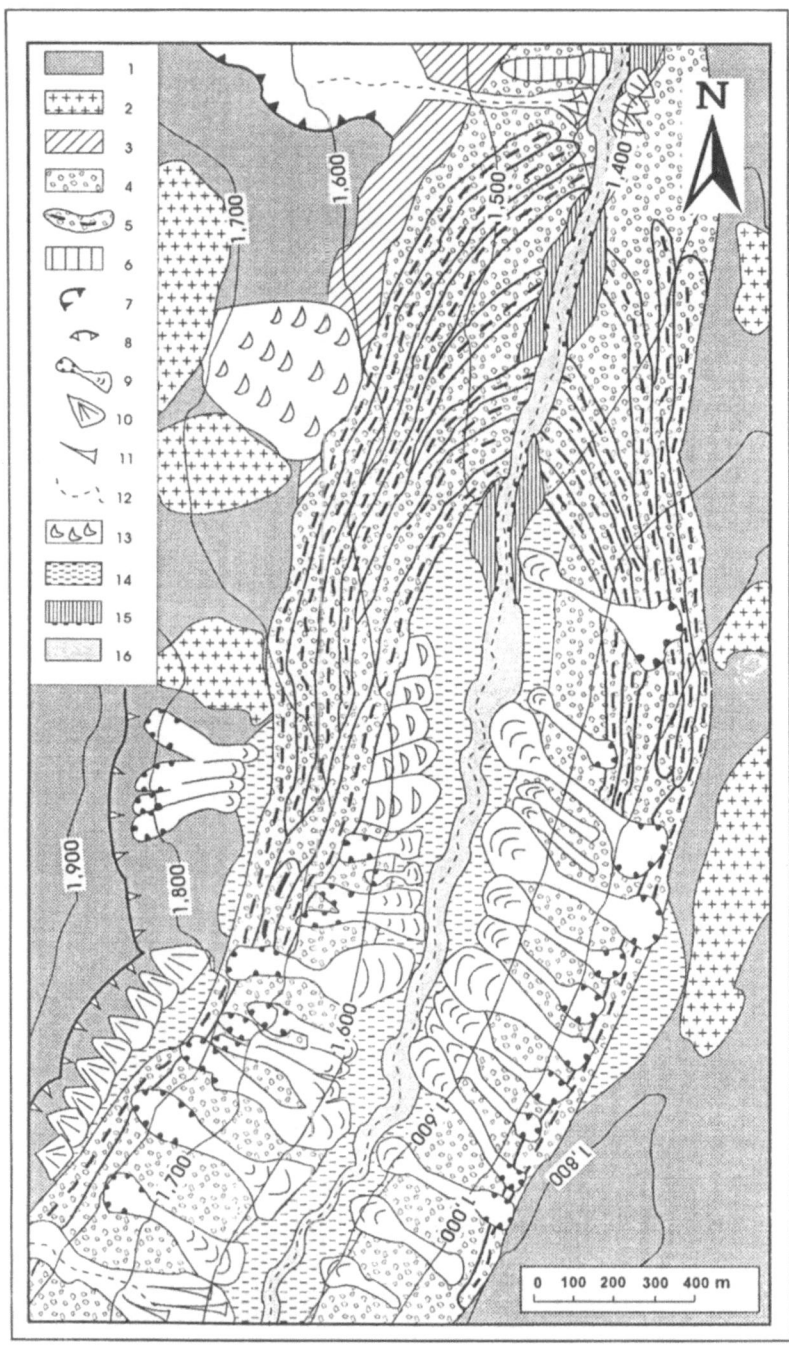

Figure 4. Geomorphologic map of Pinar Gorge.

Figure 4. (←) Key: 1. Weathered mantle area, 2. Non-glaciated rocky outcrops, 3. Tectonic step, 4. Scattered morainic blocks on slopes, 5. Morainic ridge, 6. Glaciated rocky outcrops, 7. Fluvial valleyhead, 8. Nivat on cirque, 9. Debris flows, 10. Rockfall cones, 11. Alluvial fans, 12. Water stream, 13. Solifluction lobes, 14. Lacustrine area, 15. Alluvial terraces, 17. Alluvial plain.

Debris flows may become inactive, although they can retain their peculiar morphology for a long time, or they can slowly evolve into new flows and become very hazardous.

Pinar Gorge is adjacent to and west of Gredos Gorge, and the two are geomorphologically very similar (Fig. 4). Both share the same kinds of geomorphologic risk factors (Fig. 5). In Pinar Gorge, however, the lateral moraines do not invade any tributaries, so there are fewer alluvial fans. On the other hand, the closed basins formed by the lateral moraines are well developed, and debris flows occur more often there.

3.2. THE HEADWALLS OF THE NORTHERN VALLEY: GREDOS CIRQUE

Gredos Cirque faces north and is at the head of Gredos Gorge. It is bordered to the east by Cuento Ridge, whose flat summit slopes gently from Alto de Morezón (2,349 m), to approximately 2,075m. The western edge of the cirque is formed by a sharper and more uneven ridge whose highest peak is Almanzor (2,592m). The ridge continues north and maintains elevation until it ends in two high peaks, La Galana (2,568m) and Cabeza Nevada (2,433m).

The very sharp crest of Los Hermanitos-Cuchillar de las Navajas (2,366-2,490m) forms the cirque's southern edge and separates it from the heads of Tejea, Chilla and Blanca gorges to the south.

There is also a ridge of peaks in this area with similar elevations that runs N-S, and divides Gredos Cirque into two valleys; Laguna Grande Cirque whose floor is at 1,900 and 2,000m and Garganton Cirque, at 2,000 and 2,100m.

The following geomorphologic units are found in Gredos Cirque (Fig. 6):

The flat summit, devoid of rough topographic features, is visible on Cuento, Morezon and Cabeza Nevada ridges, and is a remnant of the old pre-Alpine surface. The entire area is covered by old block fields. Below the periglacial layer there is a chemically weathered mantle that is between 0.3 and 1.7 meters thick. The periglacial deposits have been greatly worn away by chemical weathering, indicating that they are probably relatively old.

The crests on the northern side of the cirque are asymmetrical. The southern slopes are very uneven, with deep stream incisions and sharp spurs and needles. They are covered by a deep periglacial block field. On the eastern slopes, the ramps that were smoothed by the glacier reach high altitudes on the peaks. The northern and western slopes of the peaks consist of vertical rock walls that rise above glacial shoulders.

Figure 5 (→) Key: 1. No risk area on weathered mantle, 2. No risk area on the valley floor, 3. No risk area on tectonic step, 4. Risk of flooding in lacustrine area, 5. Risk of rockfall and debris flows on morainic materials. 6. Risk of flooding on alluvial plain, 7. Risk of gully erosion, 8. Risk on alluvial fan, 9. Risk of flow associated with very active solifluction lobes, 10. Risk of flow associated with not very active solifluction lobes, 11- Risk of rockfall on rockfall cones, 12. Risk on debris flows with active stream incision, 13. Risk on debris flows with no active stream incision.

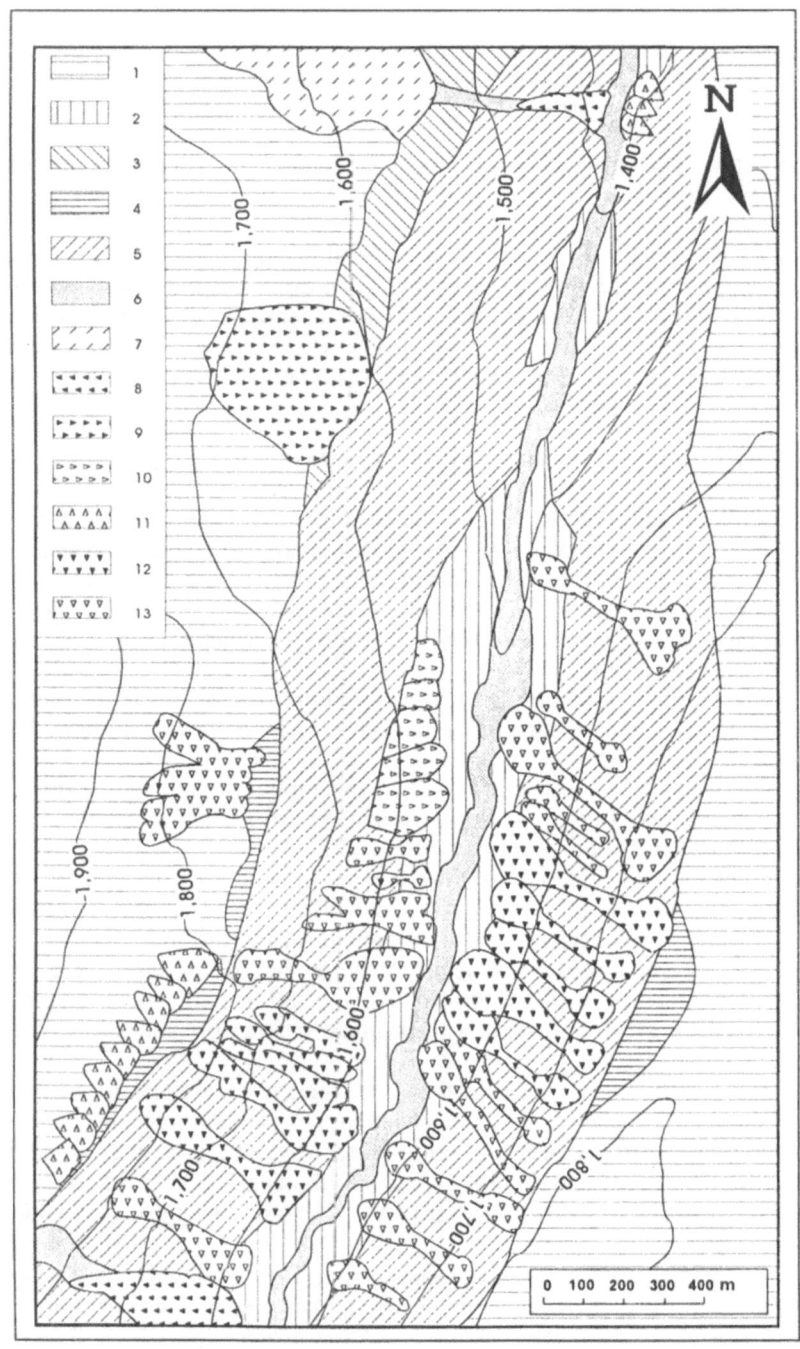

Figure 5. Risk map of Pinar Gorge.

Photo 2. The lateral moraine of Blanca Gorge is affected by erosion in the tributary valley and has formed a great alluvial fan.

Photo 3. West-facing wall of Gredos Cirque, with shoulders and crests.

Photo 4. Active channel on Gredos Cirque.

Photo 5. Overview of Blanca Gorge.

Photo 6. Recent debris flow on the lateral moraine of Blanca Gorge.

Photo 7 Recent debris flow (right of photo) on weathered mantle in Blanca Gorge.

The east-facing slopes were formed by intense downslope glacial smoothing. The channels that follow the fracture lines have been re-excavated and now cut to great depths and are free of deposits and weathered mantle.

The lower reaches of the slopes that face west and north have been modeled by the abrasive lateral action of glaciers (Photo 3). Three geomorphologic units appear on these slopes: the wall above glacier level, the glacial shoulder and the wall worn by abrasive action. The glacial shoulder is clearly discernible. Debris cones composed of material that has fallen from the near slope build up on the glacial shoulder. The surface below the shoulder has been smoothed by glacier action, and there are many striae etched perpendicular to the slope. Large gravitational talus cones are visible below the shoulder and beneath couloirs that follow the lines of weakness along faults. The cones are growing rapidly due to the amount of debris loosened by gelifraction.

The characteristic glacier morphology consisting of alternating basins and valley-steps has developed on the valley floor. Subglacial waters excavated deep incisions, and materials from the talus found at the foot of the northern and western walls were washed downslope and spread into new alluvial fans. There is some till, but it appears only on the lee side of certain valley steps.

There are two types of snow hollows on Gredos Cirque. One is found at low elevations in the depressions on the eastern side of Cuento Ridge or Cabeza Nevada. The chemically weathered layer was worn away by snow action and the eroded material was deposited at the base of the slope in a scattered manner.

The other type of snow hollow is small but noteworthy because of its climatic and geomorphologic dynamics. These hollows are located in unstable block fields on the eastern side of the ridges. There are examples of them on the eastern slopes of the spurs on the south side of Los Hermanitos-Cuchillar de las Navajas Ridge; on the eastern slope of a spur located above the northern shoulder of the ridge; and on the eastern side of Gredos Cirque where earlier deposits were dislodged by gelifraction or by erosion of the weathered mantle.

The erosion that occurs today is associated with the diversity of geomorphologic features (Figure 7). On the western and northern sides of the headwall of Gredos Cirque there is clear evidence of periglacial activity on the wall above the glaciated slope. Weathered material remains in the fissures and is easily dislodged by gelifraction (Photo 4). In addition, talus cones composed of rockfall debris from the wall, build up on the glacial shoulder. The only area where almost permanent snow patches exist is on the east face of some of the spurs on the wall. The debris cones there are remodeled and snow hollows form.

The eastern side of Gredos Cirque is virtually inactive. Where once there was a glacier, now there is deep snow cover that lasts until late summer. This inactivity is due to the previous glacial erosion in the channels that cut deeply into the pre-existing surface. Other areas affected by active processes are the edges of the flat summits, where chemical weathering generates large quantities of fine materials. This in turn, promotes gelifluction and other mass wasting processes in areas that have great snow accumulation and more snowmelt. The diffuse surface runoff on much of this chemically weathered surface is responsible for removing fine sediments and depositing isolated core-stones. The process is sometimes so severe that the slope acquires a badlands appearance.

3.3. THE SOUTHERN GORGES: BLANCA GORGE

Blanca Gorge lies to the south of Gredos Cirque (Figure 8). The head of the gorge is formed by a series of small glacier cirques that hang above a steep wall. The preglacial weathered mantle is still visible on the sides (Photo 5).

Most of this preglacial environment is dominated by stable areas located on the sides or upper reaches of shallower slopes. A few unstable areas are associated mainly with small basins on that overflow when there are torrential rains and cause severe erosion.

No evidence of mass movement of material such as debris flows was observed in the study area in Blanca Gorge. Nearby, however, this type of movement on weathered mantle is very common and is known locally as "vejigas". This phenomenon is normally associated with torrential rains (Parrillá and Palacios 1995).

Two small glaciers developed during the Pleistocene on the headwall of Blanca Gorge, one near Los Hermanitos and the other near Arroyo Chorreras. Los Hermanitos Cirque is surrounded by vertical walls that climb with an elevation gain of up to 200 m. The walls merge directly with the summit areas at 2,300 to 2,428m (Casquerazo). Their sharp-crested ridges are typical of Alpine needle morphology. There are few peaks in this area that have a massive profile or wide and flat summits like the rest of the Sistema Central.

The granite on the rocky summits has been exposed to extreme periglacial processes, mainly gelifraction. As a result, it has fractured along great fault lines and large blocks have fallen, leaving narrow geometric gaps.

There are two lateral moraines at the base of the cirque. The right one starts at about 1,700m and descends to 1,450m. The left moraine appears at 1,950m and is visible until it merges with Arroyo Chorrera at 1,600m. At this point, there is a group of scattered boulders, semi-detached from the slope, that join the terminal right moraine at about 1,450m. Evidence of the confluence of the two moraines no longer exists, and has been replaced by a proglacial terrace located at the ends of the moraines.

The Arroyo Chorrera Cirque occupies the far eastern part of the headwall of Blanca Gorge. Its morphology is completely different from that of Los Hermanitos Cirque. There are no steep sides on the headwall and it is joined to the north slope by a low ridge covered by a rock field and a few rocky outcrops.

At 1,900m the right lateral moraine of Arroyo Chorrera Cirque runs parallel to the left moraine of Los Hermanitos Cirque until it joins the latter at 1,650m. The left lateral moraine is more difficult to discern, because it is semi-detached from the wall, but at 1,850m there is a line of boulders that has no fine matrix on the upper part of the deposit.

The glacial and preglacial modeling that prevails in Blanca Gorge contributes to current erosive processes (Figure 9). Activity, however, is limited to the channels that have been worn into the granite rocks on the highest slopes. These channels are narrow and wedged into the rock, and their sides are normally steep and smooth. In some cases, the walls can reach heights of over one 100m and large blocks are usually found at their base. The length of the channels can reach up to 500m, but this varies depending on whether they cut through rocky outcrops. The channels originate mainly as

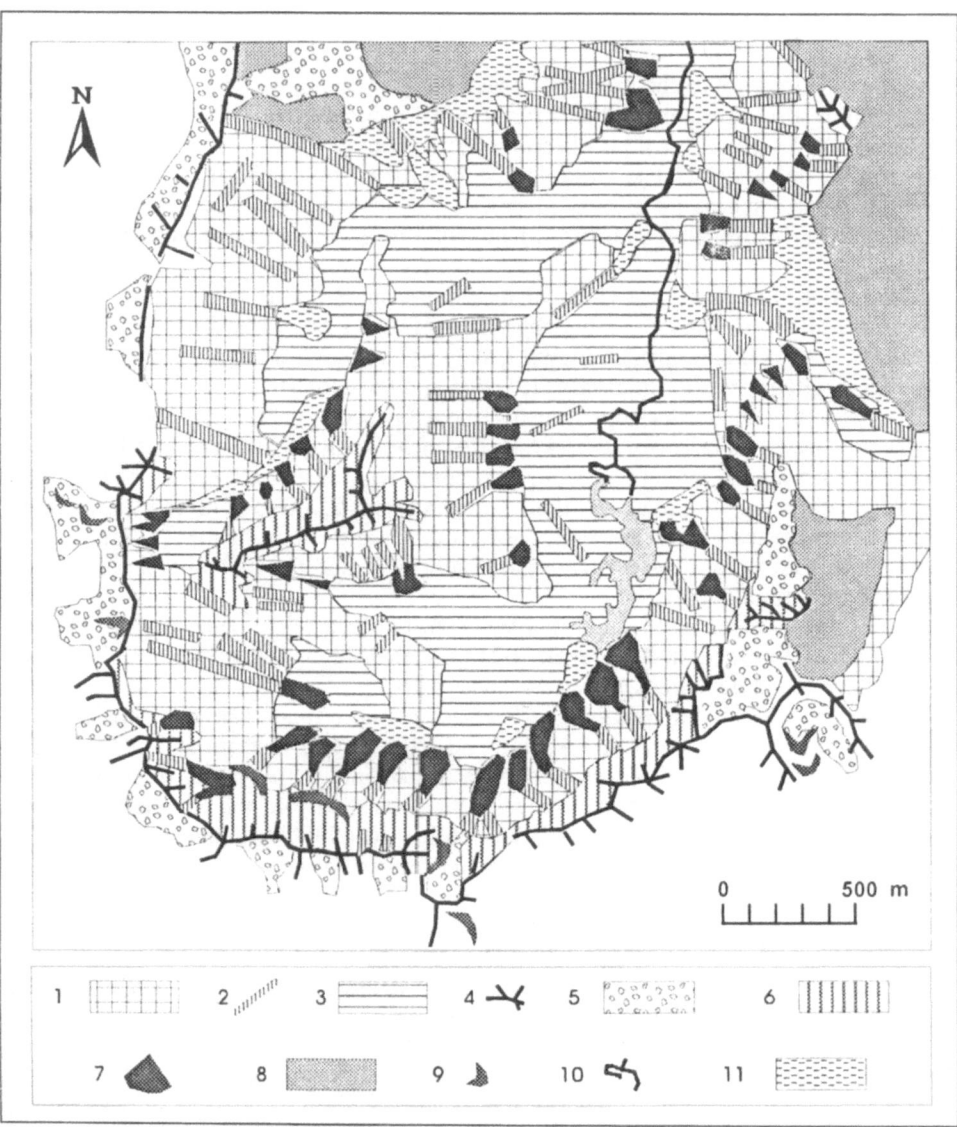

Figure 6. Geomorphologic map of Gredos Cirque (with the collaboration of Muñoz J.).
Key: 1. Glacial smoothed valley walls, 2. Channels of glacial incision, 3. Glacial valley floor, 4. Periglacial ridges, 5. Periglacial blockfields, 6. Supraglacial walls, 7. Rockfall cones talus, 8. Weathered mantle area, 9. Protalus ramparts, 10. Stream incision channels, 11. Slope with active rockfall, gelifluction and rillwashing processes.

Figure 7. Risk map of Gredos Cirque (with the collaboration of Muñoz J.).
Key: 1. Flood risk areas on alluvial plain, 2. Risk areas on alluvial fans, 3. Risk of rockslides , 4. Risk of rockfall, 5. Risk associated with permanent snow accumulation, 6. Risk determined primarily by gelifraction, 7. Risk determined primarily by gelifluction and rillwashing.

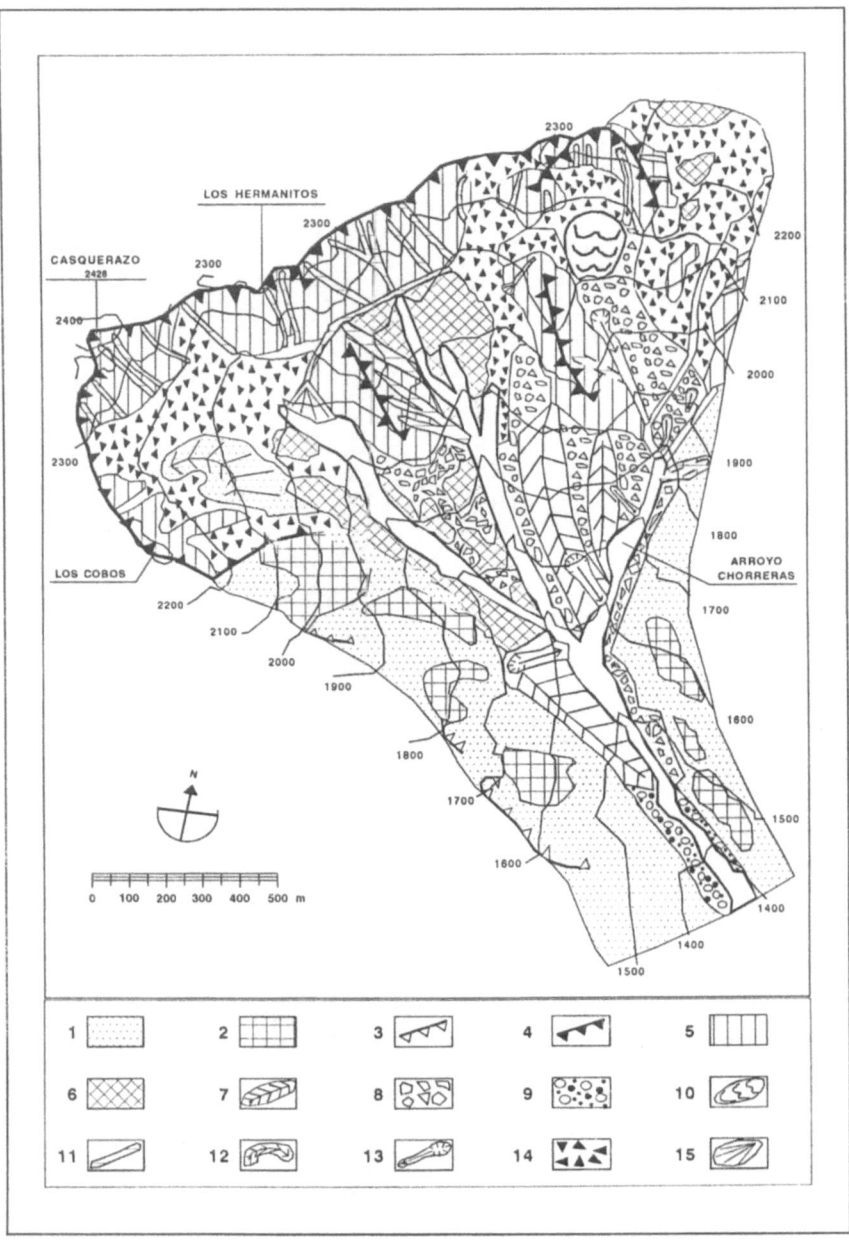

Figure 8. Geomorphologic map of Blanca Gorge.
Key: 1. Weathered mantle surface, 2. Non-glaciated rocky outcrops, 3. Fluvial valleyheads, 4. Wallhead limit of glacier cirques, 5. Rocky wallhead of glacier cirques, 6. Glacial smoothed rocky outcrops, 7. Lateral moraines, 8. Shattered morainic blocks, 9. Fluvioglacial terraces, 10. Rock glacier , 11. Glacial channels, 12. Protalus ramparts, 13. Debris flows, 14. Periglacial blockfields., 15. Rockfall cones.

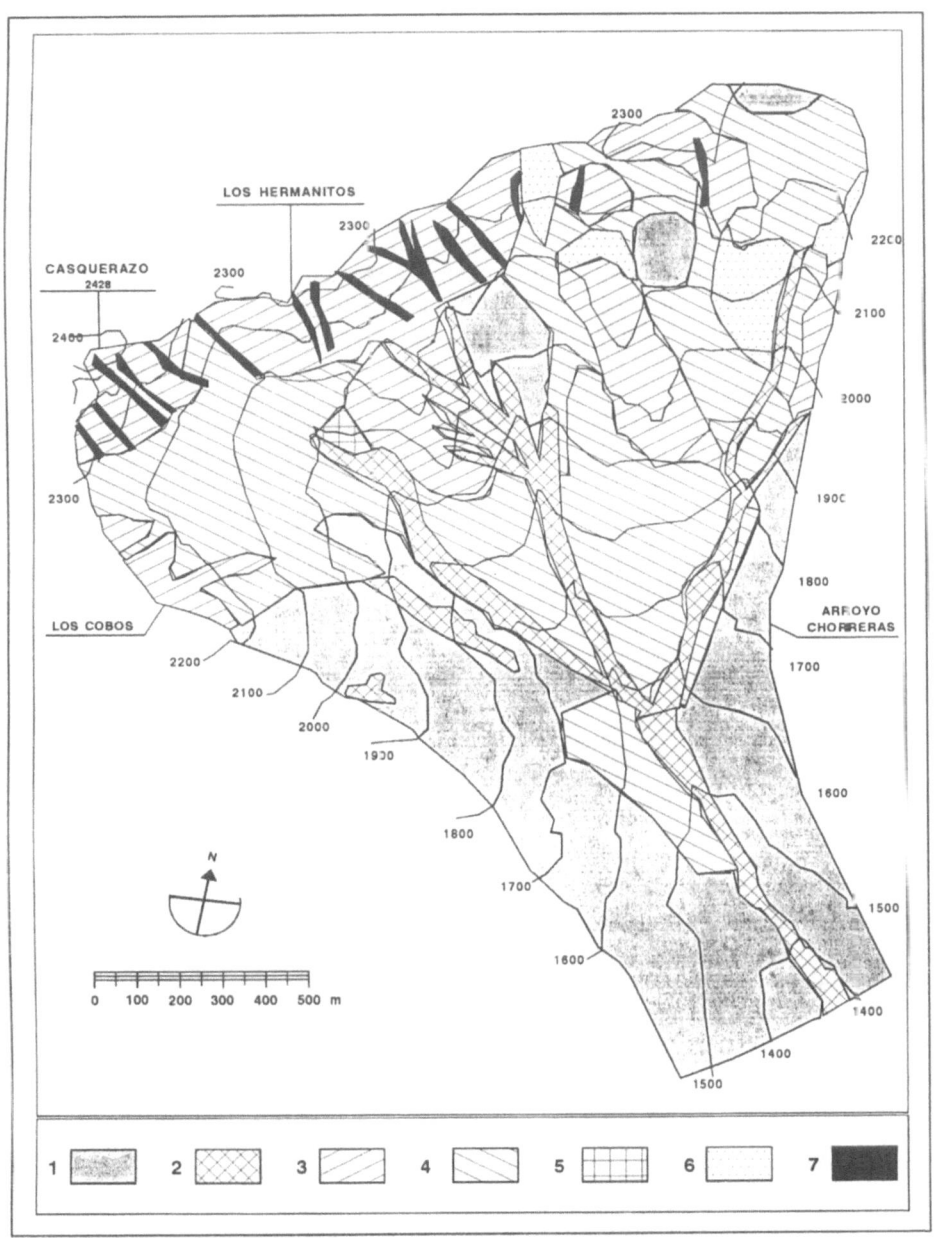

Figure 9. Risk map of Blanca Gorge.
Key: 1. No risk area, 2. Risk of rillwashing and gully erosion, 3. Risk of shattered rockfalls, 4. Risk of rockslides,
5. Risk of concentrate rockfall on cones, 5. Risk of flows in solifluction lobes, 7. Risk of avalanches.

a result of periglacial processes, especially gelifraction, followed by rockfall or snow related erosion that alter the fault lines which sometimes serve as channels for landslides or flashfloods from either permanent or spontaneous torrents.

There are also some inactive channels that have a morphology similiar to that of the active ones, except that their beds are covered with stable boulders.

The snow hollows on the headwalls of old glacial cirques and protalus ramparts are also active erosion areas.

A moraine that contains a snow field is clearly visible on the headwall of Blanca Gorge and is located on the far western side of Los Hermanitos Cirque at 2,150m at the foot of the slopes that join Casquerazo Peak and Los Cobos Peak.

The rockfall talus is also unstable and appears at the mouths of the active channels where avalanches have formed cones. Rapid snowmelt occurs on the slopes that have the greatest exposure and contribute to instability in these block talus.

The inactive rockfall talus are different from the active ones in that their morphology has undergone more alteration either through the weathering of blocks, the greater density of plant colonization or overall consolidation of the relief. There is evidence that some of the rocks are very old, and probably existed either before or during the glacier stage.

The debris flows in the study area deserve special mention. Flows are mainly found on the moraines, rockfall talus and weathered mantle (Photos 6 and 7).

4. Conclusion: Geomorphological risks in Sierra de Gredos

Snow and meltwater are agents that create hazards on the north face of Sierra de Gredos. The weathered mantle exists only on the flat summit areas, since these were virtually unaffected by glaciation. Snow easily loosens the mantle, and erosion occurs where the uplands, still covered by the weathered mantle, meet the lower walls of the glacial valleys, and where the mantle has been swept away by glacial abrasion. At this point of contact, material from the weathered mantle is loosened and moves downslope in solifluction lobes.

Snow action occurs in the channels that form in fractures. Snow provides meltwater all spring when the daily temperature contrasts are most extreme and also supplies water for gelifraction processes. Snow can destabilize gravitational talus cones caused by gelifraction and old periglacial boulder fields. The lack of precipitation on the north face is compensated for by water from snow patches that cause the distribution of torrential stream action.

The only morphologic elements that create any significant instability in the landscape on the north face are the lateral moraines that hang over the main valleys, midway up the walls. Furthermore, deforestation has affected the slopes, making them susceptible to slides, falls, debris flows and badlands-type erosion. The highest risk areas are where the moraines obstruct lateral valleys and torrents incise the moraines and create alluvial fans.

The granitic weathered mantle on the south face is better preserved, because of the low impact of glacial erosion, so in some places it reaches a depth of more than 10 m and still clings to the steep slopes. Deforestation in this area has made the weathered mantle very unstable, so there are frequent slides and debris flows. The cirques and the glaciers on the south wall left moraines that that now hang from the cliffs. These deposits are the scene of recurrent rockfall, gully erosion, rillwashing and debris flows, and constitute the second major unstable area. The greatest geomorphologic hazards on the south face are caused by torrents. Heavy rains, steep slopes and a drainage system that is deeply incised in a network of tectonic structures contribute to flooding and increased stream transport. As a result, much of the material from the weathered mantle and the moraines is washed away towards the fringes of Sierra de Gredos where they form wide and highly unstable alluvial fans.

Acknowledgements

The background information came mainly from Muñoz, Palacios and Marcos (1995), Marcos and Palacios (1995) and Martínez de Pisón and Palacios (1996). We are kindly acknowledged to Professors Eduardo Martínez de Pisón and Julio Muñoz for their collaboration.

References

1. Acaso, E. 1983. Estudio del Cuaternario en e: Macizo Central de Gredos. Ph.D. diss., Facultad de Ciencias. Universidad de Alcalá de Henares, 442 pp.

2. Alonso, F., Arenillas & Saenz, C. 1981. La morfología glaciar en las montañas de Castilla La Vieja. El espacio geográfico de Castilla y León. I Congreso de Geografía de Castilla y León, 23-43.

3. Arenillas, M. & Martínez de Pisón, E. 1977. Las gargantas meridionales del Alto Gredos, V Coloquio de Geografía. Granada, 29-33.

4. Arenillas, M. 1990. La Sierra de Gredos. En Gredos. La Sierra y su entorno. Madrid, MOPU, 49-74.

5. Asensio Amor, I. 1966. El sistema morfogenético fluvio-torrencial en la zona meridional de la Sierra de Gredos, Estudios Geográficos, 102, 53-57.

6. Carrasco, R. M. & Pedraza, J. 1992. Fenómenos gravitacionales en el Valle del Jerte: tipologías y significado morfológico. Estudios de Geomorfología en España. S E.G, 434-444.

7. Centeno J.D. 1989. Evolución cuaternaria del relieve de la vertiente Sur del Sistema Central Español. Las formas residuales como indicadoras geomorfológicas. Cuad. Lab. Xeol. Laxe 13, 79-88.

8. Franco, F. 1995. Estudio palinológico de turberas holocenas en el Sistema Central, Ph.D. diss., Departamento de Biología. Universidad Autónoma de Madrid, 392 pp.

9. García Fernández, J. 1986. El clima de Castilla y León. Edi. Ambito, Valladolid, 370 pp.

10. Hernández-Pacheco, F. 1957. Livret guide de l'excursión C-1, Gredos. 5º Congreso Internacional de INQUA, Tomo 1, 36-40.

11. Hernández-Pacheco, F. 1962. La formación de depósitos de grandes bloques de edad pliocena. Su relación con la raña, Estudios Geológicos, 18 (1-2), 75-88.

12. Huget del Villar, E. 1915. Los glaciares de Gredos, Bol. Real Soc. Española de Hist. Nat., 15, 379-390.

13. Huget del Villar, E. 1917. Nueva contribución a la glaciología 5de Gredos, las Hoyuelas del Hornillo, Bol. Soc. Española de Hist. Nat., 17, 558-567.

14. Marcos, J. & Palacios D. 1995. Evolución del relieve glaciar en la Garganta Blanca.. En Aleixandre T. y Pérez-González, A. (Edi.) Reconstrucción de paleoambientes y cambios climáticos durante el Cuaternario. Monografias 3. Centro de Ciencias Medioambientales CSIC, Madrid, 215-225.

15. Martínez de Pisón, E. & Muñoz, J. 1972. Observaciones sobre la morfología del Alto Gredos. Estudios Geográficos, 129, 3-103.

16. Martínez de Pisón, E. 1990. Unidades naturales. En Gredos. La Sierra y su entorno. Madrid, MOPU, 19-47.

17. Martínez de Pisón, E. y López, J. 1986. Las fluctuaciones glaciares pleistocenas en Guadarrama y Gredos. Libro-guía de la excursión del Simposio sobre fluctuaciones climáticas durante el Cuaternario en la regiones del Mediterráneo Occidental. Universidad Autónoma de Madrid, Madrid, 127 pp.

18. Martínez de Pisón, E. y Palacios, D. 1996. Significado del episodio glaciar en la evolución morfológica y en el paisaje de la Sierra de Gredos. Sistema Central. In "El glaciarismo en España". S.E.G. (in print).

19. Muñoz, J., Palacios, D., & Marcos, J. de 1995. The influence of the geomorphologic heritage on present slope dynamics. The Gredos Cirque, Spain". Pirineos, 145-146: 35-63.

20. Obermaier, H. & Carandell, J. 1916. Contribución al estudio del glaciarismo cuaternario de la Sierra de Gredos. Trabajos del Museo Nac. de Ciencias Naturales, Serie Geológica, 14, 54 pp.

21. Obermaier, H. & Carandell, J. 1917. Nuevos datos acerca de la extensión del glaciarismo cuaternario en la Cordillera Central española. Bol. R. S. E. Hist. Nat., 20.

22. Parrilla, G. & Palacios, D. 1995. Debris flows y cambio climático en Gredos. En Aleixandre T. y Pérez-González, A. (Edi.) Reconstrucción de paleoambientes y cambios climáticos durante el Cuaternario. Monografias 3. Centro de Ciencias Medioambientales CSIC, Madrid, 200-215.

23. Pedraza, J. & Fernández, P. 1981. Terciario y Cuaternario del Mapa geológico de Bohoyo y de arenas de san Pedro. Hojas 577 y 578. Mapa Geológico de España. I.G.M.E., Madrid.

24. Pedraza, J. & López, J. 1980. Gredos: geología y glaciarismo, Obra Social de la Caja de Ahorros de Avila, Avila, 31 pp.

25. Pedraza, J. 1993. Geomorfología del Sistema Central. En Gutiérrez, M. y Peña, J.L. (edit.) Geomorfología de España. Edit. Rueda. Madrid, 63-100.

26. Prado,C. 1862, Reseñas geológicas de la provincia de Avila y de la parte Occidental de León, Com. Nac. del Mapa Geológico, Junta General Estadística, Madrid, 260-271.

27. Rubio, J.C. 1990. Geomorfología y Cuaternario de las Sierras del Barco y de Béjar (Sistema central Español). Tesis Doctoral. Fac. C.C. geológicas, U.C.M. Madrid, 319 pp.

28. Rubio, J.C.; Pedraza, J. and Carrasco, R.M. 1992. Reconocimiento de tills primarios en el sector central y occidental de la Sierra de GredosSistema Central Español), II Reunión Nacional de Geomorfología, S.E.G., 413-422.

29. Ruiz, B. and Acaso, 1984. Clima y Vegetación durante el Cuaternario reciente en el Macizo Central de Gredos (Avila). Actas I Congreso Español de Geología. Tomo I, 723-740.

30. Sanz Donaire, J. 1977. El glaciarismo en la cara sur del macizo del El Barco de Avila, V Coloquio de Geografía, Granada. 41-47.

31. Sanz Donaire, J. 1979. El corredor de Béjar. Instituto de Geografía Aplicada, CSIC. Madrid. 195 pp.

32. Sanz Donaire, J. 1981. El macizo glaciarizado de El Barco de Avila (Provincias de Avila-Caceres). Anales de Geografía de la Universidad Complutense, 1, 184-205.

33. Schmieder, O. 1915. Die Sierra de Gredos. Mitt. d. Geogra. Ges. Müchen 10.

34. Vidal Box, C. 1932, Morfología glaciar cuaternaria del macizo Oriental de la Sierra de Gredos, Bol. Real Soc. Española de Hist. Nat., 32, 117-135.

35. Vidal Box, C. 1934. Los glaciares cuaternarios de la Sierra de Bohoyo (Avila). Bol. R. Soc. H. Nat., 34.

36. Vidal Box, C. 1936. Contribución al conocimiento morfológico del segmento occidental de la Sierra de Gredos (Bohoyo), Bol. Soc. Esp. Hist. Nat., 36, 17-31.

37. Vidal Box, C. 1948. Nuevas aportaciones al conocimiento geomorfológico de la Cordillera Central, Estudios Geográficos, 30, 5-52.

Authors

David Palacios, Javier de Marcos
Universidad Complutense
Department of Regional Geographic Analysis
28040 Madrid,
Spain

SUBJECT INDEX

The GeoJournal Library

The GeoJournal Library

42. G. Lipshitz: *Country on the Move: Migration to and within Israel, 1948–1995.*
    1998   ISBN 0-7923-4850-8
43. S. Musterd, W. Ostendorf and M. Breebaart: *Multi-Ethnic Metropolis: Patterns
    and Policies.* 1998   ISBN 0-7923-4854-0
44. B.K. Maloney (ed.): *Human Activities and the Tropical Rainforest.* Past,
    Present and Possible Future. 1998   ISBN 0-7923-4858-3
45. H. van der Wusten (ed.): *The Urban University and its Identity.* Roots,
    Location, Roles. 1998   ISBN 0-7923-4870-2
46. J. Kalvoda and C.L. Rosenfeld (eds.): *Geomorphological Hazards in High
    Mountain Areas.* 1998   ISBN 0-7923-4961-X

KLUWER ACADEMIC PUBLISHERS – DORDRECHT / BOSTON / LONDON